The Future of National Infrastructure

A System-of-Systems Approach

Infrastructure forms the economic backbone of modern society. It is a key determinant of economic competitiveness, social well-being and environmental sustainability. Yet infrastructure systems (energy, transport, water, waste and ICT), in advanced economies globally, face serious challenges.

For the first time, a leading team of researchers sets out a systematic approach to making long-term choices about national infrastructure systems. Too often in the past, decision-making has been fragmented, failing to look beyond individual projects to understand the ways in which systems function and deliver valued infrastructure services. This book shows how it is possible to devise and analyse long-term strategies for national infrastructure that are robust to uncertain futures and take full account of the interdependencies between energy, transport, water, waste and ICT. Great Britain is used as a case study to demonstrate how the methodologies and accompanying models can be effectively applied in a national infrastructure assessment. Lessons and insights for other industrialised nations and emerging economies are highlighted, demonstrating practical scenarios for delivering infrastructure services in a wide range of future socio-economic and environmental conditions.

The Future of National Infrastructure provides practitioners, policy-makers and academics with the concepts, models and tools needed to identify and test robust, sustainable and resilient strategies for the provision of national infrastructure.

Jim W. Hall is Director of the Environmental Change Institute and Professor of Climate and Environmental Risks in the University of Oxford. A civil engineer by background, Professor Hall has pioneered the use of risk analysis to inform tough decisions about the future of infrastructure systems. He advises governments, agencies and utilities globally on how to adapt their systems and plan for an uncertain future.

Martino Tran is Senior Research Fellow at the University of Oxford, and advises governments and industry on energy and environment. With a background in environmental science and mathematical modelling he applies systems thinking and decision sciences for addressing societal challenges in sustainability. He holds a doctorate from the University of Oxford and has published widely on the future impacts of technology on society and environment.

Adrian J. Hickford is Senior Research Assistant in the Transportation Research Group at the University of Southampton. As well as his recent work on possible future changes to national infrastructure systems, he has been involved in a number of projects in the U.K. and across Europe aiming to increase the use of sustainable transport, reducing the associated risks and enhanced practices of traffic accident data gathering and use.

Robert J. Nicholls, Professor of Coastal Engineering at the University of Southampton, is a coastal engineer with expertise in integrated assessment and systems perspectives and climate change impact and adaptation assessment. He was awarded the Roger Revelle Medal 2008 by the Intergovernmental Oceanographic Commission (IOC) Executive Council, which recognises outstanding contributions to the ocean sciences. He advises several national governments on coastal impacts and adaptation to climate change.

The Future of National Infrastructure

A System-of-Systems Approach

Edited by

JIM W. HALL
University of Oxford

MARTINO TRAN
University of Oxford

ADRIAN J. HICKFORD
University of Southampton

ROBERT J. NICHOLLS
University of Southampton

CAMBRIDGE
UNIVERSITY PRESS

CAMBRIDGE
UNIVERSITY PRESS

University Printing House, Cambridge CB2 8BS, United Kingdom

Cambridge University Press is part of the University of Cambridge.

It furthers the University's mission by disseminating knowledge in the pursuit of education, learning and research at the highest international levels of excellence.

www.cambridge.org
Information on this title: www.cambridge.org/9781107066021

© Cambridge University Press 2016

First published 2016

Printed in the United Kingdom by TJ International Ltd. Padstow Cornwall

A catalogue record for this publication is available from the British Library

ISBN 978-1-107-06602-1 Hardback

Contents

Contributors

David Alderson
School of Civil Engineering and Geosciences
Newcastle University
Newcastle upon Tyne
NE1 7RU

Stuart Barr
School of Civil Engineering and Geosciences
Newcastle University
Newcastle upon Tyne
NE1 7RU

Pranab Baruah
Formerly of Environmental Change Institute
Oxford University Centre for the Environment
University of Oxford
South Parks Road
Oxford
OX1 3QY

Rachel Beaven
Cambridge Econometrics
Covent Garden
Cambridge
CB1 2HT

Mark Birkin
School of Geography
University of Leeds
Leeds
LS2 9JT

Simon P. Blainey
Engineering and the Environment
Southampton Boldrewood Innovation Campus
University of Southampton

Burgess Road
Southampton
SO16 7QF

Edward A. Byers
Formerly of School of Civil Engineering and Geosciences
Newcastle University
Newcastle upon Tyne
NE1 7RU

Modassar Chaudry
Institute of Energy
Cardiff University
Queen's Building
The Parade
Cardiff
CF24 3AA

Jon Coello
Engineering and the Environment
University of Southampton
Highfield
Southampton
SO17 1BJ

Douglas Crawford-Brown
Cambridge Centre for Climate Change Mitigation Research (4CMR)
Department of Land Economy
University of Cambridge
19 Silver Street
Cambridge
CB3 9EP

Razgar Ebrahimy
School of Computing Science,
Newcastle University
Newcastle upon Tyne
NE1 7RU

Nick Eyre
Environmental Change Institute
Oxford University Centre for the Environment
University of Oxford

South Parks Road
Oxford
OX1 3QY

David W. Graham
School of Civil Engineering and Geosciences
Newcastle University
Newcastle upon Tyne
NE1 7RU

Jim W. Hall
Environmental Change Institute
Oxford University Centre for the Environment
University of Oxford
South Parks Road
Oxford
OX1 3QY

Adrian J. Hickford
Engineering and the Environment
Southampton Boldrewood Innovation Campus
University of Southampton
Burgess Road
Southampton
SO16 7QF

Ralitsa Hiteva
SPRU
University of Sussex
Brighton
BN1 9SL

Matthew C. Ives
Environmental Change Institute
Oxford University Centre for the Environment
University of Oxford
South Parks Road
Oxford
OX1 3QY

Nick Jenkins
Institute of Energy
Cardiff University

Queen's Building
The Parade
Cardiff
CF24 3AA

Cliff B. Jones
School of Computing Science
Newcastle University
Newcastle upon Tyne
NE1 7RU

Scott Kelly
Department of Land Economy
University of Cambridge
19 Silver Street
Cambridge
CB3 9EP

Chris G. Kilsby
School of Civil Engineering and Geosciences
Newcastle University
Newcastle upon Tyne
NE1 7RU

Lucy J. Manning
Formerly of School of Civil Engineering and Geosciences
Newcastle University
Newcastle upon Tyne
NE1 7RU

Robert J. Nicholls
Engineering and the Environment
University of Southampton
Highfield
Southampton
SO17 1BJ

Edward J. Oughton
Department of Land Economy
University of Cambridge
19 Silver Street
Cambridge
CB3 9EP

Alexander Otto
Formerly of Environmental Change Institute
Oxford University Centre for the Environment
University of Oxford
South Parks Road
Oxford
OX1 3QY

Raghav Pant
Environmental Change Institute
Oxford University Centre for the Environment
University of Oxford
South Parks Road
Oxford
OX1 3QY

William Powrie
Engineering and the Environment
University of Southampton
Highfield
Southampton
SO17 1BJ

John M. Preston
Engineering and the Environment
Southampton Boldrewood Innovation Campus
University of Southampton
Burgess Road
Southampton
SO16 7QF

Meysam Qadrdan
Institute of Energy
Cardiff University
Queen's Building
The Parade
Cardiff
CF24 3AA

Craig Robson
School of Civil Engineering and Geosciences
Newcastle University
Newcastle upon Tyne
NE1 7RU

Mike Simpson
Environmental Change Institute
Oxford University Centre for the Environment
University of Oxford
South Parks Road
Oxford
OX1 3QY

Anne M. Stringfellow
Engineering and the Environment
University of Southampton
Highfield
Southampton
SO17 1BJ

Scott Thacker
Environmental Change Institute
Oxford University Centre for the Environment
University of Oxford
South Parks Road
Oxford
OX1 3QY

Chris Thoung
Formerly of Cambridge Econometrics
Covent Garden
Cambridge
CB1 2HT

Martino Tran
Environmental Change Institute
Oxford University Centre for the Environment
University of Oxford
South Parks Road
Oxford
OX1 3QY

David A. Turner
Engineering and the Environment
University of Southampton
Highfield
Southampton
SO17 1BJ

Peter Tyler
Department of Land Economy
University of Cambridge
19 Silver Street
Cambridge
CB3 9EP

Geoff V. R. Watson
Engineering and the Environment
University of Southampton
Highfield
Southampton
SO17 1BJ

Jim Watson
UKERC and SPRU, University of Sussex
UKERC
11 Princes Gardens
London
SW7 1NA

SPRU
University of Sussex
Brighton
BN1 9SL

Chengchao Zuo
School of Geography
University of Leeds
Leeds
LS2 9JT

Preface

Given the importance of national infrastructure systems to societies, economies and indeed the sustainability of our planet, it is surprising how little systematic attention they receive. Individual systems, such as energy systems, transport systems and water resources systems, all have rich literatures. However, they have tended to exist in silos, without recognition of the interdependencies between them.

The need for an approach to infrastructure that looks across sectors and at a broad scale (in space and time) is beginning to be recognised around the world. In the U.K. the need for a new approach to national infrastructure was eloquently articulated in the Council for Science and Technology's 2009 report *A National Infrastructure for the 21st Century*, which called for a more strategic approach to national infrastructure provision.

This book responds to that call. We believe that a strategic approach is required because infrastructure investments are long-term commitments with high opportunity costs. The wrong decisions can lock in patterns of development in practically irreversible ways. Each individual investment should be carefully scrutinised, but we also need to understand cumulative effects of the long-term direction of travel. Given the scale of uncertainties surrounding the future and the pace of technological and societal change, any attempt at master-planning would quickly become obsolete. Yet proceeding incrementally with delivery of infrastructure projects, without rigorous analysis of the system that they will form a part of and the context that they may inhabit in the coming decades, could result in unwise choices and missed opportunities.

There have been many calls for development of a national infrastructure strategy that is long-term and cross-sectoral in its perspective. That does not yet exist in the U.K., nor in any other part of the world that we are aware of. We believe that one reason why a strategy, or indeed alternative strategies, for infrastructure provision does not exist is that the analytical approaches, models and tools that would be needed to develop such a strategy have not hitherto existed. Dealing with the complexity, uncertainty and multiple objectives of infrastructure provision has proved to be intractable in quantified analytical terms. This book aims to address that challenge, presenting concepts, a methodology and worked example of possible national infrastructure strategies in a national infrastructure assessment for Great Britain.

A critical choice in any systems problem is the definition of system boundaries. In this book the focus of our case study has been the island of Great Britain, which is occupied by the nations of England, Scotland and Wales. The geographical coherence of dealing with an island is attractive for obvious reasons, though we recognise the flows through the system boundaries, at seaports, airports and the Channel Tunnel, via electricity interconnectors and multiple telecommunications links. No choice of system boundary is ideal, so our future work is developing a multi-scale approach, which seeks explicitly to recognise interdependencies with the island of Ireland (Northern Ireland and the Irish Republic), continental Europe and the globe. Whilst the demonstration is specific to the context and issues of the infrastructure systems in Britain, we believe that the concepts and methodology are transferrable, at least to other advanced economies and to a large extent also to industrialising nations where important decisions about infrastructure development are now being made.

We have a varied readership in mind. Perhaps our most significant contribution is development of concepts for a long-term system-of-systems analysis. These concepts should be understandable to anyone sufficiently interested to have got this far into the book. The example in Britain is very important, because the concepts and methodology are best explained by example. Unless one sees what a national infrastructure assessment actually looks like in practice, it can be hard to see why, on the one hand, it is not easy, but on the other hand, it is not impossible. Having now done a comprehensive model-based national infrastructure assessment once in Britain, and built a flexible database and software architecture which makes it hugely easier to do it again, it can no longer be said that it cannot be done. Analysts and modellers will also be interested in the specifics of the implementation, so we have sought to provide them with enough information so that the technical reader can understand how we have constructed the analysis. Further technical information is provided in supporting online materials at www.itrc.org.uk and in the references cited herein.

The book is based on the work of the U.K. Infrastructure Transitions Research Consortium (ITRC), which has been funded by the U.K. Engineering and Physical Sciences Research Council (EPSRC) with a Programme Grant that began in 2011. The ITRC comprises the Universities of Oxford, Cambridge, Southampton, Newcastle, Cardiff, Leeds and Sussex, with significant additional inputs from Cranfield University, CEH Wallingford and Cambridge Econometrics. The ITRC was brought into being with the aim 'to develop and demonstrate a new generation of simulation models and tools to inform the analysis, planning and design of national infrastructure'. This book does not describe all of the ITRC's work, and further information, including a full list of publications, can be found at www.itrc.org.uk. It is also, inevitably, a snapshot on a research and development journey that is ongoing and actually now intensifying with further development and delivery of tools and databases that facilitate the analysis described in this book. That modelling and visualisation capability is enabling much more flexible generation and scrutiny of the results that are presented in this book.

The ITRC has worked closely with partners in government and industry to frame its questions and inform its analysis. We are particularly grateful to Colin Harris, formerly of Arup, who has been helping to shape ITRC since the first workshop to explore the possibility of a research programme, that we held in Cambridge in March 2009, and has since been the Chair of our Expert Advisory Group. All of the members of the Expert Advisory Group have provided tremendous support to the ITRC on its journey and have been hugely constructive in shaping this book through review and advice. They are Rosemary Albinson from BP, Garry Bowditch from the University of Wollongong, Theresa Brown from the Sandia National Laboratories, Yakov Haimes from the University of Virginia, Stephane Hallegatte from the World Bank, Geoffrey Hewings from the University of Illinois, Margot

Weijnen from Delft University of Technology, Marie-Pierre Whaley of Northumbrian Water and for the first two years of our journey Tim Broyd of Halcrow. We have also been hugely grateful for the support and enthusiasm of Caroline Batchelor, Chris White and Rob Felstead from EPSRC.

The ITRC researchers who have made substantive contributions to the research included in this book are all listed as chapter authors. We must also acknowledge the intellectual and practical contributions from all of the ITRC consortium. This has been a collective intellectual journey, wrought through a fluent discourse and a host of interactions in workshops, meetings and assemblies. It has also required disciplined management, not least to arrive at the point of delivering this manuscript, for which we thank Miriam Mendes and her predecessors in the ITRC programme manager role.

When we embarked on the ITRC journey we knew it was the most complex, ambitious and important research mission we had ever undertaken. This book represents a milestone on that journey. It is certainly not the last word, but does set out concepts, methodology and a very substantial example of how we can begin to think about the future of national infrastructure systems.

PART I

A SYSTEM-OF-SYSTEMS APPROACH

1 Introducing national infrastructure assessment

JIM W. HALL, ROBERT J. NICHOLLS, ADRIAN J. HICKFORD, MARTINO TRAN

1.1 The importance of infrastructure systems

There are many good reasons why we care about infrastructure systems. Infrastructure is a public good from which we all benefit, to a greater or lesser extent. We recognise the quality of infrastructure as being a defining feature of modern society and an advanced economy. Yet industrialised economies have a stock of ageing infrastructure, often in dire need of attention, and emerging economies need to invest heavily to provide much needed essential services. Whilst the delivery of essential services like energy, water and telecommunications is often hardly noticed, the cost certainly is, with affordability issues an important concern for householders and industries. We all notice when things go wrong, as they can during natural catastrophes, accidents or deliberate attacks, and the consequences of failure can cascade through interdependent networks representing a systemic risk at national and continental scales. Furthermore, a changing climate means that infrastructure systems are likely to be subject to more severe and uncertain future threats.

These systems have significant inertia and infrastructure investments can lock societies into patterns of development, with infrastructure systems, land-use and the natural environment co-evolving, often in practically irreversible ways. Energy and transport systems, for instance, account for nearly half of global carbon emissions, so limiting emissions to avoid dangerous climate change requires a transformation of our current infrastructure systems. The choices we make now will determine how hard it will be to limit carbon emissions in the future. For all these reasons, the choices we make about infrastructure are important and usually difficult decisions. They can involve large commitments of capital, upon which there are many competing interests and opportunity costs. Additionally, because infrastructure involves building in people's 'back yards', decisions can become intensely political with losers as well as winners.

1.2 Choosing the future of national infrastructure

Infrastructure systems in advanced economies worldwide have been created over many generations. This long legacy and their central role of underpinning economic activity, resource consumption, production processes and pollution makes infrastructure of vital importance for transitioning towards a sustainable economy through the twenty-first century. However, the management arrangements and planning tools for the different infrastructure systems have been developed in isolation and are not fit for analysing the long-term implications and impacts of infrastructure developments across multiple sectors and broad national scales.

In most advanced economies, there are established processes for making resource allocation decisions and dealing with conflicts. In particular, cost–benefit appraisal provides an explicit mechanism for weighing up benefits and impacts, but tends to exclude system-wide benefits. Yet if we take a broader perspective, through time and at a broad spatial scale, we observe systematic patterns of development, by which infrastructure projects combine to form networks and systems. To some extent these emergent networks are designed – the motorway network we have in Britain is remarkably similar to the one that was sketched out in 1946, twelve years before the first section of motorway was opened. In other cases, infrastructure systems have emerged through incremental development and interaction with patterns of human settlement and demand. An incremental approach, it could be argued, may be more adaptable and less vulnerable to future uncertainties. On the other hand, infrastructure yields its benefits to society through operation as a *system*, through the existence of networks that distribute resources and provide interconnectivity. Looking back we observe instances of the wrong choices being made, of resources being inefficiently allocated, and of negative impacts upon society and the environment. This book seeks to recognise and hence avoid those situations. It develops new concepts and methodologies to enable the construction and evaluation of long-term *strategies* for infrastructure provision.

It is certainly not appropriate for technical specialists or analysts like ourselves to propose *the* strategy for infrastructure systems. Rather, we propose credible alternatives which to some extent span the space of possibilities. Nor do we suppose that the future is fixed – far from it, we seek to identify the full range of factors in society, the economy and environment which influence infrastructure performance and may significantly change in the future. Nor do we suppose that any given strategy is cast in stone – we expect strategies to adapt as circumstances change and as we learn. However, we do need to be able to explore the implications of future uncertainties, be they in demand for infrastructure services, the policy context or the climate, all of which influence infrastructure systems performance. We therefore need to understand how the choices being made now open up, or exclude, options for systems development in the future. That requires a comprehensive systems approach, in order to construct strategies for infrastructure provision, and evaluate these strategies under a range of possible future conditions. This book presents such an approach.

1.3 Defining infrastructure systems

Before proceeding further, we will define what we mean by infrastructure systems. The analysis of infrastructure at a system level and across sector boundaries is complicated by the absence of a single comprehensive, functional and practical definition of infrastructure (Buhr, 2003; Fourie, 2006; Baldwin and Dixon, 2008; Torrisi, 2009). One approach to defining infrastructure is through enumeration of the classes of assets that are contained within the system. In this book we deal with:

- energy systems, including electricity, gas and liquid fuel networks;
- transport systems, including road, rail, ports and airports;
- water supply;
- wastewater and solid waste collection, treatment and disposal;
- digital communications networks and data storage/processing infrastructure.

This could be considered to be a fairly narrow scope. Besides energy, transport, water and communications, some studies also include emergency services, healthcare, financial services, government, food and the built environment in the list of infrastructures (Cabinet Office, 2010). Perhaps most notably, we do not deal with risk reduction infrastructures such as flood defence, coastal protection and fire services, neither is there any inclusion of social infrastructure such as schools, medical care and banking, nor do we include the built environment or natural (green) infrastructure. While there are good arguments to look at those systems as infrastructure in certain contexts (Torrisi, 2009), the focus of this book is limited to the sectors listed above because they all belong to a class of infrastructure services currently provided through networks, they belong to a list of essential human and societal needs (Buhr, 2003), and they are prone to market failure and the necessity of government intervention since they show characteristics of public goods (Baldwin and Dixon, 2008).

However, just approaching the definition of infrastructure from the perspective of classes of assets is unsatisfactory, because it focuses upon the physical assets rather than the services that they are intended to provide. Focusing upon the assets can overlook the possibility of substitution of services between networks; for example, heating services can be provided by electricity, gas or district heat networks. Thus, in this book we focus upon *infrastructure services*, acting as an enabler and catalyst for economic activity, and contributor to human well-being. Recasting the list of assets that we began with above, we deal with:

- energy services via the provision of electricity, solid, liquid or gaseous fuels to power machinery (including transport), and for heating and lighting;
- transportation services that provide mobility for passengers and freight between locations;
- water supply services, that provide water of given quality to households and industries;
- waste removal, treatment and recovery services, that remove unwanted solid waste and wastewater and, increasingly, recover resources from waste streams;
- communication, exchange and computation of information services.

Infrastructure services are provided by the operation of and complex interaction between human, economic and technical systems and the environment. Within these complex processes we identify:

(i) *providers* of infrastructure services (in the public and private sectors), who commission and operate physical facilities and accompanying human systems (these are, collectively, 'infrastructure systems');

(ii) *consumers* of infrastructure services (businesses, government and households), who demand services in order to go about their businesses or to enhance their well-being;

(iii) *externalities* to the consumer–service provider relationship, which includes people and the environment subject to the various positive and negative (e.g. pollution) side-effects of infrastructure services.

Examining the ways in which infrastructure services are provided we identify processes of conversion, storage and transmission as being inherent in the provision of infrastructure services. However, the distinction between these categories is not precise, as, for example, gas and water pipelines are a means of storage as well as transmission.

Summarising these observations leads to the following definition:

> An **infrastructure service** is the *provision* of an option for an *activity* by operating physical facilities and accompanying human systems to convert, store and transmit resources (physical and virtual).

Depending on the specific infrastructure service, the resources under consideration include different types of energy carriers, passengers, freight, water, waste products and information (data).

From this understanding of infrastructure services and the processes by which they are provided we develop a definition of infrastructure systems, recognising that the human, communications and mechanical systems that control the operation of physical infrastructure facilities are essential elements of the system:

> An **infrastructure system** is the collection and interconnection of all physical facilities and human systems that are operated in a coordinated way to provide a particular *infrastructure service*.

These definitions lead to the concept of **National Infrastructure** being defined as a *system-of-systems* of infrastructures. These systems are integral to the proper functioning of modern economies, but also face a number of serious challenges.

1.4 Challenges facing national infrastructure systems in advanced economies

In many respects the inhabitants of advanced economies (by which we are broadly thinking of countries that industrialised in the nineteenth and twentieth centuries) are in a fortunate position with respect to their infrastructure systems. They benefit from a huge legacy of infrastructure investment which in many cases, like the water supply reservoirs in the English, Welsh and Scottish hills, provides the basis for the infrastructure services upon which we now depend.

Whilst almost all of the projected population growth in the world is expected to take place in developing countries, advanced economies face challenges of ageing population, and immigration will mean that the population of some countries continues to increase. Because it is very hard to predict how much migration will take place, there are severe uncertainties surrounding any projection of population change. In the U.K., for example, the Office for National Statistics (ONS) publishes a range of long-term population projections in which the U.K. population in 2014 (64 million) is predicted to increase in 2050 to between 66 million and 85 million (ONS, 2012). Change on this scale has significant implications in terms of demand for infrastructure services. Per capita demand is modified by a range of economic and social factors. Perhaps the most important of these is the rate and structure of economic growth, with which infrastructure has a symbiotic relationship. We consider this further in the next section.

The combined effects of ageing infrastructure, demographic and economic change mean that there is a substantial requirement for global investment in infrastructure networks over the coming decades. In 2012, the OECD estimated that $53 trillion of investment, equivalent to an annual 2.5% of global GDP, will be needed to meet demand for infrastructure services over the coming decades. Over $11 trillion of that will be required for ports, airports and key rail routes alone (OECD, 2012). Standard & Poor's and McKinsey estimate that $57 trillion, or $3.2 trillion a year, will be needed to finance infrastructure development around the world over the next fifteen years (McKinsey & Co, 2013).

The fragility of infrastructure networks has been demonstrated through events like the Northeast blackout in the USA and Canada in 2003 and by major rail transport disruptions during prolonged flooding in England over the winter of 2013/2014. Both of these events reflected the ageing nature of infrastructure assets and the potential for major disruption. Furthermore, infrastructure networks have become increasingly interdependent, providing the potential for knock-on effects causing major economic and societal disruption. This shift has important implications for the resilience of infrastructure sectors, where, for example, a power failure in a major electricity exchange can result in the temporary loss of broadband service for hundreds of thousands of households and businesses (BBC, 2011). Even small, temporary failures can have significant effects on economic productivity. In the long term, these risks intensify as systems become larger and increasingly interdependent. The combined effect of ageing infrastructure, growing demand (nearing capacity limits) from social and economic pressures, interconnectivity and complexity leads to systematic weakening of the resilience of infrastructure systems (CST, 2009). These complex infrastructure networks are set to face even more severe hazards in the future, due to a changing climate that is expected to cause more severe and/or frequent extreme events (IPCC, 2013).

However, interdependence also provides opportunities when looking to the future. Many promising technological innovations, for example, in smart grids or electric vehicles, rely upon close integration between different infrastructure networks. Whilst these interdependencies potentially provide transformative opportunities, they also introduce additional layers of complexity which need to be taken into account in a consistent way in infrastructure planning and design. The framework we present in this book seeks to address that need by developing a consistent cross-sectoral approach to infrastructure assessment.

1.5 Infrastructure systems and economic growth

Infrastructure systems are integral to the proper functioning of all modern economies. Investments to provide reliable and resilient national infrastructure facilitate economic competiveness and positively impact growth (Gramlich, 1994; Lau and Sin, 1997; Sanchez-Robles, 1998; Égert et al., 2009). Infrastructure is often cited as a key ingredient for a nation's economic competiveness. The World Economic Forum (WEF), for example, lists infrastructure as the second 'pillar' in its Global Competitiveness Index, a measure of national competitiveness (WEF, 2013).

However, the link between infrastructure availability, economic growth and productivity is still the subject of much uncertainty and debate (Kessides, 1993; O'Fallon, 2003; Prud'Homme, 2004; Bourguignon and Pleskovic, 2005; Straub, 2011). Such debate exists because the shared links between infrastructure and the economy are multiple, complex and multi-directional. The economic activity enabled by infrastructure does not, in general, take place where the infrastructure is located, so the benefits can be difficult to determine. Rather, infrastructure connects resources, enabling economic activities elsewhere, for example, transmission and distribution of electricity to factories or supply of water to households.

A major shortcoming of existing literature examining the role infrastructure systems have within the economy is that its main focus has been on aggregate investment indicators and therefore ignores the heterogeneity of different infrastructure types or sectors (Serven, 1996). For example, investment in infrastructure typically enters analyses as gross public investment in fixed capital structures, machinery and equipment (Aschauer, 1989; Voss, 2002). However, large infrastructure systems such as energy and transport networks have a very different economic function. A further shortcoming is that most analyses do not adequately account for the timing of investment, which can lead to very different macroeconomic impacts. For example, public investment during a downturn may prevent a liquidity trap thus potentially saving billions. Yet another shortcoming is that most studies only consider the supply of fixed capital investment and do not consider the financial flows and transactions that occur over time as a result of infrastructure investment.

Because infrastructure differs quite fundamentally from many other forms of economic capital, the benefits of infrastructure are varied and complex, making them difficult to value. Nevertheless, it is clear that the value of infrastructure goes well beyond a simple summation of the material and labour inputs involved in their construction. The diversity of benefits (many of which are difficult to quantify) includes: increased productivity, improved communication, efficient transport systems, access to affordable energy and avoidance of natural catastrophes and systemic disruptions (Buhr, 2003; Fourie, 2006; Torrisi, 2009). Infrastructure systems have also had significant impact on non-economic indicators such as human health and well-being, and the natural environment. For example, the improvements in bathing water standards in the European Union have been a consequence of major investments in wastewater treatment infrastructure (Defra, 2012).

There is significant need for private investment into national infrastructure around the world. In order to attract these investments in an increasingly competitive global economy, it is essential to have coherent long-term goals for infrastructure and a policy and regulatory framework sufficiently stable for infrastructure providers to take investment and operational decisions consistent with these goals. The system-of-systems framework that we propose in this book is intended to demonstrate how long-term goals may be achieved and help provide policy coherence across infrastructure and between different policy objectives.

1.6 Responding to the challenges: a system-of-systems approach

The challenges we have set out are daunting: an ageing infrastructure stock and mounting investment requirements; growing demand and expectations for quality and reliability of infrastructure services; ambitious targets for carbon emissions reduction; threats from climate change and human-induced hazards; an uncertain policy landscape – all in the context of increasing system complexity and interdependence.

The motivation behind this book is that we can start to address all of these issues through the development of a system-of-systems approach for thinking about, modelling and assessing national infrastructure. Such an approach will help to provide:

- *consistency across sectors*, though the adoption of common sets of planning assumptions and metrics for evaluation of system performance;
- *robustness to uncertainty*, through quantification of the most important uncertain factors that influence infrastructure performance in the future and testing infrastructure strategies with respect to those factors;
- *a broad perspective on infrastructure performance*, through the development of multiple performance metrics that apply across infrastructure sectors;
- *identification of vulnerabilities and management of risks*, through the development of a new methodology for assessment of risks to infrastructure networks at a national scale, which targets vulnerabilities and helps to make the case for risk reduction;
- *long-term infrastructure strategies*, by looking over an extended timescale at alternative approaches to infrastructure provision;
- *reformed governance arrangements*, through analysis of current arrangements for governance of national infrastructure and exploring the challenges and opportunities provided by more integrative approaches to regulation.

To achieve these objectives, this book is structured in three parts.

Part I: a system-of-systems approach presents the conceptual framework for thinking about, modelling and assessing national infrastructure systems. The centrepiece is in Chapter 2, which sets out our approach to analysing systems and devising infrastructure strategies. Understanding national-scale infrastructure as a 'system-of-systems' of inter-connected networks, Chapter 2 recasts long-term infrastructure planning as a problem of finding strategies that deliver robust performance of infrastructure services across a wide range of future socio-economic and environmental conditions. The process for devising long-term strategies for national infrastructure starts with overarching scenarios to develop internally consistent assumptions about the willingness to invest from public and private sources, the commitment to environmental targets or the willingness to use demand management for infrastructure services.

Part II: analysing national infrastructure applies our system-of-systems approach using the island of Great Britain, comprising England, Scotland and Wales, (henceforth Britain) as a case study. We begin, in Chapter 3, by exploring the factors that will influence the future demand for infrastructure services in Britain, notably those relating to population and economic growth, pushing existing projections further into the future, to the end of the twenty-first century and disaggregating to a spatial scale that is of direct relevance to infrastructure. In Chapters 4–9 we then examine each infrastructure sector in turn (energy, transport, water supply, wastewater, solid waste and digital communications), exploring the potential future directions for infrastructure provision and appraising the options. These chapters explore the potential demand for infrastructure sectors, using the demographic and economic projections presented in Chapter 3, and assess the performance of alternative investment strategies for delivering infrastructure services in the future.

Part III: integrative perspectives for the future draws these analyses of infrastructure systems together into our system-of-systems assessment. Chapter 10 develops a set of common metrics, combining indicators for costs (including capital and operational costs), greenhouse gas emissions, security of supply and total service provision to assess total system performance. In Chapter 11, we explore particularly important interdependencies between transport and energy, and between energy and water. Chapter 12 provides an overview of our national framework for risk analysis of interdependent infrastructure networks. Chapter 13 presents the database, modelling and visualisation architecture that has been used to enable the assessment. Chapter 14 explores the governance arrangements for regulation, procurement, finance and operation of infrastructure systems, with a particular emphasis on the governance of interdependencies. Finally, Chapter 15 reflects upon the methodologies and results that have been presented and how they may be taken forwards in the future.

References

Aschauer, D. A. (1989). "Is public expenditure productive?" *Journal of Monetary Economics* 23(2): 177–200.

Baldwin, J. R. and J. Dixon (2008). "Infrastructure capital: what is it? Where is it? How much of it is there?" *The Canadian Productivity Review* No. 16.

BBC (2011). "BT suffers huge broadband failure." Retrieved from www.bbc.co.uk/news/technology-15154020.

Bourguignon, F. C. and B. Pleskovic (2005). Annual World Bank conference on development economics 2005: lessons of experience. *Annual World Bank conference on development economics 2005*, Washington, D.C., USA, World Bank Publications.

Buhr, W. (2003). What is infrastructure? University of Siegen, Economics Discussion Papers.

Cabinet Office (2010). *Sector resilience plan for critical infrastructure 2010*. London, UK, Cabinet Office.

CST (2009). A national infrastructure for the 21st century. London, UK, Council for Science and Technology.

Defra (2012). *Waste water treatment in the United Kingdom – 2012*. London, UK, Department for Environment, Food and Rural Affairs.

Égert, B., T. Kozluk and D. Sutherland (2009). Infrastructure and growth: empirical evidence. OECD Economics Department Working Paper No. 685.

Fourie, J. (2006). "Economic infrastructure: a review of definitions, theory and empirics." *South African Journal of Economics* 74(3): 530–556.

Gramlich, E. M. (1994). "Infrastructure investment: a review essay." *Journal of Economic Literature* 32(3): 1176–1196.

IPCC (2013). Climate change 2013: the physical science basis. *Working Group I contribution to the Intergovernmental Panel on Climate Change fifth assessment report (AR5).* Geneva, Switzerland, IPCC.

Kessides, C. (1993). The contributions of infrastructure to economic development: a review of experience and policy implications. World Bank – Discussion Papers.

Lau, S.-H. P. and C.-Y. Sin (1997). "Public infrastructure and economic growth: time-series properties and evidence." *Economic Record* 73(221): 125–135.

McKinsey & Co (2013). *Infrastructure productivity: how to save $1 trillion a year.* McKinsey.

O'Fallon, C. (2003). *Linkages between infrastructure and economic growth.* Prepared for Ministry of Economic Development. New Zealand, Pinnacle Research.

OECD (2012). *Strategic transport infrastructure needs to 2030.* Paris, France, Organisation for Economic Co-operation and Development.

ONS (2012). *National population projections, 2012-based.* Office for National Statistics.

Prud'Homme, R. (2004). *Infrastructure and development,* World Bank.

Sanchez-Robles, B. (1998). "Infrastructure investment and growth: some empirical evidence." *Contemporary Economic Policy* 16(1): 98–108.

Serven, L. (1996). *Does public capital crowd out private capital? Evidence from India.* World Bank Publications.

Straub, S. (2011). "Infrastructure and development: a critical appraisal of the macro-level literature." *The Journal of Development Studies* 47(5): 683–708.

Torrisi, G. (2009). "Public infrastructure: definition, classification and measurement issues." *Economics, Management, and Financial Markets* (3): 100–124.

Voss, G. M. (2002). "Public and private investment in the United States and Canada." *Economic Modelling* 19(4): 641–664.

WEF (2013). The global competitiveness report 2013–2014. *World Economic Forum.* Geneva, Switzerland.

2 A framework for analysing the long-term performance of interdependent infrastructure systems

JIM W. HALL, ALEXANDER OTTO, ADRIAN J. HICKFORD, ROBERT J. NICHOLLS, MARTINO TRAN

2.1 The nature of the infrastructure assessment problem

There are several important aspects of infrastructure services that we wish to understand when thinking about long-term planning and investment decisions. One very important set of questions relates to the vulnerability of infrastructure networks and the risks of infrastructure failure. This is the question that we address in Chapter 12. Analysing the risks of infrastructure failure involves scrutiny of the extreme states of the system: extreme or unusual loads that are placed on the system, or points of particular vulnerability. It focuses on the possible modes of large-scale (catastrophic) failure, for example, due to risk from natural or human-made hazards, the propagation of failure through (interdependent) infrastructure networks and the consequence of failure. Another set of questions relates to what one might broadly refer to as the 'business models' for infrastructure delivery: issues of governance, finance and cost recovery. The particular challenge of governing complex interdependent infrastructure systems is dealt with in Chapter 14.

Here, in all of Part II of this book, and in Chapters 10 and 11, we are concerned with the question of how we develop and analyse long-term investment strategies for infrastructure, given uncertain knowledge about future demands for infrastructure services and the future performance of infrastructure systems. We are interested in investment to provide more infrastructure capacity (supply) and also in instruments that seek to manage demand for infrastructure services (demand). We consider infrastructure performance to be related to the extent to which the infrastructure system can meet customer demands for service in a reliable way. In other words, we are addressing the question 'How can infrastructure capacity and demand be balanced in an uncertain future?'

More specifically, we are interested in comparing alternative *strategies* for providing infrastructure services in the future, taking account of the existing legacy of infrastructure stock. A strategy is a rationale for how infrastructure services may be provided. It is not a fixed plan, as we recognise that the future performance of infrastructure depends on many unpredictable factors and contingencies, to which the strategy should be adaptable. However, it is a well-defined approach that is intended to achieve certain outcomes.

There are genuine questions involved in choosing between different strategies: How much are we prepared to invest; where and when? What steps can be taken to manage demand for infrastructure services? How committed are we to reducing the environmental impacts of legacy infrastructure and of the infrastructures that will be built in the future? Modelling and analysis, which is what this book is largely about, cannot answer all of these questions. These questions imply value judgements that need to be made as part of the democratic process. However, the framework analysis that we now present is intended to help construct and analyse alternative strategies, in order to provide evidence of how they might perform in a range of different future conditions. We believe that weak strategic thinking about infrastructure has persisted because the tools have not existed to develop and analyse alternative infrastructure strategies, in the long term under a wide range of possible futures.

2.2 Overview of the assessment framework

We approach the problem of constructing, analysing and choosing between alternative strategies for infrastructure provision as being a problem of decision-making under uncertainty (Otto et al., 2014). In conventional decision analysis, a decision-maker chooses between a series of alternative *acts*, which yield some *utility* which depends upon what act is chosen, and what future *state of nature* happens to materialise. We adopt the same framework here, modifying the terminology to match our particular context. There are three important aspects of the assessment framework: (i) *Strategies* are sets of choices, or rationales for taking choices, about the provision of infrastructure services. Strategies will typically involve investment in the provision of infrastructure combined with instruments to manage infrastructure demand. They extend across infrastructure sectors (energy, transport, water, waste and ICT), so consist of strategic options from each of the sectors; (ii) *Scenarios* describe the main factors that are outside the control of decision-makers, which influence the future performance of strategies. They include changes in population, the economy, global energy prices and the climate, and are outside the scope of our sector modelling systems, so are said to be 'exogenous'. Future changes in these phenomena are uncertain, so we test a wide range of possible scenarios to analyse the sensitivity of future infrastructure performance to those scenarios; and (iii) *System models* are used to analyse the performance of strategies in the context of given scenarios. The system models take as inputs the current state of the infrastructure system and the demands placed upon it, strategies for each infrastructure sector and scenarios of exogenous changes. The model outputs are projections of infrastructure performance, given those input conditions.

This arrangement enables us to answer questions about the future performance of alternative infrastructure strategies under a range of possible futures, including: When and where are we anticipating 'breaking-points' of the current infrastructure system under changing external conditions (e.g. demographic changes)? Which infrastructure strategies provide robust infrastructure service performance across a wide range of possible future conditions? Can we identify potential multi-sector infrastructure transitions (i.e. qualitative changes in the system state), either induced by changes in demand or by proactive planning interventions?

Our approach seeks to provide evidence to inform these and other 'what if' questions, from the perspective of a decision-maker who is interested in choosing between alternative strategies. There are many people who might be interested in this evidence: governments, regulators, infrastructure owners/operators, customers/passengers and so on. Our framework is neutral as to who the decision-maker actually is. We are simply seeking to provide evidence to inform strategic choices. One might think of a grand infrastructure 'master planner', but such a construct would be flawed because the notion of master-planning neglects the profound uncertainties that face provision of infrastructure services in the future, which means that strategies will need to be adaptive and robust to a wide range of future conditions. Furthermore, in practice in the context of the complex governance arrangements of infrastructure in Britain and other advanced economies (see Chapter 14) there is no single agent or agency that makes strategic choices within and across infrastructure sectors. There are multiple agents making choices with a variety of different objectives. Nonetheless, we believe that a modelling and decision analysis framework, broadly constructed around a view of the public interest, can provide a normative perspective and common platform for developing a shared understanding about future challenges and trade-offs, and for testing alternative strategies for national infrastructure provision.

In practice, the partitioning between factors under the control of decision-makers (strategies), and external changes over which they have no control (scenarios) is not clear-cut. Behavioural change and technological change in particular lie on the boundary between controllable and uncontrollable factors. We broadly assume that choices about the uptake of technology lie within the strategic choices available to decision-makers, whilst acknowledging that technological development, more broadly, is a global process, and is therefore out of the control of a national decision-maker. Nonetheless, the large-scale nation-wide application of a specific technology (e.g. shale-gas extraction, carbon capture and storage, desalination and autonomous vehicles) is within the authority of national legislation and regulation, whereas some external technology-related factors, such as global energy prices, are considered to be exogenous.

Similarly, there are many processes of behavioural change that are largely outside the control of decision-makers. For example, we assume demographic changes like household size or internal migration to be outside the control of infrastructure decision-makers, even though these issues will profoundly change the demand for future infrastructure services. These are exogenous variables in our analysis. On the other hand, instruments that are intended to directly influence the demand for infrastructure services, such as road-user charging, are taken as being part of the infrastructure strategy and have an assumed effectiveness.

We also recognise that there are feedbacks between the provision of infrastructure services and the factors that are taken as being exogenous to the modelling framework. The choices that are made about infrastructure will influence the economy, demography and future demand for infrastructure services. At a global level, the choices made about energy infrastructure will influence global energy prices and the climate. In our assessment framework we choose to neglect these feedbacks. Doing otherwise would involve constructing a fully dynamic coupled model of infrastructure systems and the economy, which adds considerable complexity and could detract from the clarity of insights provided by the current analysis process.

Figure 2.1 provides an overview of the analysis process that we have just outlined and present in more detail in the remainder of this chapter. The analysis is designed to provide

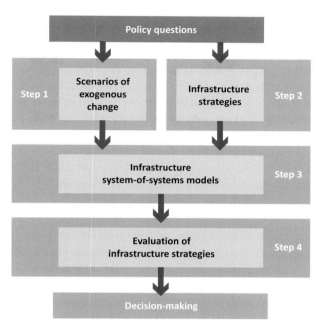

Figure 2.1   Overview of the analysis framework

evidence to inform particular questions about the future performance of national infrastructure under a range of uncertain conditions, as follows: Step 1 involves the construction of
quantified scenarios of change in the exogenous factors that influence future infrastructure
system performance; Step 2 involves the formulation of strategies for national infrastructure
provision; Step 3 uses system models to simulate those strategies and generate metrics of
system performance; and Step 4 involves evaluation of infrastructure strategies to inform
decision-making.

Step 1: Scenario generation

The first step of the modelling framework is to develop scenarios of future socio-economic
and environmental change. We capture uncertainties arising from these processes by including a wide range of different future demographic, economic and climatic scenarios, which
determine demand for infrastructure services, as well as a comprehensive set of climate
change projections, primarily impacting the water sector. We define a *scenario* of future
conditions as a complete set (time series) of external parameters defining the boundary
conditions in which infrastructure systems operate. Four components of the environment
in which infrastructure systems operate are taken to be completely exogenous to the analysis: (i) *Demographic change* which largely affects demand for infrastructure services. See
Chapter 3 for high resolution demographic projections for Britain; (ii) *Economic change*
which affects the population's ability to utilise infrastructure services, where higher GDP
is likely to result in higher demand for infrastructure services, as people tend to consume more energy and travel further and more often. Economic growth may also affect

governmental and infrastructure owners' ability to invest in infrastructure systems. Chapter 3 also provides multi-sectoral economic projections for Britain; (iii) *Global fossil-fuel costs* affect operating costs and transport costs in particular. National policy measures such as carbon taxes may affect these costs, but these are generally assumed to be exogenous to the models; and (iv) *Climate change* will affect the availability of water resources and the potential for extreme loads on infrastructure systems. Scenarios for future projected U.K. climate change (seasonal temperatures, rainfall and sea-level rise) are discussed in Chapter 6 focusing on water supply systems.

In total we are able to generate eight demographic scenarios, eight economic scenarios, three sets of global energy costs (oil, gas and coal), and eleven climate change scenarios. This potentially provides a total of 2,112 combinations, which is a large but not infeasible number of conditions to test. In practice, however, many of these combinations result in very similar results, so we have been able to filter the number of scenarios to a more manageable number. Moreover, we recognise that certain scenario combinations are more likely than others, so, for example, population growth and low energy prices are likely to be accompanied by high economic growth. We therefore primarily report three scenario combinations, as presented in Table 3.2.

Step 2: Strategy generation

The next step in the modelling framework is to develop alternative long-term strategies for infrastructure provision. These need to be distinctive and recognisable as national policy alternatives. They also need to be resolved in detail in each sector, so that they can then be tested in our simulation models. This leads to our approach to strategy generation that combines a top–down approach of exploring the main policy choices facing decision-makers, with a bottom–up approach to testing the options for infrastructure provision within different sectors (Figure 2.2). The bottom–up perspective ensures that strategies are described in sufficient detail for it to then be possible to test them in our infrastructure system models.

From a top–down perspective, we identify three dimensions of high-level policy commitment associated with infrastructure provision (Figure 2.2): the overall willingness to invest in new infrastructure assets, the environmental ambition (in terms of greenhouse gas emissions and other environmental impacts of infrastructure operation), and the commitment to demand management (e.g. strong price signals, promotion of behavioural and technological change). These, we argue, represent fundamental high-level political commitments, which define the overall direction of infrastructure provision. Whilst there are many different levels of commitment on these three main policy axes, in practice there is a limit to the extent to which it is possible to distinguish between them. We therefore limit the number of possible combinations, for example, by restricting the environmental and investment policy dimensions to two levels each: an Ambitious versus Neutral environmental focus and High versus Low investment levels. An ambitious environmental focus would entail building on the current policies aimed at decarbonising the energy and transport system (through reducing energy consumption, improving energy efficiency, emphasis on renewable energy and sequestration of carbon into carbon sinks). Less ambition may entail more focus on

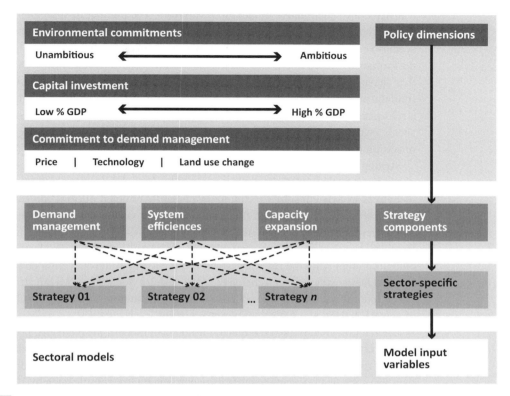

Figure 2.2   The top–down strategy generation process, from high-level policy dimensions (environmental ambition, capital investment level, demand-related policy focus), through components of infrastructure strategies (demand management, system efficiencies, capacity expansion), to implementation of these strategies as inputs in the system models

future development, resilience and growth regardless of environmental impacts. An additional strategic dimension is the rate at which investment, environmental improvements or demand management measures are implemented; the changes can be implemented urgently, or the approach can be steadier.

From a bottom–up perspective, we have conducted a review of the different strategic options that exist in each infrastructure sector: energy, transport, water, waste and ICT (Hickford et al., 2015). These include the following sub-strategy components that are used to derive sector-specific strategies: (i) *Demand management* instruments which seek to modify the scale and timing of demand for infrastructure services, for example via prices and regulation; (ii) *System efficiency* measures which change the technology, the modes and practices of operation and management of infrastructure systems. This may be achieved by using more efficient plant or utilising infrastructure capacity (e.g. road-space) more efficiently; and (iii) *Capacity expansion* changes, which modify the number, capacity and connectivity of physical infrastructure assets.

A qualitative description of each of the sector strategies is based on choosing the extent and timing of using specific single-sector policy dimensions. The strategies are then translated into quantitative model input variables. The number of strategies for each sector has been deliberately limited to between four and eight, which can be combined with the scenario combinations discussed above to provide extensive sensitivity analysis on strategy performance. While this is more than traditional scenario studies, it is still much smaller than the number of conceivable strategies, but for computational feasibility, it is desirable to limit the number of combinations of strategies. The emphasis has been put upon developing a reasonably concise set of distinct strategies spanning the range of possibilities for each sector, which broadly influence (i) demand management and (ii) capacity utilisation and provision discussed next.

Strategies for demand management: demand for infrastructure services can be influenced by changing consumer behaviour, either through education, pricing structures and incentives or new consumer technologies. Specific targeted information can affect overall use of infrastructure services: for example, energy saving schemes can reduce domestic energy requirements; societal and environmental pressure can promote modal shift away from private vehicles to more sustainable modes, as well as influencing change in waste recycling and other resource recovery; local levels of grey water recycling could be increased by introducing water usage schemes (e.g. metering). Such measures are attractive in terms of the likelihood of lower overall costs of implementation, especially if subsequent levels of societal behaviour change are not insignificant.

Behavioural change can be effected through a number of mechanisms: pricing measures, such as taxation and financial incentive policies, could have greater influence on behaviour change; road user charging measures, or other regulations or taxes designed to reduce fossil fuel use, such as preferential tariffs for electric vehicles, could change travel behaviour as well as the overall make-up of the vehicle fleet; tax incentives could encourage investment in new technologies; and volume-based tariffs for water consumption or waste generation could reduce the requirements for water and waste treatment. Demand can also be influenced by technological changes which alter the way in which a system is used: for example, increased energy efficiency in domestic appliances, alongside the national roll-out of smart meters is likely to influence energy demand, and increased use of ICT could result in changes in travel habits.

Strategies enabling change in capacity utilisation and provision: technological advances and different approaches to capacity utilisation can affect the overall efficiency of an infrastructure system. For example, efficiencies in road transport can be achieved through increased fuel economy, optimised route planning or vehicle-to-vehicle (V2V) and vehicle-to-grid (V2G) interactions. Losses in energy and water distribution could be reduced through new technologies enabling real-time monitoring of the distribution systems. Structural changes to the infrastructure system itself will be achieved through new-build (such as new rail and road links, power stations, reservoirs, water treatment works or recycling provision), and adaptation of existing infrastructure by replacing out-dated infrastructure with modern materials or incorporating new technologies. The transition to renewable energy generation is one example of how the physical infrastructure required for distribution of energy may remain relatively unchanged while the landscape of options for energy generation changes significantly. Additionally, these sector strategies can be combined to create contrasting cross-sectoral infrastructure strategies that integrate strategies from each sector, discussed next.

Developing long-term strategies for national infrastructure systems-of-systems: in order to assess the impact across all sectors of different approaches to infrastructure provision, cross-sectoral infrastructure strategies can be developed by combining strategies from different sectors with similar underlying assumptions. In order to test a wide range of possibilities, whilst still dealing with a manageable number, we have developed four contrasting cross-sectoral infrastructure strategies, as follows:

Minimum intervention (MI): takes a general approach of minimal intervention, reflecting historical levels of investment, continued maintenance and incremental change in the performance of the current system. There is no long-term vision to reduce future demand or implement more stringent commitments to environmental policies; instead, the focus is on short-term incremental improvements at a sector level, and thus fails to account for sectoral interdependencies.

Capacity expansion (CE): focuses on planning for the long term by increasing investment in infrastructure capacity. Priority is given to the expansion of physical capacity to alleviate pinch-points and bottlenecks. There is no long-term vision to reduce future demand. However, provision of new capacity will provide opportunities for introduction of some more environmentally benign technologies.

System efficiency (SE): focuses on deploying the full range of technological and policy interventions to optimise the performance and efficiency of the current system, targeting both supply and demand. This implies targeted investments to increase capacity at severe bottlenecks in the shorter term, but the medium- to long-term vision is to invest heavily in maximising the throughput of the current system, without high investments in capacity expansion. Improvements in the efficiency of plant and vehicles enable significant increases in the throughput of the system with relatively modest capacity investments. There may be a limit to the extent to which these efficiency improvements can address expected demand pressures from population and the economy.

System restructuring (SR): focuses on fundamentally restructuring and redesigning the current mode of infrastructure service provision, deploying a combination of targeted centralisation and decentralisation approaches. This will require long-term investments aiming to utilise a wide range of technological innovation, incorporating policy incentives and integrated planning and design.

Subsequent assessment of these infrastructure strategies can be used to compare the total systems performance. The strategies are designed to be contrasting and distinct, although there will inevitably be some overlap between them. Figure 2.3 illustrates how each of these strategies are positioned with respect to the high-level policy dimensions illustrated in Figure 2.2. Other sectoral and cross-sectoral strategies can be developed as needed, including strategies developed as a response to policy needs, and learning from the initial analysis in an iterative manner. Chapter 10 synthesises the results of our quantified assessment of these national infrastructure strategies.

Step 3: Infrastructure system-of-systems models

The performance of infrastructure sectors is analysed by a suite of coupled system models, which estimate capacity of and demand for infrastructure services. The models that have been developed to assess Britain's national infrastructure system are described in Part II

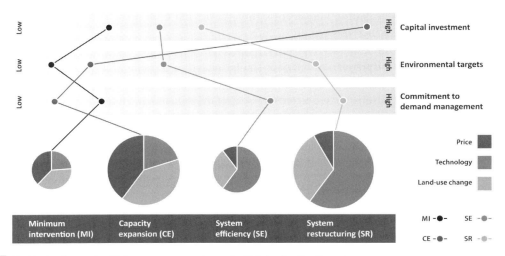

Figure 2.3 Positioning four national infrastructure strategies on the high-level policy dimensions

of this book. The coupling allows an iterative approach where the input variables for one sector model are dependent on the outputs from another sector. The network representations of the systems of physical infrastructure assets used to derive capacities are common to all the sector models, while the method used to derive demand for infrastructure services is sector-specific. The system-of-systems perspective enables us to address the strong interconnections between the infrastructure sectors, originating from a series of cross-sector demands and feedbacks, most notably:

- **Energy–transport links:** electric vehicles, smart grids and the structure of the power grid need to be co-designed and impact upon one another via a range of feedbacks (prices, demand).
- **Energy–water, energy–waste links:** the quest for energy efficiency and decarbonisation links energy supply, wastewater and solid waste treatment, via potential schemes of energy recovery, circular economy or co-emissions. Meanwhile, energy generation requires water resources for cooling and hydropower.
- **ICT–infrastructure links:** the rise of integrated ICT systems changes demand patterns for classical infrastructure, and also allows new connections between modes of transport and, for example, demand for energy or water.

A system-of-systems modelling framework (shown in Figure 2.4) is required to represent the role of these interdependencies in determining the impact of strategies for infrastructure service provision across sectors, by allowing rapid communication between a family of models, that is, cross-sector model integration, rather than just running a series of specialised single sector models. Even more importantly, a system-of-systems modelling framework allows implementation of a cross-sectoral decision-making process that can determine performance trade-offs between different sectors at each step in time in order to allocate scarce resources.

While demand for infrastructure services in most cases is directly derived via elasticity functions from changes in external variables, the capacity of the national infrastructure subsystems to provide infrastructure services is a quality emerging from the state of the physical (or network) representations of the infrastructure assets. In particular, we are concerned about whether the possibility of future capacity limitations means that investment or other policy interventions will be required to ensure that demand does not exceed capacity of supply.

Capacity for infrastructure service provision requires fine-grained consideration of the performance of individual assets, for example, a specific road or a power plant. The system-wide capacity for service provision is an emergent property, depending on the asset base, the functional, physical and socio-economic networks and processes operating the assets. Here again, for the long-term assessment of national infrastructure, we assume that this system-wide capacity can adequately be captured by a network of nodes and links with specified capacities that are based on an aggregation of the assets and the functional spatial networks and contains some element of operational and management arrangements.

Framing the modelling architecture around modules of capacity and demand for infrastructure services within and across sectors simplifies the interfaces between sector models, allows repeated inter-model data exchange, and makes the whole modelling architecture itself modular. The implementation of each of these modules in Britain is described in Part II this book. Chapter 4 describes a model of Britain's electricity and gas generation and supply networks (Chaudry et al., 2014) and a disaggregated energy demand module (Baruah and Eyre, 2014); Chapter 5 describes a national strategic model of trunk road, rail, port and airport infrastructure, with regional resolution and an elasticity-based demand module (Blainey and Preston, 2013); Chapters 6 and 7 describe a national water resources system model, coupled to a model of wastewater treatment systems (Tran et al., 2014); Chapter 8 describes a national solid waste assessment model and Chapter 9 describes the technologies and networks for digital communications (ICT).

A schematic representation of the linking architecture is shown in Figure 2.4 which depicts our system-of-systems modelling framework. This coupling of each of the system models enables a centralised sampling of model runs as well as a collection of model results and centralised post-processing and visualisation of complex cross-sector simulation outputs on top of a single database structure. The framework also shows the different stages at which engagement with stakeholders enables feedback on framing the strategies and validating model inputs and outputs.

The most important interdependencies between the different infrastructure sectors originate from correlated demand (e.g. demographic change as a common driver of increases in household energy demand and transport demand for commuting) and from cross-sector demand from infrastructure service provision (e.g. provision of energy services requires water for cooling, pumping of water requires energy, etc.). The integrated model architecture, combining the different infrastructure sector capacity and demand models, allows addressing these demand interdependencies through harmonised strategy and scenario assumptions and through iterative solving within the modelling framework.

With respect to the representation of the infrastructure sectors we are constrained by our long-term and national perspective. In the time dimension, the long-term perspective

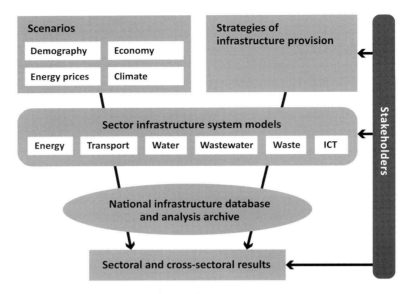

Figure 2.4   Overview of the system-of-systems modelling framework

means that we are covering annual changes to 2050, and beyond in some cases, and representing shorter timescales only in a time-slice approach (peak vs. non-peak demand, seasonal variation). We recognise the severe uncertainties associated with modelling on these timescales, and emphasise the central role of our uncertainty framework. We represent sub-annual processes by a time-slice approach that recognises changes over daily, weekly and seasonal scales as a parameterisation derived from annual mean values.

Concerning the spatial scale of investigation, our focus on networks already implies some spatial extent of analysis, but the system boundaries are a matter of judgment, and of course there will always be boundary issues – unless one deals with the whole world (and space). Our choice of a national perspective is justified (i) because it informs national policy (in federal governments where infrastructure is vested at the state level that might be a more appropriate scale); (ii) it is the most common scale at which data are available (e.g. economic input-output data); and (iii) for an island of nations like Great Britain the physical boundary with a number of well-defined ports, airports, tunnels, interconnectors and cables provides a well-defined geographical unit.

The focus on the national, long-term and capacity/demand perspective leads to a specific choice of spatial and functional resolution of the system representation. Whilst a very detailed description of the infrastructure networks at an asset level is essential for day-to-day infrastructure operation and for failure analysis, the average infrastructure system performance is the focus for capacity considerations. Hence, whilst the performance indicators we develop here include aspects of reliability, peak versus off-peak behaviour, and security of supply, we do not deal explicitly with issues of (optimal) infrastructure operation or the impact of infrastructure failure, although the latter is dealt with in Chapter 12 and in related work elsewhere (Pant et al., 2014). As these aspects of system performance are supposed to change significantly over longer time horizons (e.g. through changes in technology, demography and socio-economic conditions, as well as in the topology and functionality of the national infrastructure system), a too detailed representation of the current infrastructure system would be overconfident, over-complex and, consequently, unhelpful. Thus representing infrastructure on a regional scale (local authorities up to larger subnational regions), with parameterisations of smaller scale processes, is most appropriate.

With respect to demand, all the sector models take a similar approach of combining information about elastic behavioural reaction to changes in macro-economic variables (e.g. modal shifts depending on delays and transportation costs) with a description of the activities that are enabled by the infrastructure service (e.g. space heating, appliances).

Policy decisions that might change the system state are introduced by strategy input variables representing societal change (incorporated as changes in demand), technological change (incorporated as changes in efficiencies and technology-cost parameters) and systemic change within the physical system of infrastructure assets (incorporated in the configuration and capacity of infrastructure networks). Changes in exogenous variables that influence the performance of the infrastructure system are represented by scenarios of socio-economic variables, climate variables and other technological variables, which are outside the influence of infrastructure planners and investors, and which are assumed to be independent of the performance of infrastructure itself.

Step 4: Evaluating infrastructure strategies

In long-term infrastructure assessment we are concerned about the quality of present and future infrastructure provision and the interventions required to ensure the requisite quality of infrastructure service provision. The performance of national infrastructure systems in providing infrastructure services to customers (e.g. households, businesses and the public sector) is a multivariate construct as performance can be defined and measured at different locations, points in time, and with respect to a number of priorities and metrics. As a general principle we refrain from aggregating different dimensions of performance by weighting into unit-free composite indicators as we are especially interested in gaining an intuition into cross-sectoral comparisons of single dimensions of performance under different possible future conditions.

The performance indicators we propose vary in space and time. We deal with temporal and spatial aggregation by applying appropriate aggregations up to a characteristic scale and additional statistics that report about changes on smaller scales. A list of generic, non-sector-specific performance indicators is given below:

Capacity of infrastructure services defines the extent and amount of activities that may be enabled by operating the infrastructure system. While representing capacity as average over a region, we do acknowledge the difficulty of interpreting a regional capacity in terms of a range of proxies, ranging from asset numbers, single asset capacities or capacities of known bottlenecks.

Demand for infrastructure services is defined as the amount and extent of actions enabled by infrastructure services that consumers seek to conduct. Demand can be reported in spatial aggregates relatively easily.

Actual supply of infrastructure services (service delivered) is defined as the amount and extent of actions that are actually enabled. This might be equivalent to demand in some sectors, as they are legally bound to deliver sufficient quantities of service. As this setting excludes the possibility of supply shortages occurring at large scales, infeasibility of a certain infrastructure strategy under a certain external condition would then be indicated not by an actual supply shortage but by astronomical costs for service provision.

The **capacity utilisation (capacity margin)** is defined as the part of the available capacity that is used for providing the actual level of supply. While this measure of capacity utilisation can be applied on any time scale (from hours to annual) it is most useful in application to shorter timescales (hours to daily) to identify hotspots of capacity shortages. The aggregation to an annual timescale can then be achieved by looking at the level of supply reliability, defined as the probability of occurrence of a failure of supply to meet demand over the whole year.

Supplementary indicators of infrastructure service quality measure attributes of infrastructure service provision beyond availability, for example, electricity frequency fluctuations, passenger comfort, road safety and water quality.

Indicators of cost and efficiency of infrastructure service provision measure the cost of infrastructure services from the perspective of (i) consumers and (ii) service providers. Costs for consumers are in terms of units of service provision, as are many of the operating costs for service providers. These cost indicators are the reciprocal indicators of efficiency (service provision per unit cost input), so in the same category we also include other efficiency indicators, notably service provision per unit of energy input. Service providers will also be interested in other cost elements, including annual maintenance costs and capital costs of new infrastructure provision.

Indicators of externalities of infrastructure service provision measure the extent of a number of 'side effects' of infrastructure service provision, such as greenhouse gas emissions and effluent water quality, toxin emissions, rate of accidents, etc. Each of these generic indicators is interpreted in natural units of the corresponding sector.

These sector-specific indicators provide the basis for a cross-sector comparison. Our central tool for this analysis is an interactive, web-based data-viewer (see Chapter 13) that provides a flexible 'dashboard view' to visually combine and compare performance across sectors, across time, across regions and across future conditions. Browsing through the ensemble of infrastructure strategies using this view allows identifying stress-points in the system and relationships between strategy variables and performance outcomes. More generally, this also allows clustering the high-dimensional and complex solution space of our robust decision analysis into sections defined by relative sensitivity of the policy relevant metrics to the strategies (Lempert et al., 2009).

Rather than reporting on optimal strategies, the approach allows investigation of the robustness of infrastructure strategy performance across the range of uncertain future conditions. This approach is equivalent to a robust decision analysis of these geographically explicit national-scale model results (Lempert, 2002). It not only enables the identification of capacity constraints in space and time, but also of levers and links for critical investments to ensure robust and secure supply of infrastructure services. With this capability, our national system-of-systems analysis tools can provide the basis for developing national infrastructure strategies for the twenty-first century.

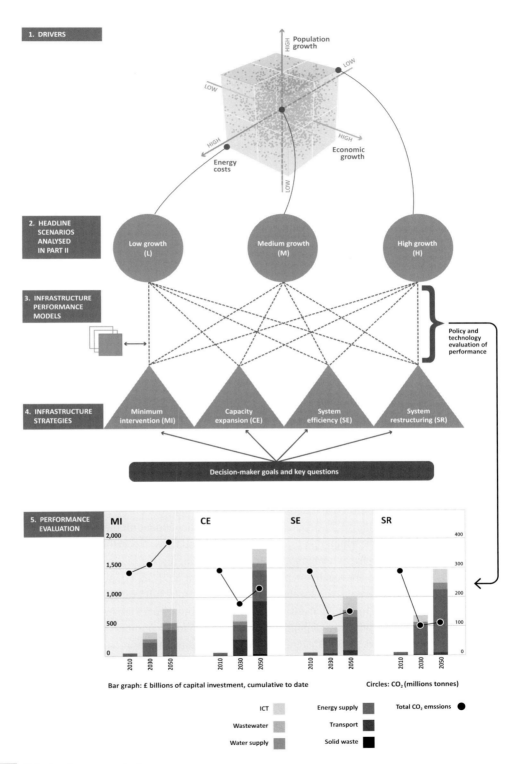

Figure 2.5 Robust performance evaluation

2.3 Conclusions

We have now set out the framework and the main components in our system-of-systems assessment of infrastructure performance. The main steps within that framework are:

1. **Scenario generation** in which a set of key exogenous factors are sampled, in order to test the performance of infrastructure with respect to a wide range of possible future changes.
2. **Strategy generation** in which strategies for national infrastructure provision are developed from a combination (and iteration) of top–down and bottom–up approaches, with a strong involvement of national infrastructure stakeholders. The strategies have an explicit connection with policy-relevant commitments (e.g. investments and the approach to demand management and environmental issues), spanning a wide spectrum of policy instruments (from demand to technology and major investments). These strategies span infrastructure sectors in a coherent way.
3. **Infrastructure system performance evaluation** using an integrated suite of infrastructure sector models around a central database. The system models simulate strategies and generate metrics of system performance.
4. **Evaluating infrastructure strategies** according to a range of metrics which can be explored through visualisation tools that enable a comparison and integration of single sector performance indicators across time, regions, future conditions and sectors and the identification of robust strategies to inform decision-making.

The framework combines the methods of robust decision-making, large-scale ensemble-based simulations, bottom–up engineering-based infrastructure system modelling and database supported post-processing and interactive data visualisation to provide novel analysis capability. Our approach is complementary and compatible to risk-based analysis of the failure of complex, interconnected infrastructure systems described in Chapter 12.

We should be aware of the limitations of the modelling framework. There is an assumption that the socio-economic system is not dependent on the performance of infrastructure, a feedback which this framework does not address. Nevertheless, treating the economy and demography as exogenous and uncertain inputs we believe offers a more readily understandable approach for policy and decision analysis (IEA, 2012; IPCC, 2013). Another, related issue is the limitation with respect to a full equilibrium treatment of cross-sector demand. In our feed-forward simulation framework, cross-sector feedbacks can be modelled as one-way dependencies and via a small number of iteration steps of the whole framework. Notwithstanding these limitations, our approach does build towards an integrated capability of long-term policy evaluation in national infrastructure provision that is needed to develop a vision of the future of national infrastructure.

References

Baruah, P. and N. Eyre (2014). "Simulation of residential energy demand in Great Britain under a range of demographic change and energy system transition pathways." *Energy Policy*. In preparation.

Blainey, S. P. and J. M. Preston (2013). Assessing long term capacity and demand in the rail sector. *13th World conference on transport research*. Rio de Janeiro, Brazil.

Chaudry, M., N. Jenkins, M. Qadrdan and J. Wu (2014). "Combined gas and electricity network expansion planning." *Applied Energy* 113: 1171–1187.

Hickford, A. J., R. J. Nicholls, A. Otto, J. W. Hall, S. P. Blainey, M. Tran and P. Baruah (2015). "Creating an ensemble of future strategies for national infrastructure provision." *Futures* 66: 13–24.

IEA (2012). *Energy technology perspectives 2012: pathways to a clean energy system*. Paris, France, International Energy Agency.

IPCC (2013). Climate change 2013: the physical science basis. *Working Group I contribution to the Intergovernmental Panel on Climate Change fifth assessment report (AR5)*. IPCC, Geneva, Switzerland.

Lempert, R. J. (2002). "A new decision sciences for complex systems." *Proceedings of the National Academy of Sciences of the United States of America* 99(Suppl 3): 7309–7313.

Lempert, R. J., J. Scheffran and D. F. Sprinz (2009). "Methods for long-term environmental policy challenges." *Global Environment Politics* 9(3): 106–133.

Otto, A., J. W. Hall, A. J. Hickford, R. J. Nicholls, D. Alderson and S. Barr (2014). "A quantified system-of-systems modelling framework for robust national infrastructure planning." *IEEE Systems Journal* 99: 1–12.

Pant, R., J. W. Hall, S. L. Barr and D. Alderson (2014). Spatial risk analysis of interdependent infrastructures subjected to extreme hazards. *Second international conference on vulnerability and risk analysis and management (ICVRAM) and the sixth international symposium on uncertainty, modeling, and analysis (ISUMA)*, Liverpool, UK.

Tran, M., J. Hall, A. Hickford, R. Nicholls, D. Alderson, S. Barr, P. Baruah, R. Beavan, M. Birkin, S. Blainey, E. Byers, M. Chaudry, T. Curtis, R. Ebrahimy, N. Eyre, R. Hiteva, N. Jenkins, C. Jones, C. Kilsby, A. Leathard, L. Manning, A. Otto, E. Oughton, W. Powrie, J. Preston, M. Qadrdan, C. Thoung, P. Tyler, J. Watson, G. Watson and C. Zuo (2014). *National infrastructure assessment: analysis of options for infrastructure provision in Great Britain, interim results*, Environmental Change Institute, University of Oxford.

PART II

ANALYSING NATIONAL INFRASTRUCTURE

3   Future demand for infrastructure services

CHRIS THOUNG, RACHEL BEAVEN, CHENGCHAO ZUO, MARK BIRKIN, PETER TYLER, DOUGLAS CRAWFORD-BROWN, EDWARD J. OUGHTON, SCOTT KELLY

3.1 Introduction

The nature and scale of demand for national infrastructure services is driven by long-term changes in population, the economy, technology, society and the environment. However, how those factors affect the longer-term demand for infrastructure is not straightforward. Investment in infrastructure will be challenged by a number of fundamental long-term trends that include demographic developments (e.g. ageing and urbanisation), increasing constraints on public finances, environmental factors (climate change), technological progress (especially in the area of information and communication technology), trends in governance (particularly decentralisation), an expanding role for the private sector and an increasing need to maintain and upgrade existing infrastructures (OECD, 2007).

Chapter 1 highlighted the relationship between infrastructure and the economy. However, the causation between infrastructure availability, economic growth and productivity remains subject to much uncertainty because the relationships between infrastructure and the economy are multiple and complex (Kessides, 1993; Serven, 1996; O'Fallon, 2003; Prud'Homme, 2004; Bourguignon and Pleskovic, 2005; Straub, 2011). Moreover, it is not necessarily generally the case that growth in the economy and population results in increasing demand for infrastructure services. For example, in the transport sector the notion of 'peak car' (Le Vine et al., 2009; Millard-Ball and Schipper, 2010), asserts that car ownership and usage may have plateaued (Chapter 5); and, the quantity of solid waste which needs to be dealt with by infrastructure has been decoupled from economic growth over time (Chapter 8). Nonetheless, it is clear that a growing population and economy are important factors that influence demand for infrastructure services, even while the relationship between these underlying factors and demand is changing. Moreover, it is not just aggregate demand that is relevant when we are thinking about the future of infrastructure systems. The spatial patterns of demand are highly relevant to infrastructure needs at particular locations thereby influencing network configuration. The structure of the economy also influences the nature of demand, as does the demographic structure of the population, for example, age and gender distribution.

Some insight into how the demand for infrastructure has changed in the past can be gained by considering the experience of Britain's road and energy networks. There has been a dramatic increase in user demand for road space over the last sixty years: in the mid-1950s road travel was around 60 billion vehicle kilometres per year, but by 2010 this had increased to nearly 500 billion vehicle kilometres per year (DfT, 2010). In contrast, rail travel averaged 40 billion passenger kilometres per year until the end of the 1990s, but this has increased to the current average of around 60 billion passenger kilometres per year (DfT, 2013). This increase has prompted an urgent reappraisal of how much more capacity might be required. The U.K. electricity system has undergone somewhat more incremental development over the last sixty years (Devine-Wright et al., 2009). Over the period a high voltage transmission system has been installed to meet a demand that has increased by over 70% since the early 1980s. The generation of electricity was historically dominated by the use of coal, but gas has now become the dominant source.

In some cases the growth of demand and subsequent supply has reflected significant technological advances that have enabled a dramatic change in how people and businesses interact. The striking example is digital communications where Britain has become one of the world leaders in the provision of new infrastructure to meet an ever expanding demand by consumers. The U.K. currently has a very advanced information and communications technologies (ICT) infrastructure comprising communication (fixed and mobile telephony, broadband, television and navigation systems) and computation (data and processing hubs). Much of this infrastructure has been provided in a relatively short space of time compared to other more traditional forms of infrastructure.

Previous studies of infrastructure demand take a broadly similar approach to ours in defining the primary drivers of demand. Consideration is routinely given to infrastructure as meeting both final demand and intermediate production capacity (Fay and Yepes, 2003). The literature, however, suffers from both theoretical and empirical limitations summarised by Romp and De Haan (2007). These limitations stem from several issues, and infrastructure growth is often seen to be exogenous to the economic analysis which then assesses the costs of infrastructure provision, whereas the infrastructure may itself generate economic growth that will have economic value and generate social benefits (Cashman and Ashley, 2008; Hayhoe et al., 2010). In other cases, particular infrastructure components are examined in isolation from others despite these being increasingly interconnected both technologically and economically (Santos and Haimes, 2004). Sometimes, it is assumed that infrastructure provision is driven by optimisation of utilities of actors rather than examining the more dynamic nature of such investments (Partridge and Rickman, 2008) and/or in examining specific nations or regions in isolation of the global economy (Heintz et al., 2009).

Infrastructure assets are long-lived, so decisions today should recognise future needs in addition to issues of resilience, interdependence and flexibility. The nature of these long-term trends is uncertain and hard to predict. It is impossible to forecast precisely how the population and economy will change in the future. The notion that we can 'predict and provide' infrastructure services has been discredited. However, we do wish to use the best evidence that we have to understand the *range* of possible futures that infrastructure systems will inhabit. The framework set out in Chapter 2 is designed to incorporate the uncertainties in the exogenous factors that will influence future demand for infrastructure services. This chapter presents approaches for analysing and modelling those factors, most notably population and economic activity. The projections from this work

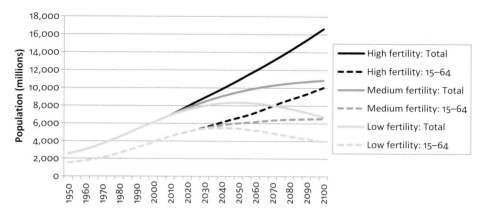

Figure 3.1 World population projections. *Source*: United Nations, 2012

feed into the infrastructure analysis in later chapters. We begin by exploring these issues from a global perspective and the challenges of projecting them into the future. We then set out the modelling and analysis framework that has been developed as part of our national infrastructure assessment.

3.2 Trends in population and demography

The current global population exceeds seven billion people, compared to around two billion in the mid-twentieth century. The UN's latest 'Medium fertility' projection (United Nations, 2012) suggests that the global population could rise to almost ten billion by 2050 (see Figure 3.1). The range of the UN's projections is wide, with the 'High fertility' projection indicating a population of almost eleven billion in 2050, growing to almost seventeen billion in 2100. Conversely, the 'Low fertility' projection represents slower growth to 2050 (to around eight billion) followed by a decline in the global population, to a level below that seen now. In all cases, the proportion of the population that will be of working age is projected to decline over time, from 61–65% in 2050 to 58–60% in 2100. The majority of the future-expected population growth will be in the developing world.

Many developing countries are seeing rapid economic growth and are now also undergoing the transition from agrarian to more manufacturing and services-based economies. There is a substantial socio-economic transition underway. The 2011 Revision of the UN's *World Urbanization Prospects* (United Nations, 2012) projects the population of the developed world to grow by a further seventy-five million in the next forty years, with a mild continued shift from rural to urban areas (see Figure 3.2). In contrast, the developing world (whose population today is almost five times that of the developed world) is expected to grow by over two billion people over 2010–2050. This will raise the demand for infrastructure services, from the simple growth in population, but also as the per capita incomes of these people increase, leading them to demand more services.

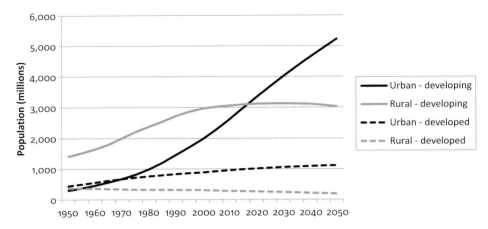

Figure 3.2   Projected world population to 2050 by developing/developed and rural/urban. *Source*: United Nations, 2012

In the developing world, the UN (United Nations, 2012, 2013) projects the urban population to grow rapidly over that period, while the rural population will plateau. This will further alter the future pattern of demand for infrastructure, requiring a different set of planning decisions in order to accommodate this urbanisation, for example, in the form of transport within cities; greater electricity provision; waste and water collection. Depending on activity patterns (location of production and offices), there may also be requirements for higher-capacity links between cities.

As is clear from Figure 3.2, the transition from rural to urban areas has largely been completed in developed economies with urban agglomerations a key feature in their spatial make-up. In the U.K. for example, cities now account for less than 10% of the landmass but more than 50% of businesses, population and jobs; and almost three-quarters of the country's highly skilled workers (Centre for Cities, 2013). With so much economic activity concentrated in such areas, and with it increasing (possibly faster than expected or planned for) cities' infrastructure assets are likely to struggle to meet the needs of the population, if they are not struggling already. Given the age of much of the existing infrastructure assets, which were designed to accommodate an earlier set of socio-economic circumstances, a change in how infrastructure is planned for and provided may be necessary in order to support the economic-spatial configuration that has since evolved.

3.3 Economic growth

There is a general tendency over the long term for economies to grow over time (as measured by their GDP). Living standards have increased markedly in the last 200 years (since the Industrial Revolution) and global GDP per capita has increased particularly rapidly from the 1950s onwards. Across countries, however, the pattern of growth has differed, initially led by what are now the developed economies and, more recently, driven by emerging (developing) economies such as China and India. There is a multitude of analyses of the long-term future of the global economy undertaken by a variety of organisations (e.g. PwC Economics (PWC, 2013), Carnegie Endowment for International Peace (Dadush and Stancil, 2010)). Such analyses typically focus on the influence of 'supply-side' factors on an economy's productive capacity. This is done on the basis that this productive capacity represents the long-term constraint on future growth. Viewed in this way, productive capacity (potential output) depends on the availability of inputs (labour and capital) and the efficiency with which these inputs can be transformed into outputs. This efficiency depends on the state of technology which, in this context, is a broad concept that encompasses physical assets but also intangible processes and knowledge involved in the transformation of inputs into outputs. Observed fluctuations in economic growth are then taken to represent variations in the utilisation of productive capacity.

Set out this way, the long-term productive capability of economies and their consequent requirements for infrastructure services can be broken down into distinct sets of components, with particularly important productive factors being the availability of labour (i.e. the working-age population and its productivity) and technological innovation and adoption. One can then consider how socio-economic trends such as demography might affect long-term growth by creating increased demand but also creating the potential for infrastructure provision to be inadequate in meeting that demand, thus creating bottlenecks in production.

Demography is an important factor in long-term economic growth since households are both *consumers* of goods and services, and *suppliers* of labour inputs to production. Larger populations will consume more, generating demand for more goods and services. As the composition of the population changes, patterns of consumption will change, since younger people will have different consumption preferences to older people. The size and composition of the population will also affect the supply of workers (labour) to contribute to economic activities: for a given population (and productivity per worker), a higher proportion of working-age persons provide a larger supply of potential labour in the economy. Moreover, the scarcity or abundance of labour (and/or skills) may also affect wage rates in the economy and, in turn, household incomes (feeding back into expenditure and consumption effects). One further potential effect of demographic change relates to the nature of government provision of services such as the welfare system and the funding mechanisms for healthcare. In both cases, differing demographic trends place different demands on these systems, with implications for economies' fiscal positions, that is, the level and sustainability of public deficits/surpluses and debt.

The scope for growth in labour and technology (whether adoption or innovation) is what motivates the majority of analyses to conclude that the nature and scale of change in the developing world will be a key driver of future growth in the global economy (and, conversely, why one might expect growth in the developed world to slow somewhat over that same period). In particular, most projections expect continued strong growth in the populations of developing countries, accompanied by a shift towards greater urbanisation and more manufacturing/services-oriented economies. This corresponds to, for example, expansion of the burgeoning middle classes of these countries and changes in consumption patterns as a result of rising incomes.

3.4 Challenges in projecting future demand for infrastructure

Population growth leads to increased demand for infrastructure services, but better infrastructure services also attract population to a region. This is true in regard to both final and intermediate demand within the economy. The result is a feedback loop with non-linear dynamics. In contrast, our approach is based on a feed-forward analysis that starts with demographic and economic scenarios, which drive infrastructure demand providing inputs to individual infrastructure sector models. There is some feedback between transport infrastructure capacity and the actual use of these services; however, our analysis is not fully endogenous, so does not fully capture the effect of new infrastructure provision on the demand for infrastructure services.

Whilst the demographic projections we have developed are highly granular in space, the spatial scale of the macroeconomic model is limited by the availability of reliable data. We deal with that issue in this chapter by aggregating the more spatially resolved demographic projections upwards to the spatial resolution of the macroeconomic model. Implicit in this approach is the assumption that economic change in a region is approximated by scaling the spatial pattern of demographic change uniformly while maintaining the relative amounts of population from the more spatially resolved projections.

Inevitably, there is considerable uncertainty in our demographic and economic projections, which we address through provision of a wide range of possible scenarios. It is important to be aware that, while economies do tend to grow over the long term, economic growth is not a continuous process, with few periods in world history of truly sustained economic growth over long periods of time. Instead, economies exhibit 'business cycles'. While economies *tend* to grow over time, as a trend, there are periods of decline (recession/depression). Unexpected events and discontinuities are not necessarily negative in terms of economic growth and welfare. Indeed, technological innovation has an important role to play in economic growth, with information technologies enabling new methods of commerce and communication at speeds and distances not previously possible, at least where adequate infrastructure is available.

For long-lived assets like infrastructure, performance in terms of service provision can be affected dramatically by changes in future conditions. By extension, high uncertainty about future conditions in the long term makes 'optimisation-oriented' infrastructure planning extremely difficult, if not impossible. This puts forward a case for a more 'robustness-oriented' planning approach to identify strategies that perform 'well' over a range of possible future conditions, rather than ones that perform 'best', but under much narrower ranges of conditions. The aims of a robustness-oriented approach should be both to limit the adverse consequences of unexpected future conditions (e.g. disruption/damage to infrastructure networks or unexpectedly high demand/loads), and to maximise the opportunities to capitalise on the benefits of future social and technological developments. An historical example of this would be the internet, which has facilitated a range of innovations far beyond those originally envisaged by its inventors.

The discussion above highlights the difficulties of long-term forecasting. Over longer periods of time, an understanding and acceptance of uncertainty is critical. As such, the approach followed in the analysis presented in the following sections should be interpreted as scenarios of what could happen under defined conditions, rather than attempts to produce point forecasts of what will happen. In that sense, the scenarios are projections contingent on their input assumptions, in order to examine the implications of different future states of the world.

3.5 Modelling future demand for infrastructure services

The modelling framework set out in Chapter 2 starts with projections of the main factors that influence demand for infrastructure services, most notably population and economy. These projections need to be on spatial and temporal scales that are of relevance to the infrastructure systems analysis, that is, at a suitably high spatial resolution and over timescales extending at least over the next forty years (to the 2050s) and beyond towards 2100. Here we present the modelling methodology that has been used to generate the demographic and economic scenarios for Britain.

3.5.1 Generating the demographic projections

Attempts to model demographic change over such a long period of time are unusual. Interesting work has explored variations between EU regions but at a level of spatial aggregation which is at least an order of magnitude less detailed than is used here (de Beer et al., 2010). The idea of a set of demographic scenarios has been adapted and extended from the two-dimensional matrix of social and political futures adopted in that work. Research by Rees and colleagues has examined the components of population change with particular respect to ethnic sub-populations using detailed and highly disaggregated parametric estimation (Rees et al., 2011). However, the scenario-based elements of that work are relatively underdeveloped and less ambitious in their temporal extent. The demographic projections provided here are based on a more straightforward yet robust projection mechanism which allows the model to be extended further into the future, at a refined spatial level, and within a framework of multiple future scenario paths.

Our approach models demography as the outcome of changes in three components: fertility, mortality and migration. In order to establish a baseline, the model was calibrated to U.K. data. The process established the current rates of fertility and mortality (for each subnational area) and then reproduced the baseline trends for these components. The baseline projections to 2033 were used to estimate net migration rates for each geographical area, to approximate the local migration trends which are embedded in official projections. The calibrated rates of fertility, mortality and migration were then extrapolated through to 2100, subject to a series of scenario multipliers being applied: these multipliers adjust the projected components (fertility, mortality and migration) (Figure 3.3).

Since 1992, the Office for National Statistics (ONS) has produced spatially disaggregated Sub-National Population Projections (SNPP) which run over a twenty-five-year time horizon, and National Population Projections (NPP) which extend seventy-five years into the future. These projections are updated every two years. The latest available ONS projections

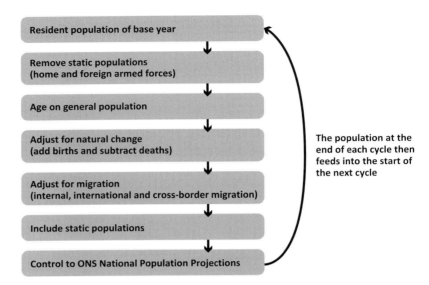

Figure 3.3   Steps in generating the demographic scenarios

were for the base year 2008 which means neither of the projections can reach the horizon of 2100. Therefore, projections were introduced to expand and extrapolate the existing ONS projections from three perspectives: (i) our projections run till 2100 rather than the year 2083 of ONS projections; (ii) our projections are spatially disaggregated to the local authority district (LAD) level whereas the ONS NPP Projections are spatially aggregated; and (iii) we provide an extended portfolio of demographic scenarios which allows users and policy-makers to trace future developments in relation to background parameters of social, political and economic change, while the ONS projections are represented as three variant projections.

In view of the uncertainties associated with the longer timescales, we have developed an extended set of demographic scenarios. The scenario framework is illustrated as a three-dimensional space, in which three broad drivers of long-term demographic change are characterised as the level of economic prosperity, social attitudes to sustainability and the political effects of isolation (in particular the importance of spatial political integration and international migration policies). In what follows the eight vertices of the scenario box (Figure 3.4) will typically be regarded as discrete scenarios (denoted scenario A–H). Table 3.1 shows the model components associated with each scenario.

This eight-scenario framework represents a substantially larger space of potential outcomes than those produced by the ONS, ranging from a mild contraction of the population (scenario F) to sustained, rapid growth (scenario D) (see Figure 3.5). We consider these eight scenarios to lie at the extremes of the possible future outcomes: in all probability the future pattern of demographic change lies somewhere within the box.

The population projections for Britain are shown in Figure 3.5. An important trend in the demographic projections is simply 'growth'. For example, in the baseline all regions will experience growth in the range of 20–35% before 2100, which has significant implications

Table 3.1 Numerical values for the model components under different scenario combinations

Scenario	Prosperity	Sustainability	Isolation	Fertility multiplier	Mortality multiplier	Migration multiplier
A	H	H	H	1.15	0.975	0.01
B	H	H	L	0.95	0.965	1.33
C	H	L	H	1.15	0.995	1.33
D	H	L	L	0.95	0.985	2.65
E	L	H	H	1.05	1.015	0.00
F	L	H	L	0.85	1.005	0.67
G	L	L	H	1.05	1.035	0.67
H	L	L	L	0.85	1.025	1.99

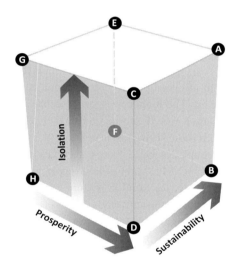

Figure 3.4   Scenario space of population projections

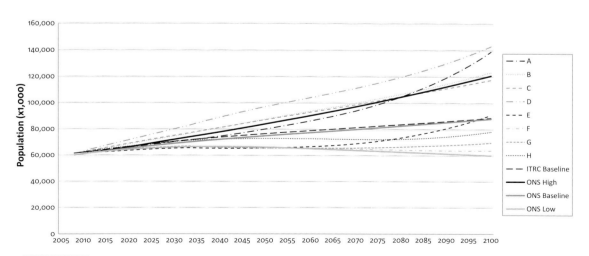

Figure 3.5   Total ITRC population projections for Britain compared with ONS 2012-based projections

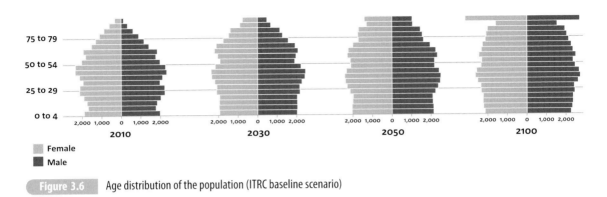

Female
Male

Age distribution of the population (ITRC baseline scenario)

for infrastructure. The ageing of the population is a process that has been well documented, and this trend continues in all of the scenarios (Figure 3.6). An elderly population has important policy impacts in terms of the kinds of transport services that need to be provided (there is certainly food for thought here regarding investment in public transport) but also in relation to varying patterns of consumption for energy, water and waste. The relatively slow uptake of information technology amongst more elderly consumers also renders this trend significant. The higher prosperity scenarios (scenario A, B, C and D) will lead to a higher population growth (compared to E, F, G and H), due to the lower mortality rates associated with these scenarios.

The demographic profile varies significantly both geographically (Figure 3.7) and between scenarios. Migration policy has a considerable impact on the regional population trend especially for London (Figure 3.8). The higher population growth can be found in the Northwest than the surrounding areas (i.e. Yorkshire and the Humber, and the Northeast). Scenarios involving high migration and relatively slow growth in life expectancy (e.g. scenario H) will give rise to much younger age pyramids. These impacts will be most serious in the Southeast, because it is here again that growth is mostly strongly influenced by migration. There seems a mismatch between population growth and infrastructure availability/pressure on existing facilities – for example, strong growth in the Southeast where transport congestion is highest; modest growth in Scotland where resources such as water are most abundant.

The model that has been described here provides spatially explicit demographic projections of the British population over the next century. Whilst extreme values have been explored in the scenarios, and it has been noted that future expectations probably lie somewhere in the interior of this range of possibilities, one of the assumptions of this model has been of a continuous future trend. In practice, economic prosperity is generally observed to be cyclical (Kondriatev, 1935) and there is evidence that variations in fertility are similar in nature (Easterlin, 1975). Fluctuations in these elements are not represented in the current version of the model, although there is no reason why hybridisation of the paths and trends might not be accommodated in the application of the model. This in itself raises a further important question of how confident one can actually be about future demographic paths. The obvious response to this is that a scenario-based approach may be the

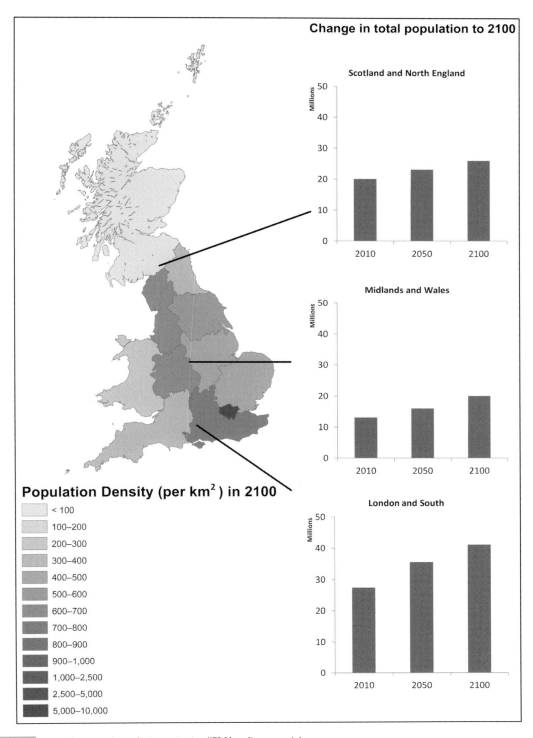

Change in total population to 2100

Scotland and North England

Midlands and Wales

London and South

Population Density (per km²) in 2100

< 100
100–200
200–300
300–400
400–500
500–600
600–700
700–800
800–900
900–1,000
1,000–2,500
2,500–5,000
5,000–10,000

Figure 3.7   Example regional population projection (ITRC baseline scenario)

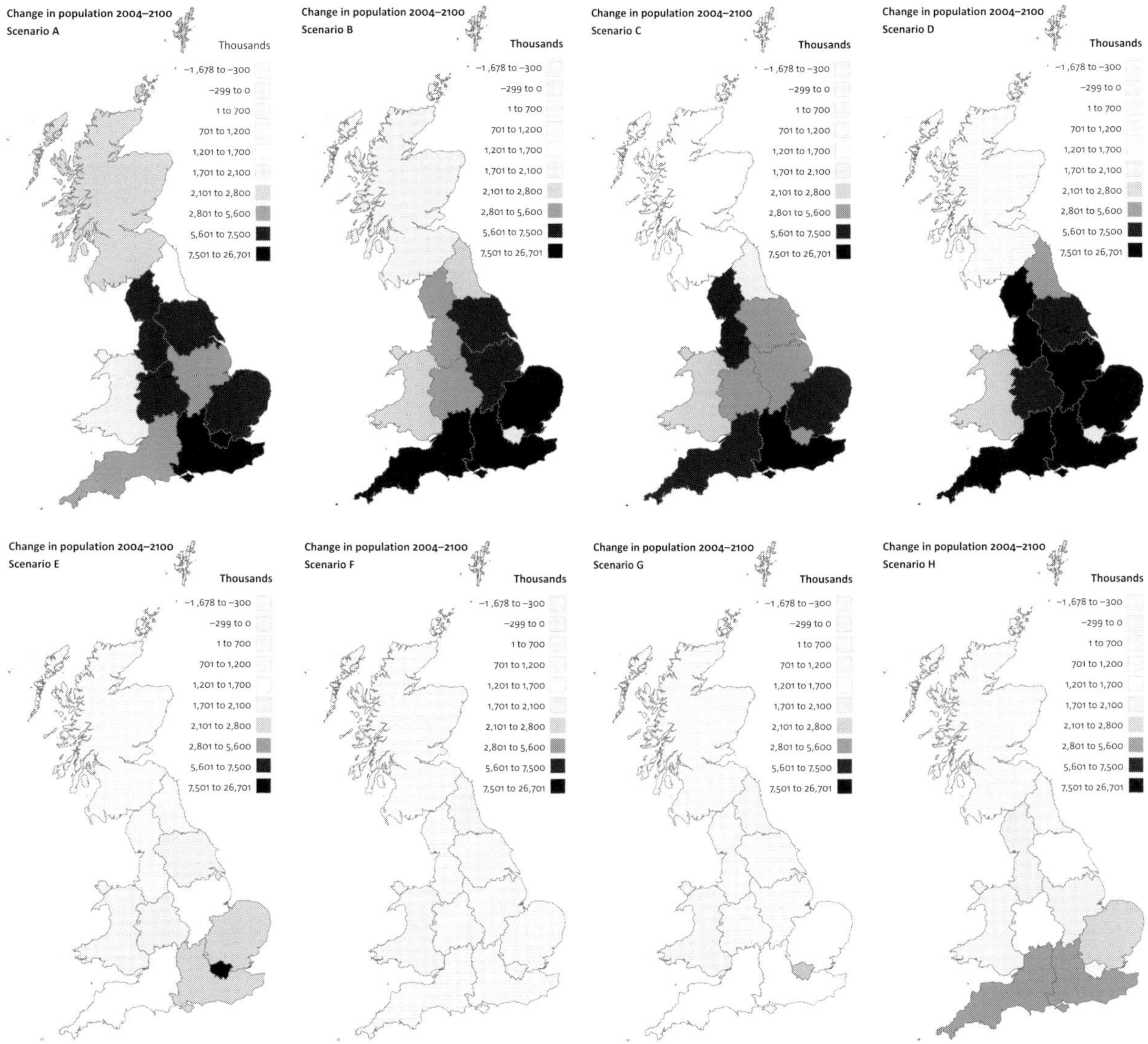

Figure 3.8   Projected changes of population by region (ITRC scenarios A–H)

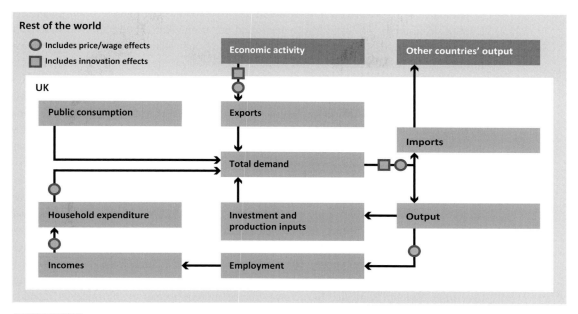

Figure 3.9 Overview of the MDM-E3 model

best one can do in exploring the implications of various potential futures. Furthermore, these fluctuations and alternative scenarios may themselves be related to potential cycles of growth and investment. This raises a related issue of whether it is necessary to close the loop between infrastructure and population – hence it is not just population change that influences infrastructure provision but the reverse may also be true.

3.5.2 Generating the economic projections

Analysing the drivers of demand for infrastructure services in the economy requires an approach with a coherent treatment of detailed industrial and spatial demands. The approach we have adopted uses the Multisectoral Dynamic Energy–Environment–Economy (MDM-E3) model, which projects key indicators (such as output, prices, employment and components of demand) separately for each industry sector and region, building up the macroeconomic picture by aggregating this detail. Key features of the model are: the high level of disaggregation in terms of industrial sectors and categories of final expenditure; and the regional disaggregation of key economic variables. The model's structure mirrors the organising principles of official U.K. national statistics and embeds an essentially Keynesian logic for determining final expenditure, output and employment. Importantly, this framework recognises interdependencies between sectors (i.e. supply chains) and how these might change over time through technological change, relative price movements and changes in the composition of each industry's purchases and outputs (Figure 3.9).

The purpose of MDM-E3 is to simulate the dynamics of Britain's economy based on a set of econometrically estimated behavioural equations that capture short-term impacts followed by medium-term adjustments to a long-run steady state, underpinned by economic fundamentals. Relationships are not generally imposed by theory, although the ranges of some coefficients are limited to prevent implausible or perverse simulation properties. The model draws from a range of official, publically available data, integrating them in a consistent manner: the Office for National Statistics (ONS) for the national and regional economic accounts and labour market data; the Digest of U.K. Energy Statistics (DUKES) published by the Department of Energy and Climate Change (DECC) for energy data; and the National Air Emissions Inventory (NAEI) for emissions data.

A principal uncertainty in projecting into the future is the nature of structural and technological change. Successfully predicting such changes in future economic structure becomes more difficult as the projection horizon (2050 in this economic analysis) extends and implicitly requires some view on the nature of future goods and services, including how (the nature of the inputs and production process) and where (the, possibly increasingly global, geography of the supply chain) they are produced. Such trends are captured to some degree in the MDM-E3 framework, tracking the general pattern of structural change over time. Examples of such structural change include trends in the material intensity of production and the productivity of labour. Because the demographic and economic projections serve as inputs to the individual infrastructure models, changes in infrastructure-service demand per unit of population/activity ('infrastructure intensity') are dealt with in the individual infrastructure models. In MDM-E3 these and other trends are estimated econometrically, based upon historical data. A criticism of this econometric approach is that, especially in the longer term, future relationships may deviate from those observed in the past. Our scenario approach explores possible future outcomes within an envelope defined by past economic relationships; it does not project the consequences of, as yet, unforeseen technologies.

In order to generate a projection, MDM-E3 requires a number of exogenous assumptions: natural resources (production and prices), government expenditure, tax rates and allowances, global economic conditions (activity, prices, interest and exchange rates) and population. The effects of population change captured within MDM-E3 are: that a larger population tends to drive up household expenditure in the economy, that changes in the demographic structure of the population alter consumption patterns in the economy, and that a larger population represents a larger workforce, allowing for higher employment. The larger workforce potentially has further labour-supply implications as the higher availability of labour may curb wage growth. Effects not captured endogenously within MDM-E3 are those that arise because a larger population puts greater demand on healthcare and education services, requiring additional government expenditure and investment to provide them. These effects are not endogenous in the model and revised exogenous assumptions for government expenditure must be developed off model to be consistent with the demographic assumptions. Estimates were made of the historical relationships between population (by region and children/working-age/old age pensioners (OAPs)) and government consumption of goods and services (education, health and social work and public administration); and

Table 3.2 Variants of the three assumptions sets		
Variant	Description	
POPULATION BY REGION		
D	Highest population variant, with a population approaching 100 m people by 2050	
B and C	Medium-high population variants, with the population reaching around 87 m people by 2050	
A	Broadly, a 'central' population projection, with the population growing to almost 80 m in 2050	
H	Medium-low population projection with a similar profile to scenario A until 2035, when growth plateaus at almost 73 m people	
E, F and G	Low population variants with initially slow growth followed by either flat population or mild decline	
WORLD ECONOMIC GROWTH		
High	Average global (non-British) economic growth of 4% pa over 2010–2020, rising to 5–6% over 2020–2050	
Central	Baseline view with global economic growth averaging 3.5% pa over 2010–2020 and 4–5% pa over 2020–2050	
Low	Global economic growth of 2% pa over 2010–2020 and growth of 3–3.5% pa to 2050	
FOSSIL FUEL PRICES		
	Oil price 2050 ($/bbl)	Gas price 2050 (p/therm)	Coal price 2050 (£/tonne)
High	486	263	313
Central	345	183	199
Low	204	105	126

government investment (education, health and social work and housing). These estimates were used to derive alternative sets of exogenous assumptions for government expenditure given the alternative projections of demographic change.

The space of outcomes from the economic projections are defined by the following three dimensions: (i) *British population by region* (the eight variants generated by the demography model); (ii) *World economic growth* (three variants) developed to represent a range of world economic conditions that would in turn affect British international trade with the rest of the world; and (iii) *Fossil fuel prices* (three variants) based on the DECC fossil fuel-price assumptions from its most recent Updated Energy and Emissions Projections publication (DECC, 2012) extended to 2050. These three were chosen as the primary drivers common to all the infrastructure sectors to be analysed. It is intentional that these scenarios do not assess the impact of, for example, specific policies or legislation, rather they provide a consistent context in which each infrastructure sector is to be analysed.

The variation in the assumption sets characterises a range of conditions within the bounds of trends observed over recent decades, that is, the scenarios do not attempt to predict tipping points or regime changes. The variants are summarised in Table 3.2 and yield seventy-two scenarios once all possible combinations are considered. Strictly, there is no 'Central' or

'Baseline' scenario but, for the purposes of comparison, we consider the 'A' population variant and the Central world economic growth and fossil fuel price variants as the inputs that constitute a 'Reference' scenario. We summarise the results separately for each of the three assumption sets. When the assumption sets are varied in combination, while there are some interactions between them these are relatively small and the impacts of different scenario combinations are largely additive in nature. For example, the population variant D results in a 21% higher GDP than the reference case by 2050; the High world economic growth variant gives a 10% higher GDP, and a variant with both population D and High world economic growth (with the reference fossil-fuel price) gives a 32% higher GDP than the reference case.

The population variants in combination with the Central world economic growth and fossil fuel prices variants in all cases generate GDP growth over 2010–2050 in the range of 1.7% pa (variants E and G) to 2.4% pa (D). The range of GDP results in 2050 is around 33% while the range of employment results is around half that size at 16%. The reference population variant (A) gives GDP towards the lower end of the results (GDP growth 1.9% pa). The largest GDP and employment effects are seen in the scenario with the highest population (D), driving higher household expenditure and government expenditure to the benefit principally of activity in both private and public services. Conversely, the smallest GDP effects are in variants E and G which are characterised by weaker population growth which results in a relatively smaller boost to public services. The population variants cause some shift in the sectoral structure of the British economy (principally higher growth leads to a greater share of public services). However, this does little to change the regional distribution of economic activity with London and the surrounding regions continuing to dominate. This is shown in Figure 3.10 which illustrates how the regional contributions to British gross value added (GVA) output by 2050 vary across the different population variants.

Although relatively faster growth in public services tends to favour the spatial 'rebalancing' of activity, because public services account for a larger proportion of GVA outside of the Greater Southeast, the scale of structural change is not sufficiently large to stem the relative expansion of London and the Southeast. London and the Southeast currently account for around 36% of GVA, and this is projected to rise in *all* of the population variants (41% by 2050 in the central A variant, versus 39% in variants B (medium-high population growth) and D (highest population growth), and 41% in variant E (low population growth)).

Although there is no mechanism to ensure that the detailed assumptions underpinning the demographic projections are consistent with the results of the macroeconomic model, the results of the macroeconomic model show broad consistency regarding economic *prosperity*: the demographic model assumes that greater economic prosperity will foster stronger population growth and the results of the economic model show that stronger population growth drives faster economic growth.

The world economic growth variants (in combination with population variant A and the Central fossil fuel prices variants in all cases) generate results ranging by 29% for GDP and 11% for employment in 2050 shown in Figure 3.11. The differences in GDP are driven directly by the impact of world economic growth on demand for U.K. exports, and indirectly

(£bn 2009 prices)

0 - 100	100 - 200	200 - 300	300 - 400	400 - 500	500 - 600	600 - 700	700 - 800

Figure 3.10   Regional gross value added (GVA) by 2050 for ITRC population growth variants

Figure 3.11 Composition of regional gross value added (GVA) by 2050 for world economic growth variants

by household expenditure and imports (from changes in national income) and investment (to support higher/lower levels of U.K. output). Consequently both manufacturing and private services see variation in growth prospects in these scenarios. The High world economic growth variant generates a U.K. economy by 2050 that is more rebalanced than any of the population variants with exports and investment accounting for a greater proportion of GDP, and household and government consumption accounting for a smaller proportion, with the reverse being true for the low world growth variant. Manufacturing output (and employment) is more responsive to the world economic growth variants than to the population variants. However, the other broad sectors of the economy are generally less responsive. With the exception of manufacturing, which accounts for a comparatively small proportion of output in all U.K. regions, the shift in the sectoral structure of the U.K. is less pronounced, and consequently the world economic growth variants result in even less of a spatial rebalancing of activity. By 2050, London plus the Southeast is projected to account for nearly 41% of U.K. GVA in the Low, Central and High variants.

The range of fossil fuel-price variants is relatively narrow in macroeconomic terms because energy-intensive industry represents a small share of U.K. economic activity and, even then, energy is but one component of production costs, further diluting the impact of price increases. While the differences in nominal wholesale prices are large (around 40% either side of the baseline), after factoring in other components of the price, the difference in the final retail prices is smaller. This leads to relatively small changes in energy demand in the scenarios, which feed through into relatively small changes in industrial cost structures. Consequently, the economic impacts of the fossil fuel-price variants are relatively small in comparison to the population and world growth variants. The fossil fuel-price variants generate results ranging by less than 5% for GDP in 2050.

3.6 Implications for infrastructure services

The scenarios generated using MDM-E3 provide a consistent and cross-cutting framework for the detailed analyses, in the subsequent chapters of this book, of demand, supply and performance of the different infrastructure sectors. All of the variants of demography, world growth and energy prices that have been considered project that both U.K. GDP and population will tend to grow over the period to 2050, indicating that, unless substantial efficiency gains are made, demand for infrastructure services will continue to grow. Although the variants of population and world economic growth drive different patterns of expenditure that generate alternative economic structures (internal/population/services-led versus external/world/manufacturing-led), a number of trends persist across many of them. The contribution of manufacturing to total output and employment will continue to shrink, though at a slower pace than in recent decades. The contribution of service activities, in particular financial and supporting business services, will continue to expand. In addition, the population will be larger and older, leading to a larger proportion of people working to an older age and greater demand for health and social care services.

Demand for energy will be shaped to some extent by these structural changes, along with measures in place to target the U.K.'s greenhouse gas emissions reduction commitments. The U.K.'s manufacturing activity will shift from energy-intensive basic manufactures, to higher value (lower energy intensity) manufactures and this, in addition to measures to improve energy efficiency, will drive down energy use from industry. In general, electricity used in the expanding services sector, and the increase in appliances in the home as well as the increasing consumption of consumer electronics, will drive an increase in electricity consumption over the long term, despite likely improvements in the efficiency of energy-using technologies (if measures such as EU Products Policy are assumed to be successful). Trends in the way we work will shape demand for infrastructure services. A larger proportion of the workforce will have jobs in services, in professional and managerial occupations, in which there is greater potential to telework and for flexible working. This may ease transport congestion at peak commute times, but change the pattern of energy demand away from shared offices to (potentially less efficient) individual homes. However, some work may become more transport intensive, requiring more journeys: a larger proportion of the workforce will have more than one job; and the number of some peripatetic jobs, such as social care, is projected to increase as the population ages.

A key issue for the provision of all infrastructure services will be the location of economic activity – where people will work and where they will live. Our results show that as might be expected structural change usually occurs slowly, especially at the regional level within the U.K. Even when significant shifts in industrial structure occur, which has certainly been the case in the U.K. over the last forty years, the impact on the distribution of activity spatially tends to be smaller because of the existing balance of economic activity in the U.K. London and the Southeast already account for one-third of U.K. output; in order to 'catch up', the other nations and regions would need to achieve considerably faster than average growth sustained over a number of decades. This has not happened over the last fifty years (Gardiner et al., 2013).

Our projections indicate that, under all variants, the concentration of economic activity in London and the surrounding regions will increase further, increasing pressure on transport infrastructure. Congestion on transport networks has the potential to hinder economic growth, either through delays caused or if higher prices are charged to ration demand. Information and communication technologies (ICT) may offer potential technological solutions to help manage or control demand for infrastructure services, for example, through increased use of ICT to enable teleworking. Concentrations of economic activity and population also present challenges and opportunities to the water supply, wastewater and solid waste sectors. Increased demand for water is expected to grow most rapidly in regions of the U.K. where collection will be most limited, so for the water supply sector redistribution will continue to present a key challenge, along with the ageing and deteriorating distribution network. Illustrating the interdependencies between the infrastructure sectors, there is potential for further expansion of facilities and networks located to serve densely populated areas where the management of solid waste recovers resources that can be used for energy generation.

The model projections show that depending on the scenario adopted it is possible that the pattern of economic growth observed by region and sector could be quite different.

Aggregate growth trajectories can be achieved through a wide range of alternative economic structures. As might be expected, the scenarios have different implications for infrastructure-service provision and future network configuration. The composition of the growth trajectory is clearly of significant importance in determining the parts of the infrastructure system that will be challenged the most.

3.7 Conclusions

The approach adopted to assess the future demand for infrastructure services has enabled a number of demographic and economic growth scenarios to be generated that are based on plausible assumptions of global and national factors that influence these changes.

There are a number of limitations to the approach adopted that should be considered in future work. One criticism is the 'feed-forward' approach whereby demographic and economic projections are exogenous inputs to the infrastructure systems analysis that is described in the chapters that follow. There is no feedback from changes in the stock of infrastructure back to the economy. In addition, demographic trends are an exogenous input to the economic model and there is thus no feedback from changes in the economy to the demographic projections.

A key assumption of the approach (and of econometric modelling in general) is that the past is a useful guide to the nature of relationships in the future, that is, that agents respond in a similar way to future developments and policies as they have in the past. The possibility that this might not hold is the essence of the 'Lucas Critique' in economics (Lucas, 1976): behavioural parameters estimated on past data do not necessarily ensure the validity of a model applied for future forecasting and policy analysis if there is the possibility that agents' behaviour might change in the future. This possibility has clear theoretical importance and should be acknowledged by modellers but, in practice, the empirical evidence in support of the Lucas Critique is relatively weak. Thus, it is not unreasonable to engage in model-based projections of this kind; indeed, such modelling represents accepted practice in the field.

It is also important to emphasise the purpose of these projections, which is to examine a range of possible outcomes under alternative future conditions ('what could happen') rather than forecasts ('what will happen'). This work has illustrated how important advances can be made building on established models and techniques that begin to recognise key infrastructure interdependencies in the future demand for infrastructure services.

References

Bourguignon, F. C. and B. Pleskovic (2005). Annual World Bank conference on development economics 2005: lessons of experience. *Annual World Bank conference on development economics 2005*, Washington, DC, USA, World Bank Publications.

Cashman, A. and R. Ashley (2008). "Costing the long-term demand for water sector infrastructure." *Foresight* 10(3): 9–26.

Centre for Cities (2013). *Cities outlook 2013*. London, UK, *Centre for Cities*.

Dadush, U. and B. Stancil (2010). The world order in 2050. Policy Outlook, Carnegie Endowment for International Peace.

de Beer et al. (2010). DEMIFER: demographic and migratory flows affecting European regions and cities. Applied Research 2013/1/3, Final Report, ESPON.

DECC (2012). *Updated energy and emissions projections 2012*. London, UK, Department of Energy and Climate Change.

Devine-Wright, P., Y. Rydin, S. Guy, L. Hunt, L. Walker, J. Watson, J. Loughead and M. Ince (2009). "Powering our lives: sustainable energy management and the built environment." *Final project report*. London, Government Office for Science.

DfT (2010). *Transport Statistics for Great Britain*. London, UK, *Department for Transport*.

DfT (2013). *Rail trends, Great Britain 2012/13. DfT transport statistics factsheet*. London, UK, Department for Transport.

Easterlin, R. A. (1975). "An economic framework for fertility analysis." *Studies in Family Planning* 6(3): 54–63.

Fay, M. and T. Yepes (2003). *Investing in infrastructure: what is needed from 2000 to 2010?* Washington, DC, USA, World Bank Publications.

Gardiner, B., R. Martin, P. Sunley and P. Tyler (2013). "Spatially unbalanced growth in the British economy." *Journal of Economic Geography* 13(6): 889–928.

Hayhoe, K., M. Robson, J. Rogula, M. Auffhammer, N. Miller, J. VanDorn and D. Wuebbles (2010). "An integrated framework for quantifying and valuing climate change impacts on urban energy and infrastructure: a Chicago case study." *Journal of Great Lakes Research* 36: 94–105.

Heintz, J., R. Pollin and H. Garrett-Peltier (2009). "How infrastructure investments support the US economy: employment, productivity and growth." Political Economy Research Institute, University of Massachusetts at Amherst.

Kessides, C. (1993). The contributions of infrastructure to economic development: a review of experience and policy implications. World Bank – Discussion papers.

Kondriatev, N. (1935). "The long waves in economic life." *The Review of Economic Statistics* 18(6): 105–115.

Le Vine, S., P. M. Jones and J. W. Polak (2009). Has the historical growth in car use come to an end in Great Britain? *European Transport Conference*, Frankfurt, Germany.

Lucas, R. E. (1976). "Econometric policy evaluation: a critique." *Carnegie-Rochester Conference Series on Public Policy* 1: 19–46.

Millard-Ball, A. and L. Schipper (2010). "Are we reaching peak travel? Trends in passenger transport in eight industrialized countries." *Transport Reviews* 31(3): 357–378.

O'Fallon, C. (2003). *Linkages between infrastructure and economic growth*. New Zealand, Pinnacle Research, prepared for Ministry of Economic Development.

OECD (2007). *Infrastructure to 2030, Volume 2 – mapping policy for electricity, water and transport*. Paris, France, Organisation for Economic Co-operation and Development.

Partridge, M. D. and D. S. Rickman (2008). "Computable General Equilibrium (CGE) modelling for regional economic development analysis." *Regional Studies* 44(10): 1311–1328.

Prud'Homme, R. (2004). *Infrastructure and development*, World Bank.

PWC (2013). World in 2050 – The BRICS and beyond: prospects, challenges and opportunities. PriceWaterhouseCoopers.

Rees, P., P. Wohland, P. Norman and P. Boden (2011). "A local analysis of ethnic group population trends and projections for the UK." *Journal of Population Research* 28(2–3): 149–183.

Romp, W. and J. De Haan (2007). "Public capital and economic growth: a critical survey." *Perspektiven der Wirtschaftspolitik* 8(S1): 6–52.

Santos, J. R. and Y. Y. Haimes (2004). "Modeling the demand reduction Input-Output (I-O) inoperability due to terrorism of interconnected infrastructures." *Risk Analysis* 24(6): 1437–1451.

Serven, L. (1996). *Does public capital crowd out private capital? Evidence from India.* World Bank Publications.

Straub, S. (2011). "Infrastructure and development: a critical appraisal of the macro-level literature." *The Journal of Development Studies* 47(5): 683–708.

United Nations (2012). *World urbanization prospects: the 2011 revision.* New York, USA, United Nations Department of Economic and Social Affairs.

United Nations (2013). *World population prospects: the 2012 revision.* New York, USA, United Nations.

4 Energy systems assessment

PRANAB BARUAH, MODASSAR CHAUDRY, MEYSAM QADRDAN,
NICK EYRE, NICK JENKINS

4.1 Introduction

Universal, affordable and clean energy is a prerequisite for economic and social development. Addressing global challenges from poverty eradication to food security, essential healthcare to peace, and from ecosystems conservation to security requires greater access to modern energy. Almost all goods and services have energy implications: from the foods we consume, the buildings we live and work in, vehicles we use and the day-to-day consumables at home and work – all require energy. With the current fossil fuel dominant global energy mix, energy use worldwide is responsible for $\sim 60\%$ of global CO_2 emissions. This also makes the energy system central to mitigation and management of anthropogenic climate change. It is estimated that in order to transform the energy system to address its current challenges with existing and emerging technologies, investments in the order of $\sim\$45$ trillion will be required by 2050 (OECD, 2012). Given the reliance of long-lived assets, path dependency and the size and complexity of the energy system, acting earlier and with a long-term and holistic approach will ensure that these investments will be cheaper and more effective to deliver long-term sustainable, effective and affordable energy systems.

Now, with industrialisation and economic development occurring at break-neck speed in high-population emerging economies, global energy demand hotspots and trends are once again changing rapidly, with the share of non-OECD countries' energy consumption surpassing OECD countries' in recent years (from 36% in 1973 to 57% in 2011). Non-OECD countries' share is set to grow further to $\sim65\%$ by 2035 (IEA, 2013) driven by increasing population, urbanisation and income, key drivers of energy demand in that part of the world. Global population is expected to increase from the current 7 billion to 8.3 billion by 2030, with 90% of the growth in low and medium income countries outside the OECD. The current trend of industrialisation, urbanisation and motor vehicle growth in developing and emerging economies will continue, leading to more than an 80% increase in energy use worldwide in a business-as-usual scenario (IEA, 2012). Demands in power generation and industry are projected to dominate, accounting for 57% and 25% growth in primary energy consumption by 2030, respectively (BP, 2010). Between 2011 and 2035, China is projected to account for $\sim30\%$ of net demand growth followed by India (18%), SE Asia (11%), Mid-East (10%) and Africa (8%), with demand likely to increase slightly in the US (1%) and decrease in the EU and Japan.

Today, fossil fuel still dominates the global energy system accounting for 82% of total primary energy use (2011 data). Ten years earlier, it was 80%, and without ambitious energy reduction measures it is projected to remain around that level by 2035 (IEA, 2013). After the oil shock of the 1970s, the world has seen a gradual increase of energy supply, mainly in electricity, from nuclear and hydropower, the overall share of which remains still considerably less at 8% in comparison to ~81% from incumbent fossil fuels (in 2008) (IEA, 2008). Climate change and supply security concerns have pushed development of renewable energy dramatically in the last two decades, yet their share in the energy system remains below policy aspirations and the requirements to keep global rise in temperatures below 2 °C by the end of the century, the threshold to avoid dangerous climate change. On the development front, energy access remains a key issue for much of the developing world with ~1.3 billion people without access to electricity (in 2009), most of whom are from developing countries in Asia and Africa (IEA, 2013).

4.2 Energy systems in Britain

4.2.1 System description

The analysis here focuses on Britain, but given that Britain has 97.5% of the U.K.'s population and is responsible for ~97% of annual U.K. energy consumption in 2012 (DECC, 2013a), we can also use national (U.K.) statistics as being relevant to Britain where subnational data are not readily available. The main energy networks in the U.K. are the gas and electricity systems. Liquid fuels are important for transportation (so are analysed in Chapter 5), whilst solid fuel (coal, biomass) is primarily used as an input for electricity generation. Both networks are designed to transport energy from remote locations to demand centres, often a considerable distance apart. It is this geographical separation that results in the energy transmission system being of such high national importance. Generally, both gas and electricity energy systems can be structured into the following categories:

- fuel source (coal, gas oil, uranium, etc.) and power generation;
- transmission (high voltage power network; high pressure gas network);
- distribution (medium/low voltage power network; medium/low pressure gas network);
- consumers (electricity/gas demand).

There are also important differences between these two networks. Natural gas constitutes a primary form of energy that comes from gas fields, while electrical energy is a secondary form of energy which is formed by the transformation of primary energy (fuel) in a power plant. Gas is transported from the gas fields (suppliers) to customers through pipelines while electricity is transmitted through power circuits. Additionally, gas networks can store natural gas to be used at peak load periods while electricity cannot yet be stored efficiently (although future electricity storage technologies may change this).

For the electricity system, the transmission network transports electricity from generation plants to distribution companies and large industrial customers. The distribution companies then deliver the electricity to the majority of customers through lower voltage networks. There are undersea interconnections to northern France (HVDC Cross-Channel), Northern Ireland (HVDC Moyle), Isle of Man (Isle of Man to England Interconnector), Netherlands (Britned) and the Republic of Ireland (EirGrid). There are also plans to lay cables to link the U.K. with Iceland and Norway (Scotland–Norway interconnector) in the future. For the gas system, the U.K. has nine gas terminals. The largest two are St Fergus (indigenous and Norwegian gas supplies) and Bacton (indigenous and Belgian/Continental gas supplies) gas terminals. The U.K. has previously been self-sufficient in gas supplies, but by 2005 was a net gas importer. To meet the challenge of import dependency the gas system has seen developments in three key areas including (i) pipeline imports, (ii) liquified natural gas (LNG) facilities and (iii) gas storage.

Total energy consumption in the U.K. has been decreasing over the past eight years. In 2012, primary energy consumption increased by 2.1% from the previous year, owing to a 1°C cooler weather and consumption increase in domestic (11%) and services (2.7%) sectors. Correcting for temperature, demand was 0.6% lower than the previous year. The government forecasts that the industrial sector will have a larger share of electricity demand by 2020, with an outlook of lower share in the residential sector.

4.2.2 Challenges and opportunities

With mounting evidence of dangerous anthropogenic climate change, the U.K. has taken a leadership role in setting a unilateral legally binding target of an 80% greenhouse gas reduction by 2050 from 1990 levels under the Climate Change Act (2008). This requires moving away from the incumbent fossil fuel-based energy system with dramatic transformations in order to meet the triple objectives of reducing greenhouse gas emissions while maintaining energy security and providing affordable energy to end-users.

A continued dependence on fossil fuels would imply a long-term trend of rising imports, due to falling domestic production of North Sea gas and oil, which would leave the U.K. more exposed than historically to international fossil fuel markets. These are notoriously volatile, so that in periods of rising international energy prices, there would tend to be widening trade and budget deficits. For example, until the recent fall in world oil prices, increasing wholesale prices put upward pressure on retail energy prices. Retail prices for electricity and gas increased by 75% and 122%, respectively, during 2004–2009, much faster than incomes, increasing the number of households in fuel poverty from 2 million in 2002 to 5.5 million in 2009 (DECC, 2013b). Under the EU Renewable Energy Directive, the U.K. also has a target of 15% renewable energy in final energy consumption by 2020. With the likelihood of modest amounts of renewables in heating and transport, a significant proportion of electricity supply is expected from renewable sources by 2020.

The U.K., however, is not new to managing transformative change in its energy system. In the past 150 years or so, it has successfully transformed its energy system on at least three occasions. The first was during the industrial revolution when its energy system transformed from a largely biofuel based system to an industrial economy heavily dependent on coal. In the early twentieth century, local systems of power generation and gas production, both largely using coal were developed, with increasing scale of production and the development of national systems after World War II. With the rise of oil and onset of mass mobility, the U.K. moved to a more oil-dependent energy system. After the oil-shocks and discovery of the North Sea oil and gas in the 1960s/1970s, the energy system again went through transformation to the current state, with natural gas increasingly playing a dominant role, first in heating and later in electricity generation. Privatisation of the energy market in the 1990s saw large shifts in the way the energy system is owned and largely drove the shift to gas-fired power generation. Transitioning the current system to meet new policy objectives will involve new risks and challenges, as it requires moving away from the current highly (technically) reliable energy system. Uncertain types and levels of distributed generation, the possible electrification of transport and heating, ICT-enabled options and variable renewables bring additional complexities (as well as potential benefits) for technology, policy and governance.

To provide certainty and direction to this challenging transition, particularly to meet the carbon goal while satisfying security and affordability objectives, in 2008 the U.K. Government established an independent advisory body, the Committee on Climate Change, responsible for recommending evidence-based five-year carbon budgets. Particularly over the last fifteen years, successive Governments have published a number of reviews and strategy documents and enacted policies both to improve energy efficiency and begin the decarbonisation of energy supply. Key policies for energy efficiency include the Energy Company Obligation, building standards, the Carbon Reduction Commitment and Climate Change Agreements, renewable energy (e.g. EU renewables target), and structural reforms of the market, while renewable energy uptake has been largely driven by the Renewables Obligation, Feed-in-tariff and the Renewable Heat Incentive. All of these operate within the framework of EU energy and climate policy measures, of which the EU Emissions Trading is the best known, but vehicle and appliance standards are also important. An overview and historical list of U.K. energy policy instruments and related strategy documents can be found in IEA (2012) and the IEA country policy and measures database (IEA, 2014).

Appendix A (available online at itrc.org.uk) provides an overview of the evolution of the U.K. energy system within a global context. A summary of the current state and future outlook for energy demand and supply infrastructure in the U.K. energy system is also provided. Governance issues for energy systems, in particular concerning the interdependence between energy and other networks, are dealt with in Chapter 14.

The U.K. now faces the 'energy trilemma', which also materialises in many different settings globally, namely balancing energy security, environmental goals and affordability. Energy security is now threatened by life-expired generation plant and declining domestic gas supplies. Environmental goals, in particular relating to carbon emissions reduction, imply a major reconfiguration of the energy supply infrastructure, whilst this is challenged by affordability, in particular for low income ('fuel poor') households.

4.3 Assessment of national energy systems

The Committee on Climate Change (CCC), the government's independent advisory body on meeting the carbon target, has set out five-year carbon budgets leading to year 2050 and has published quantitative pathways to a lower carbon energy system (Committee on Climate Change, 2008) adopted by successive governments (HM Government, 2009; 2011). Academic studies (Skea et al., 2010) have also contributed to understanding the range of technological, social, governance and policy issues involved in meeting the transition goals. The work described in this chapter contributes to this body of work, with a specific goal of providing strategic foresight from an infrastructure perspective. With a radically different energy system, infrastructure challenges and opportunities are likely to be significant in three areas: (i) implementing new technologies and infrastructure; (ii) managing the transition and decommissioning of incumbent energy infrastructure; and (iii) exploiting opportunities from the interdependencies between different infrastructure systems (e.g. transport electrification and energy infrastructure, energy mix and water use, energy from waste). Additionally, these have to be implemented on time and in a cost-effective way. The combination of triple objectives, an evolving energy market and a largely globalised technology market means the long-term transition of a complex national energy system is fraught with uncertainty. This can be further exacerbated by an uncertain socio-economic landscape in a country like the U.K. that is highly linked internationally. With these underlying uncertainties, strategic foresight on risks and opportunities for a robust transition pathway is valuable to avoid lock-in to costly and undesirable pathways, especially in the light of the long lifespan and the high costs of energy system investment.

To carry out the analysis, coupled spatially explicit models of demand and supply in the energy system, its drivers and key transition measures, are employed. A number of plausible yet distinct futures for energy infrastructure and patterns of energy demand are developed, modelled and analysed in a wider infrastructure context. The analyses focus on the two major energy carriers, electricity and gas, with significant network infrastructures both in the context of current and future U.K. energy systems. Both demand and supply-side transition aspects and energy system-wide impacts are investigated. This enables investigation of challenges and opportunities at both energy end-use and supply levels as well as from energy system interdependencies with other infrastructure sectors, notably with the transport and water sectors (Chapter 11). The quantitative analysis methodology and results are presented next, followed by analysis of strategy performance, critical insights and policy recommendations. Institutional and governance implications are discussed in Chapter 14.

4.3.1 Modelling approach

The modelling framework (Baruah et al., 2014) couples eight separate models to provide an analysis of the GB energy system and its alternative futures. Residential, services and industry sector energy consumption, as well as electricity and gas peak loads, are estimated by bespoke models. Transport energy consumption is modelled from transport services demand estimated by the transport model described in Chapter 5. Electricity supply analysis is carried out with the cost optimisation model, CGEN+ (Combined Gas and Electricity Network model) (Chaudry et al., 2014). Figure 4.1 shows the modelling framework with key inter-model data links (note that 'Transport model' in this figure refers to an internal model of transport energy use, rather than the model described in Chapter 5). Appendix B (available online at itrc.org.uk) provides an overview of different energy modelling approaches and details of the demand and supply models employed for this analysis.

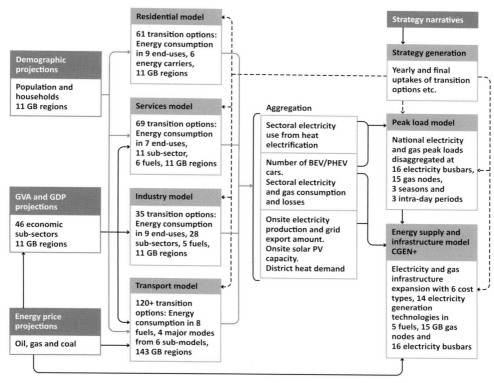

Figure 4.1   Schematic of the modelling framework and inter-model data-links

4.3.2 Strategy description

The four chosen energy strategies and their variants are briefly described below. Detailed narratives of these strategies are given in Appendix C (available online at itrc.org.uk).

(1) Minimum Policy Intervention (MPI) is the reference strategy and assumes minimal energy efficiency, conservation and fuel switching in the energy system. The MPI energy demand estimates are used with three variants of supply-side mix – with the Government's announced carbon price floor enforced, without any carbon price (termed MPI-No CO_2 cost) and gas demand met with domestic shale gas production (GasWorld).

(2) Electrification of Heat and Transport (EHT) envisages a highly electrified future, comprising three different variants of electricity generation mixes – nuclear power centric (EHT-Nuclear), offshore wind centric (EHT-Offshore) and a third dominated by a fossil-fuel mix with carbon capture and storage (EHT-CCS).

(3) Local Energy and Biomass (LEB) envisages a future with uptake of distributed generation technologies, especially solar- and biomass-based systems, on the demand-side, and a cost-optimised expansion of the supply-side infrastructure with a carbon price floor enforced.

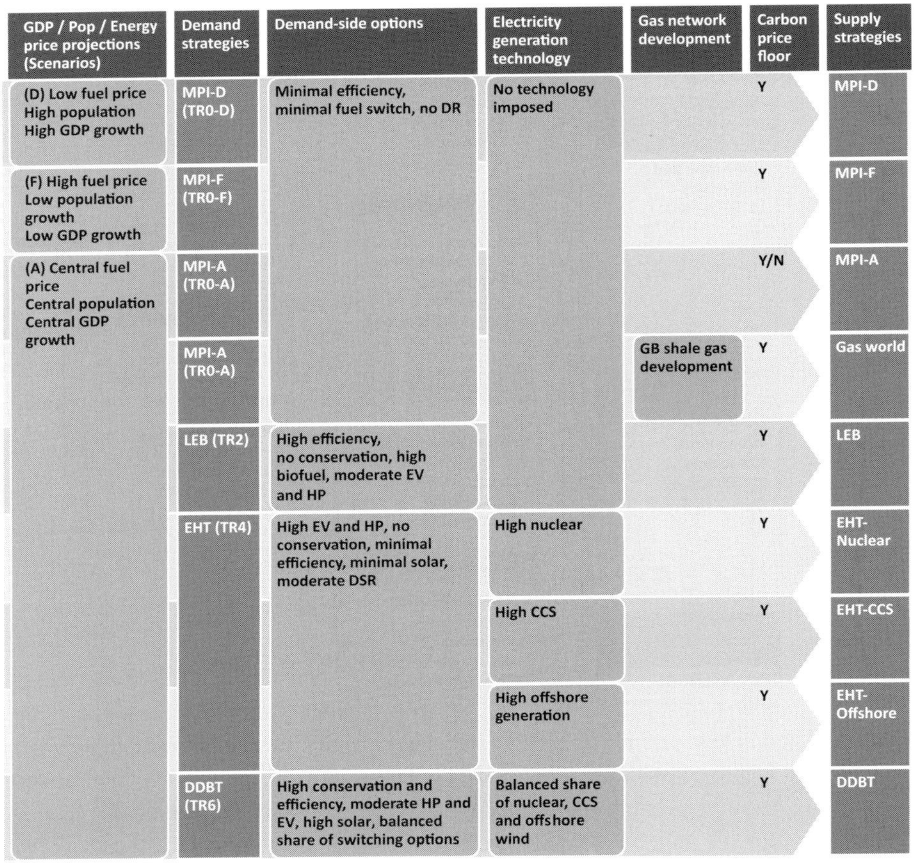

GDP / Pop / Energy price projections (Scenarios)	Demand strategies	Demand-side options	Electricity generation technology	Gas network development	Carbon price floor	Supply strategies
(D) Low fuel price High population High GDP growth	MPI-D (TR0-D)	Minimal efficiency, minimal fuel switch, no DR	No technology imposed		Y	MPI-D
(F) High fuel price Low population growth Low GDP growth	MPI-F (TR0-F)				Y	MPI-F
(A) Central fuel price Central population Central GDP growth	MPI-A (TR0-A)				Y/N	MPI-A
	MPI-A (TR0-A)			GB shale gas development	Y	Gas world
	LEB (TR2)	High efficiency, no conservation, high biofuel, moderate EV and HP			Y	LEB
	EHT (TR4)	High EV and HP, no conservation, minimal efficiency, minimal solar, moderate DSR	High nuclear		Y	EHT-Nuclear
			High CCS		Y	EHT-CCS
			High offshore generation		Y	EHT-Offshore
	DDBT (TR6)	High conservation and efficiency, moderate HP and EV, high solar, balanced share of switching options	Balanced share of nuclear, CCS and offshore wind		Y	DDBT

Figure 4.2   Schematic of the high-level assumptions that are analysed for the energy sector

(4) Deep Decarbonisation with Balanced Transition (DDBT) envisages a future with ambitious demand reduction measures to meet decarbonisation, supply security and affordability goals. It assumes that there is no policy preference for a single or a set of technologies and all major options, both on the demand and supply side, are used by 2050.

Each of these four strategies is modelled for the high (D), central (A) and low (F) socio-economic growth projections (described in Chapter 3) to quantify uncertainties in demand and to investigate potential for peak demand response. All energy supply modelling analyses (except for MPI) are based on energy consumption from central (A) projections. Strategy names with the socio-economic projection extension '-A' or no extension (e.g. MPI and MPI-A) denote the central socio-economic projection as an input.

Figure 4.2 shows the high-level assumptions in the strategies for major transition drivers and options. Transport strategies that correspond to the energy-side strategies are also shown.

Table 4.1 Salient demand-side features in the transition strategies	
Transition strategy	Salient features
Minimum policy intervention (MPI)	Achieve 10% of efficiency and fuel switching potential by 2050.
Local energy and biomass (LEB)	Moderate heat pumps, moderate micro-combined heat and power (CHPs), high solar photovoltaic (PV) and biomass systems.
Electrification of heat and transport (EHT)	High heat pumps, high electric cooking, high electric Arc furnace, low micro-CHPs, low solar, minimal efficiency.
Deep decarbonisation with balanced transition (DDBT)	Efficiency (housing and appliances) at the full potential level, high solar PV/thermal, high fuel switching with a balanced uptake of heat pumps, micro-CHPs, biomass and district heating systems.

Demand-side assumptions

Table 4.1 lists the salient features of the demand-side options in the four strategies. Uptake levels of the major transition options in these four strategies are given in Table 4.2.

Energy supply strategies

The CGEN+ model plans the energy supply system to minimise the costs, subject to any constraints imposed (such as capacity margins). Different strategies are therefore implemented through the model by exogenous choices, such as carbon prices, or minimum/maximum requirements for a specific technology. All strategies assume the Government's announced carbon price floor (i.e. a minimum U.K. carbon price for U.K. power generators, even when the EU-ETS price is lower), rising from £16 per tonne of CO_2 in 2013 to £70 in 2030 unless stated otherwise (at the time of writing it is capped at £18 until 2018/2019 (HMRC, 2014)).

Minimal policy intervention (MPI) – no particular generation technology is imposed. Emission targets in 2050 as well as the renewable target for 2020 are not enforced. Therefore the generation mix and expansion of the gas and electricity infrastructure is determined by cost minimisation, subject to limits on resources and meeting gas and electricity demands. The impact of imposing a carbon price floor on the generation mix was investigated through two variants, with and without a carbon price floor.

Gas world (with MPI demand regime) – the Gas World strategy is very similar to MPI-A, the only difference is that exploitation of GB shale gas was considered. All the other assumptions regarding energy demand and generation mix are the same. In the Gas World strategy, it was assumed that shale gas exploitation in Bowland-Hodder (Northwest England) will be started in 2020. According to the BGS (Andrews, 2013) shale gas resources (Gas in Place) in Bowland-Hodder are between 23.3 tcm and 64.6 tcm. Assuming a recovery factor of 8–20%, similar to that achieved in the US, potentially recoverable resources of shale gas in Bowland-Hodder are estimated to be between 1,800 bcm and 13,000 bcm. In this study, the lower estimates of shale resource (23.3 tcm) and recovery factor (8%) were used. It was

Table 4.2 Uptake levels of major demand-side energy transition options

Efficiency and electrification in residential, services and industry sectors

Major transition options	MPI	EHT	LEB	DDBT
Energy conservation: internal temperature change [degree C]	None	Same as MPI	Same as MPI	Residential: 1 Services: 1
Building fabric performance [% improvement in average leakage rate of building stock]	Residential and services: 5	Same as MPI	30	50
Lighting efficiency: efficiency/stock change [% of savings and/or technology uptake potential achieved]	Residential and services: 10 Industry: 20	Same as MPI	Residential and services: 50–80 Industry: 25	Residential and services: 100 Industry: 25
Appliance efficiency [% of savings potential achieved]	Residential: 10 Services: 10 Industry: 20	Same as MPI	Residential: 57–65 Services: 75–80 Industry: 100	Residential: 100 Services: 100 Industry: 100
Onsite solar thermal installation [% of water heating energy consumption met]	Residential: 6 Services: 5 Industry: 0	Same as MPI	Residential: 40 Services: 30 Industry: 0	Residential: 60 Services: 50 Industry: 0
Heating demand met by recovered heat in specified end-uses in sub-sectors [% of sub-sectoral heating demand met]	Services: 0–4 Industry: 4–8	Same as MPI	Services: 0–20 Industry: 5	Services: 0–40 Industry: 20–40
Heating fuel (technology) replaced by alternative technologies in specified end-uses [total % replacement range of incumbent fuel consumption]	Residential: 9–13 Services: 7–10 Industry: 9–12	Residential: 86–91 Services: 71–83 Industry: 21–49	Residential: 70–100 Services: 87–89 Industry: 40–42	Residential: 85–93 Services: 80–100 Industry: 65–100
Heating fuel demand replaced with electric heat pumps in specified end-uses [% replacement of gas/electric/oil/solid fuel consumption]	Residential: [3/3/3/3] Services: [3/3/3/3] Industry: [3/3/3/3]	Residential: [80/80/80/80] Services: [65/75/60/60] Industry: low temp. process: [15/15/15/–]; space heating: [40/40/40/–]	Residential: [40/40/40/40] Services: [30/30/35/30] Industry: low temp. process: [7.5/7.5/7.5/–]; space heating: [7.5/7.5/7.5/–]	Residential: [40/70/40/50] Services: [35/40/40/30] Industry: low temp. process: [15/15/15/–]; space heating: [20/40/20/–]

Parameter				
Share of electric arc furnace in iron and steel sector	16	80	40	80
Relative ground source heat pump installations in residential and services sectors [%] (ground source + air source heat pumps = 100%)	6	50	25	60
Onsite solar PV installation [Watt-peak per person]	43	Same as MPI	344	430

Efficiency and electrification in transport sector

Parameter				
Road transport fuel efficiency improvement [% in ICE / BEV / PHEV / Hybrid / Hydrogen fuel cell vehicle] (ICE: Internal Combustion Engine)	Cars: [0 / 0 / 0 / 0 / 0] LGV: [0 / 0 / 0 / 0 / 0] HGV: [0 / 0 / 0 / 0 / 0] PSV: [0 / 0 / 0 / 0 / 0]	Cars: [18 / 23 / 27 / 18 / 24] LGV: [19 / 11 / 18 / 18 / 0] HGV: [19 / 0 / 0 / 0 / 25] PSV: [18 / 9 / 24 / 16 / 22]	Cars: [18 / 23 / 27 / 18 / 24] LGV: [19 / 11 / 18 / 18 / 0] HGV: [19 / 0 / 0 / 0 / 26] PSV: [18 / 9 / 24 / 16 / 22]	Cars: [18 / 23 / 27 / 18 / 24] LGV: [19 / 11 / 18 / 18 / 0] HGV: [19 / 0 / 0 / 0 / 26] PSV: [18 / 9 / 24 / 16 / 22]
Rail, air and sea transport fuel efficiency improvement in [%]	Rail: 0 Air: 0 Sea: 0	Rail-electric: 58; Rail-diesel: 42 Air: 30 Sea: 35	Rail-electric: 58; Rail-diesel: 42 Air: 30 Sea: 35	Rail-electric: 58; Rail-diesel: 42 Air: 30 Sea: 35
Road transport electrification rate [% of BEV / PHEV / Hybrid / Hydrogen fuel cell vehicle]	Cars: [0 / 0 / 0 / 0] LGV: [0 / 0 / 0 / 0] HGV: [0 / 0 / 0 / 0] PSV: [0 / 0 / 0 / 0]	Cars: [50 / 30 / 10 / 0] LGV: [50 / 30 / 10 / 0] HGV: [40 / 0 / 40 / 0] PSV: [40 / 30 / 10 / 0]	Cars: [50 / 30 / 10 / 0] LGV: [50 / 30 / 10 / 0] HGV: [40 / 0 / 40 / 0] PSV: [40 / 30 / 10 / 0]	Cars: [45 / 20 / 5 / 20] LGV: [50 / 25 / 5 / 0] HGV: [40 / 0 / 40 / 5] PSV: [40 / 30 / 10 / 5]
Rail electrification rate [km per year]	None beyond existing projects	300	300	200
Number of EV/PHEV cars [million]	0	26	13	17

Smart meter uptake and use of vehicle to grid (V2G) and grid to vehicle (G2V)

Parameter				
Share of end-users connected to grid through smart meters	2010: 0%; 2020: 100%	Same as MPI	Same as MPI	Same as MPI
BEV/PHEV cars connected to the grid for G2V/V2G during peak hours [%]	NA	G2V: 20 / V2G: 10	G2V: 20 / V2G: 10	G2V: 20 / V2G: 10

Table 4.3 Total cumulative generation capacity for coal and CCGT equipped with CCS at each planning milestone in EHT-CCS (Source: DECC, 2010)

Year	Coal CCS (GW)	CCGT CCS (GW)
2020	2.5	0
2030	10	7
2040	25	11
2050	40	17

Table 4.4 Total cumulative generation capacity for nuclear at each planning milestone in EHT-Nuclear (Source: DECC, 2010)

Year	Nuclear (GW)
2020	13
2030	30
2040	60
2050	90

assumed that only 30% of the potentially recoverable resources are economically viable. The cost of shale gas extraction was assumed to be £4.40 per MBTU (the minimum quoted by BNEF (2013)). Maximum shale gas production rate that can be achieved was assumed to be approximately 32 (bcm) per annum (Taylor and Lewis, 2013).

Electrification of heat and transport with large capacity of CCS (EHT-CCS) – it was assumed that carbon capture and storage is demonstrated on both coal and gas power stations and rapidly becomes the preferred form of generation investment. There is rapid investment after 2030 when the technology is projected to become mature. The total capacity of CCS-equipped power plants at each year was taken from DECC 2050 Pathways – level 3 (DECC, 2010) and is presented in Table 4.3.

Electrification of heat and transport with large capacity of nuclear (EHT-Nuclear) – it was assumed that there is successful investment in nuclear power in the 2020s and a steady growth in investment in the subsequent decade, followed by new generation IV technologies after 2030. Investment is confined to existing coastal nuclear sites. The total capacity of nuclear power plants at each year was taken from DECC 2050 Pathways – level 3 (DECC, 2010) and is presented in Table 4.4.

Electrification of heat and transport with large capacity of offshore (EHT-Offshore) – there is early and rapid investment in offshore wind, primarily in the North Sea, followed by wave and tidal flow investment, mainly in the Atlantic, after 2030. The total capacity of offshore generation at each year was taken from DECC 2050 Pathways – level 3 (DECC, 2010) and is presented in Table 4.5.

Table 4.5 Total cumulative generation capacity for offshore technologies at each planning milestone in EHT-Offshore (Source: DECC, 2010)			
Year	Wind offshore (GW)	Tidal range (GW)	Tidal stream (GW)	Wave (GW)
2020	25	0	0	0
2030	70	4	0	0
2040	95	13	4.5	10
2050	100	13	9	20

Table 4.6 Total cumulative generation capacity at each planning milestone in DDBT (Source: DECC, 2010)		
Year	Wind offshore (GW)	Coal CCS (GW)	CCGT CCS (GW)
2020	25	–	–
2030	25	2.5	–
2040	25	5	2.5
2050	25	5	2.5

Local energy and biomass (LEB) – expansion of supply infrastructure in electricity and gas sectors was determined based on the least cost optimisation approach. No generation technology was enforced on the model.

Deep decarbonisation with balanced transition (DDBT) – a balanced contribution from a number of low carbon generation technologies (including wind offshore, coal with CCS and CCGT with CCS) was assumed. Table 4.6 shows capacities of the generation technologies imposed at each planning milestone in DDBT strategy.

4.4 Strategy analysis and discussion

4.4.1 Exploring energy demand across strategies

Demand evolution in the MPI strategy – no energy conservation, management or reuse measures are applied. Uptake of efficiency measures is $\sim$10% of assumed potential level by 2050. Similarly, fuel switching by 2050 is $\sim$10% of assumed switching potential. Each of heat pumps, micro-CHP and biomass boilers replace gas boilers for space heating by 3% by 2050. Uptake of solar PV is minimal and solar thermal heating does not contribute to meeting hot water demand.

Figure 4.3 shows annual energy consumption (excluding transport) under high (D), central (A) and low (F) socio-economic projections with a MPI strategy. By 2050 annual

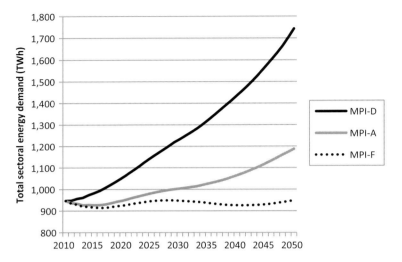

Figure 4.3 Total sectoral energy consumption (excluding transport) in the MPI strategy under high (D), central (A) and low (F) socio-economic growth scenarios

consumption can vary by ∼1.8 times base year demand. High (D) and low (F) estimates are 46% above and 25% below the central estimate (A).

Results in Figure 4.4 show similar trends in annual energy consumption. The higher divergence from central estimates in solid fuel is primarily driven by metals and coke production in the industry sector. About half of biomass consumption in the MPI-D strategy by 2050 is in industry, and is driven by rapidly expanding industry sub-sectors with large low temperature process heat loads (e.g. food and beverage, pulp and paper, chemicals) and replacement of gas and oil in low temperature processes and space heating with biomass CHP systems (∼3% replacement in both end-uses). About 30% of demand is from fuel switching for space and water heating in the residential sector. Two-thirds of the heat sold by 2050 in the MPI-D strategy comes from switching to district heating in the residential sector.

Half-hourly electricity peak load and daily gas peak loads are estimated from annual demands, using current data on the profiles of different energy services (e.g. heating). Further description of peak load estimation methods are given in Appendix B (available online at itrc.org.uk). Figure 4.5 shows the estimated national electricity and gas peak load evolution in the MPI strategy.

Figure 4.6 shows disaggregated total energy, electricity and gas consumption in each strategy. All the three non-MPI strategies have lower annual energy consumption than the base year and reference strategy in 2050. Heating energy consumption (which consists of space heating, water heating, cooking, catering, process use and drying/separation) is responsible for ∼80% of the total (excluding transport). Fuel switching measures, in conjunction with energy reuse, energy efficiency and solar thermal water heating, results in lower heating energy consumption in the DDBT, LEB and EHT strategies by about 60%, 40% and 40%, respectively, below reference strategy by 2050. In the electricity specific

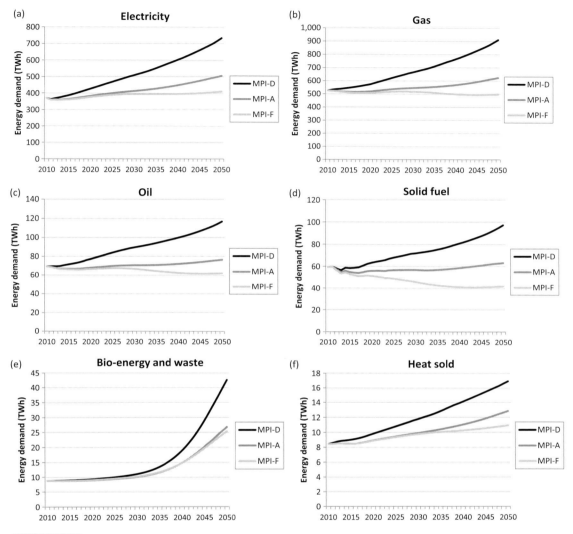

Figure 4.4 Total sectoral energy consumption (excluding transport) by fuel type in three MPI strategy variants using high (D), central (A) and low (F) socio-economic growth scenarios

end-uses, energy conservation, management, efficiency and reuse measures resulted in significantly lower electricity consumption in computing (~60% lower), cooling and ventilation (~59%), lighting and appliances (~58%) and motors and compressed air (~40%) by 2050 compared to the reference strategy.

As the majority of heating services are currently met with gas-based technologies, switching to efficient alternative technologies reduces annual gas consumption significantly. Switching to alternative gas-based systems, such as gas CHP, could potentially increase annual gas consumption. Uptake of gas CHP is not as high as that of heat pumps in any of the strategies investigated here.

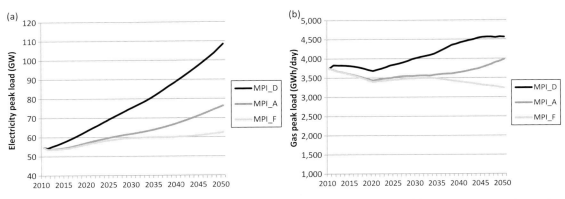

Figure 4.5   Electricity and gas peak loads in MPI strategy under high (D), central (A) and low (F) socio-economic growth scenarios

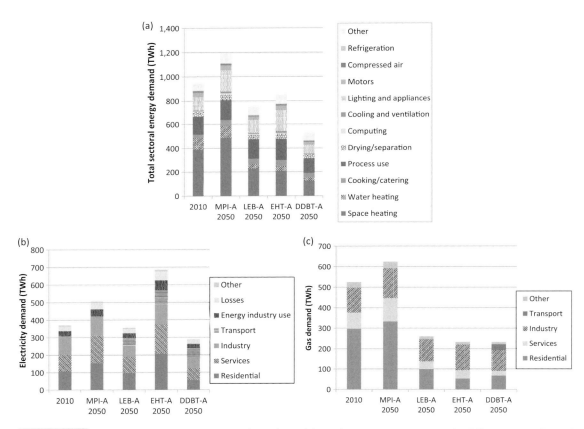

Figure 4.6   Total energy demand, electricity and gas demand (central socio-economic scenario for different energy demand strategies)

Minimum demand for electricity is achieved in the DDBT strategy despite a significant increase in electricity consumption (over the MPI reference strategy) in transport and heating, with total consumption 22% lower than the base year and 43% below the reference strategy in 2050. This is achieved by energy conservation, energy efficiency and fuel switching in key end-uses and use of electricity from onsite solar PV and CHP systems. In the LEB strategy, lower uptake of conservation, efficiency measures, solar PV and fuel switching result in electricity consumption higher than the DDBT strategy despite less transport electrification. The largest savings in the DDBT strategy are achieved in the residential sector, followed by services and industry.

Among the four strategies, the EHT strategy envisages the highest level of electricity use for transport and heating services. Transport electrification rises to ~81 TWh (~12% of total) by 2050, about twice the level in the LEB strategy. High levels of heat pump uptake (in the residential and services sectors) and 80% conversion to electric arc furnaces (EAF) for steelmaking (from the current ~25%) further increase electricity consumption by ~61 TWh. All residential gas cooking is also replaced with efficient electric hobs and ovens. Heat pumps drastically reduce overall heating energy consumption, mainly by replacement of gas boilers in the residential sector. By 2050, electricity consumption in the EHT strategy is ~35% more than the reference strategy with implementation of minimal efficiency and conservation measures and uptake of minimal distributed generation.

Compared with the MPI strategy, the reduction in electricity consumption by 2050 in the LEB and DDBT strategies are significantly higher (about 12 and 18 times, respectively) than Britain's electricity import in 2012 (12 TWh). In contrast, consumption increased by ~14 times this amount in the EHT strategy. Plausible supply strategies and energy infrastructure implications for meeting this demand in the EHT strategy are analysed in Section 4.3.2. The results indicate significant levels of bi-directional electricity flows in all three non-MPI strategies from vehicle-to-grid use and grid export of electricity generated onsite – under a central socio-economic growth scenario, exported electricity to the grid was about 3.7, 17.0 and 20.8 TWh by 2050 in the EHT, LEB and DDBT strategies, respectively.

Figure 4.6 also shows the annual gas consumption across the strategies. In contrast to electricity, gas consumption in all three non-MPI strategies has similar levels, about 60% below today's consumption by 2050. In the EHT strategy, this reduction is achieved primarily by gas boiler replacement with heat pumps. In the LEB strategy, efficiency improvements (in housing fabric and heating systems) are deployed along with fuel switching (heat pumps and biomass-boilers). Despite efficiency measures, gas consumption in the LEB strategy is higher than the EHT strategy, highlighting the importance of heat pumps in reduction of gas consumption – in the LEB strategy, heat pump uptake by 2050 is at half the level of EHT strategy. Biomass CHP systems replace low temperature process and space heating systems to a greater level in the LEB strategy than heat pumps do in the EHT strategy, resulting in lower gas consumption in LEB compared to EHT strategy. The DDBT strategy produced the lowest annual gas consumption through a multi-pronged approach of energy conservation, ambitious building fabric and appliance efficiency and higher uptake of heat pumps than the LEB strategy.

Figure 4.6 does not include gas consumption from gas-fired electricity generation. The energy supply modelling and analysis described in Section 4.3 takes this into account to provide total gas consumption.

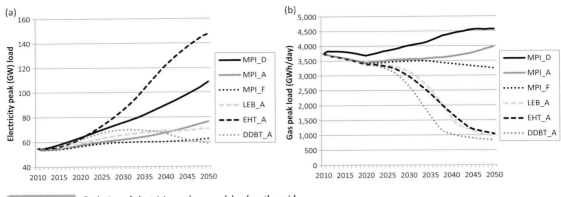

Figure 4.7   Evolution of electricity and gas peak load on the grid

4.4.2 Peak electricity and gas demand across strategies

Figure 4.7 shows electricity peak load evolution in the different strategies. Peak electricity demand is impacted by a number of factors, including change in current demands and the level of electrification of heat and transport (refer to Appendix B for methodology and details). In the EHT strategy, a large number of Electric (EV) and Plug-in Hybrid (PHEV) cars charging from the grid (G2V) and electric heat pumps running at peak hours result in significant increase in the peak load. In a business-as-usual type scenario (MPI strategy) with no new electrification of heat and transport, residential and services sectors' demands drive peak electricity load. In the MPI-D strategy (MPI with high socio-economic growth assumptions) peak demand increases by about 40% with a ~45% increase in annual electricity consumption above the MPI-A strategy (MPI with central socio-economic growth). The level of electrification and peak hour usage regimes determine the level and shape of the peak load. In the EHT strategy, peak demand nearly doubles (~93% over MPI-A) and reaches ~2.7 times that of base year level, even though annual electricity consumption increases by only ~35%. In the EHT strategy, transport and heat electrification's share of total demand are 12% and 9%, respectively.

Peak electricity demand has attracted greater attention in Britain in recent years in light of the retirement of a large capacity of power generation plants and the lack of investment in new gas generation capacity due to an unattractive 'spark-spread' for gas-based generation and uncertain policy signals. Even small reductions in generation capacity margin from the current levels would result in a significant increase in security supply risks (Ofgem, 2013).

Gas peak demand evolution across the strategies follows similar trends to annual gas consumption. As seen in Figure 4.7, in the LEB, EHT and DDBT strategies, gas peak load decreases by more than 70% compared to the MPI strategy in 2050. The MPI-D strategy (high socio-economic growth) has the highest peak load of about ~14% over the MPI-A strategy (central socio-economic growth).

This contrasting evolution of electricity and gas peak loads highlights differing challenges for energy supply infrastructure across the strategies. In the EHT strategy, there are significant implications for both electricity infrastructure (from expansion) and existing gas

Table 4.7 Assumptions on peak hour usage of G2V, V2G and back-up systems in the electricity peak load sensitivity analysis based on EHT-A demand			
Sensitivity (S)	% of PHEV/BEV cars on G2V	% of PHEV/BEV cars on V2G	Is back-up heating system used with heat pump?
S-1	20	0	Yes
S-2	20	10	Yes
S-3	10	20	Yes
S-4	0	20	Yes
S-5	20	0	No
S-6	20	10	No
S-7	10	20	No
S-8	0	20	No

Note: S-6 is the reference case used in all energy supply analysis.

infrastructure (from disuse and decommissioning). Depending on the supply-side strategy, cost, environmental, public acceptance and governance implications can be significantly different, as discussed below. In the LEB and DDBT strategies, where electricity consumption is significantly lower, implications are less pronounced. Both sectoral annual and peak demands show that in all non-MPI strategies, gas infrastructure implications would be at a comparable level.

4.4.3 Impact of demand response on peak electricity demand

The EHT, LEB and DDBT strategies assume that, during peak hours, 20% of EV and PHEV cars are charging from the grid (G2V) and 10% are providing vehicle-to-grid (V2G) demand response (enabled by a smart grid). Heat pumps are assumed to be operating and meeting full heating services demand during the peak hours. A sensitivity analysis is carried out to assess the implications of different levels of V2G, G2V and heat pump back-up systems (e.g. gas boilers in hybrid systems or heat storage) during peak hours.

Eight simulations to analyse sensitivity (as shown in Table 4.7), with different G2V, V2G and back-up system usage levels, are modelled. As shown in the Figure 4.8, peak hour usage of back-up systems impacts peak load significantly. Back-up systems have the potential to reduce the electricity peak load by ~28% (S-2 compared to reference S-6). However, without V2G demand response, peak load will increase. The sensitivity analysis shows that, under current EV and PHEV cars' charging and discharging assumptions, a 10% increase in peak hour V2G demand response provides between 7% and 9% reduction in electricity peak load (S-2 compared to S-1 and S-6 compared to S-5). On the other hand, a 10% increase in peak hour G2V would result in between 25% and 33% increase in the peak load. Factors contributing to differences in the peak load include charging and discharging power characteristics – in our analysis we assume G2V charging power at twice the level that of V2G discharging power.

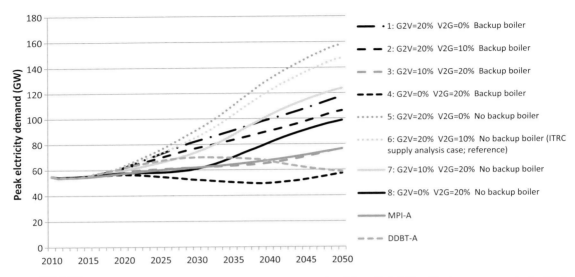

Figure 4.8 Evolution of electricity peak load under different levels of peak hour V2G, G2V and backup system use in the EHT-A strategy. Peak loads in the MPI-A and DDBT-A strategies are shown for comparison

The peak demand management benefits from V2G and back-up systems in an EHT-like strategy are illustrated by S-3. With back-up systems for heat pumps, a V2G level of 20% and G2V level of 10%, it is possible to keep peak load at MPI-A levels (i.e. more than 45% below reference S-6) even with a much higher annual electricity demand. If no EV and PHEV cars are charged during the peak hours (S-4), peak demand can be further reduced to the DDBT-strategy level by 2050 (this is ~60% below reference S-6). However, it should be noted that the infrastructure planning implications are likely to be significantly different as peak load evolution paths for these two cases are different despite similar levels of peak load by 2050. Also, with similar levels of V2G, G2V and back-system usage, as in the case of S-4, peak demand in the DDBT strategy would be lower than currently.

4.4.4 Impact of uncertainty in socio-economic projections on annual electricity and gas consumption

Energy system transition modelling studies may focus on a limited number of trajectories of scenarios and strategies. However, there can be a wide range of plausible future consumption levels driven by varying socio-economic scenarios, new energy service demands and/or technologies (e.g. electric vehicles and heat pumps) and new efficiency measures. Socio-economic drivers are often the primary causes of consumption change and can offset emissions reductions from elsewhere in the energy system transition. Sensitivity of energy consumption to different plausible levels of socio-economic growth projections (GDP and population) is carried out for the four strategies and is shown in Figure 4.9.

Our analysis shows an electricity consumption range from 250 TWh to 900 TWh in 2050. The highest (EHT-D) and lowest (DDBT-F) estimates are 35% above and 52%

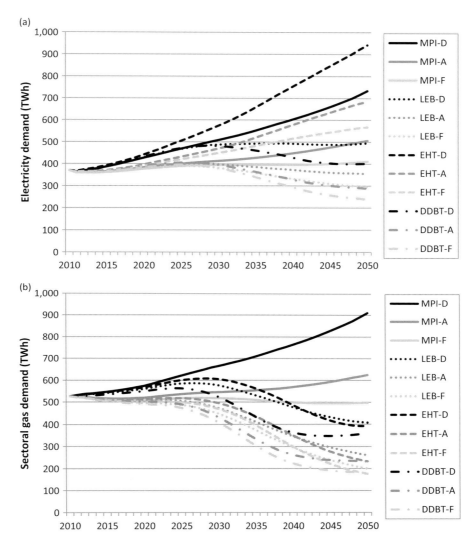

Figure 4.9   Annual electricity and sectoral final gas consumption evolution in the strategies with high (D), central (A) and low (F) socio-economic growth scenarios (electricity includes consumption from transport and other sources as shown in Figure 4.6, gas excludes gas in electricity generation)

below MPI-A, respectively. Within each transition strategy, the socio-economic growth sensitivities investigated resulted in 17–19% decreases and 39–45% increases in electricity consumption.

For gas, the highest and lowest consumption levels are 45% above (MPI-D) and 71% below (DDBT-F) the central strategy (MPI-A) in 2050, about 40% of base year gas consumption. Within each strategy, the socio-economic sensitivity analysis shows consumption 20–23% below and 30–45% above the central estimates in 2050.

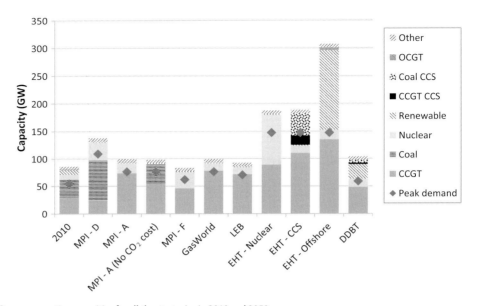

Figure 4.10 Power generation capacities for all the strategies in 2010 and 2050

4.4.5 Strategies for electricity generation

The electricity supply analysis explores the infrastructure required to meet the various demand scenarios that have just been described. The generation capacity mix for all the strategies is shown in Figure 4.10. The MPI strategy uses three different socio-economic scenarios (D, A and F) to show sensitivity around the reference case. All other strategies use the central scenario A as an input. In MPI-A strategy with no carbon floor price imposed, combined cycle gas turbine (CCGT) capacity increases from under 30 GW in 2010 to 54.9 GW by 2050 and accounts for ~60% of total capacity. CCGT capacity is built mainly due to low capital and fixed operating and maintenance (O&M) costs and also because of favourable fuel price variations (lower gas price in summer) compared with other fuels. New coal plants are built to compensate for decommissioning of old coal plants, and this maintains coal capacity at around 35 GW (36% of total capacity in 2050). Coal plants have an overall load factor of 80% by 2050 while the CCGT load factor is 39%. Due to gas price variations in 2050, the CCGT load factor during summer/intermediate periods is much higher than the annual value of 39%. Additionally, CCGTs provide the reserve capacity required to maintain the security of the power system.

Imposing a carbon price floor (£16/tonne in 2014, £30/tonne in 2020 and £70/tonne in 2030 and beyond) makes coal power plants more expensive to run, and therefore less economic compared to CCGTs and even nuclear plants. CCGT capacity increases from under 30 GW in 2010 to 73.5 GW by 2050 and accounts for ~70% of total capacity. New nuclear plants become economically attractive from 2030s onwards as the carbon price floor increases to £70 per tonne.

Total generation capacity in MPI-D in the period between 2020 and 2050 is higher than the generation capacity in MPI-A, due to a rise in the electricity demand from higher GDP and population growth, alongside a fall in fuel prices. The capacity of CCGTs increases from roughly 30 GW in 2010 to 52 GW in 2020 and then drops to 24 GW by 2050. Despite the adverse impact of the carbon price floor on the competitiveness of coal plants, the capacity of coal plants reaches 73 GW by 2050. This is due to a decrease in coal price that makes generation from coal economically viable.

In MPI-F, there is lower electricity demand due to lower GDP and population growth alongside a rise in fuel prices. CCGT capacity at 46 GW and nuclear with 30 GW are the predominant generation technologies in 2050. High coal prices, in addition to the carbon price floor, make it expensive for coal plants to be built after existing plants are decommissioned by 2030.

In the EHT-Nuclear and the EHT-Offshore strategies, total CCGT capacity is exceptionally high. This is due to high electricity peak demand and, in part, to deal with inflexible (nuclear) and variable (wind, wave, etc.) generation. The large amount of capacity in the EHT strategies (in comparison with the other strategies) is due to higher electrification of heat and transport. These results reflect the considerably higher annual energy and peak demand requirements (almost two times larger than in the MPI-A strategy). Generation capacity is even higher (>300 GW) in EHT-Offshore strategy mainly due to the variable output of wind turbines and other renewables.

In the LEB and DDBT strategies, efficiency improvements to end-use devices, in addition to an increase in micro-generation technologies such as PV and CHP, results in an annual electricity demand of 350 TWh in LEB and 270 TWh in DDBT by 2050. This reduction in demand, coupled with decommissioning of wind farms installed before 2030, results in generation capacity (connected to the transmission network) declining post 2030. In the Gas World strategy, the generation mix is relatively similar to the generation mix in MPI-A with a slightly larger capacity of CCGT plants (4.3 GW more by 2050), due to the availability and price competiveness of shale gas.

CCGT capacity factors across the supply strategies – the capacity factor of CCGT plants drops significantly in EHT and DDBT strategies, mainly due to the use of CCGT plants to back up variable and inflexible generation from wind and nuclear. The analysis illustrated that in the EHT-Offshore strategy, provision of reserve through 12 GW of interconnection between Britain and Western European countries reduces the need for CCGTs by the same capacity and leads to slightly higher capacity factor for these plants by 2050 (16% vs. 11%). In the MPI-A strategy, CCGT power plants on average maintain a capacity factor of almost 50% throughout the period from 2010 to 2050 (Figure 4.11).

The capacity factor of CCGT plants in the Gas World strategy is higher than all the other strategies (46% in 2010 rising to 58% in 2050), as more competitively priced domestic gas (shale gas) is available and therefore CCGT operation is more economical in meeting the base load.

Generation capacity margins across the supply strategies – the derated generation capacity margins for all the strategies are shown in Figure 4.12. The derated capacity margin is the percentage by which generation capacity exceeds likely peak demand, taking into account the estimated availability of generation at that time. A gradual decrease in the

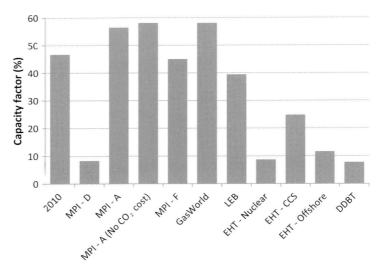

Figure 4.11   Capacity factor for CCGT plants in 2010 and across all the strategies in 2050

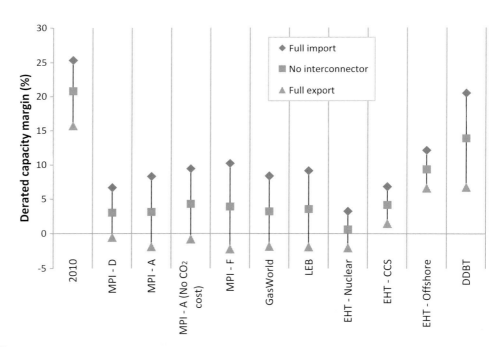

Figure 4.12   Generation capacity margins in 2010 and across all the strategies in 2050

capacity margin in all strategies is seen by 2050. The results show the capacity margin in the EHT-Offshore strategy by 2050 is higher than most other strategies. This is mainly due to a large amount of investment in back-up CCGT power plants. The impact of interconnectors on the capacity margins is also shown in Figure 4.12 (some capacity margins drift

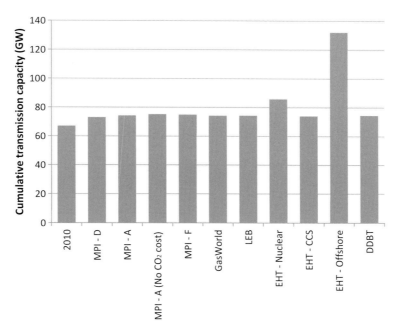

Figure 4.13 Cumulative power transmission capacities in 2010 and across all the strategies in 2050

into negative territory in full export mode, this is due to the model only calculating the generation capacity required for Britain's electricity demand). Interestingly the CEGB (former nationalised owner of the England and Wales electricity system) planned the system to operate with 20% gross capacity margin (equivalent to a derated capacity margin of ~5%) (RAE, 2013). Therefore figures in the vicinity of 3–5% do not imply a security of supply risk. A caveat is that an optimisation model was used. Hence, the results show that existing generating assets are used to their maximum potential and generating plants are only built to meet demand (the model takes account of ACS – Average Cold Spell peak electricity demand and only builds capacity if it is economically viable and not due to any capacity margin target).

4.4.6 Investment in power transmission infrastructure

The CGEN+ model, which is a spatial model of electricity supply/demand and the electricity and gas transmission network enables analysis of the new transmission infrastructure requirements associated with the various strategies. The cumulative power transmission capacity in 2050 for different strategies is compared to the value in 2010 in Figure 4.13. There is a direct correlation between the transmission capacities added and the amount of new generation capacity connected to the electricity network. In all MPI strategies, transmission expansion takes place in the 2020s (approximately 8 GW); this is mostly due to connection of onshore wind in Scotland and offshore wind in the northeast and west of England and Wales.

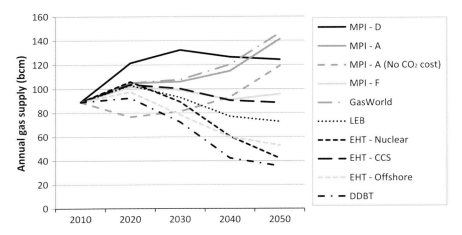

Figure 4.14 Evolution of gas supplies across the supply strategies

In EHT-Nuclear, transmission investment takes place from 2020 to 2050 (~18 GW additional by 2050) to deliver power generated by nuclear plants that are mainly located on the east and west coasts of Britain to the demand centres. For the EHT-Offshore strategy, transmission investment takes place from 2020 to 2050 (~65 GW additional by 2050) to accommodate large amounts of renewable/offshore technologies on the system. Transmission expansion for the Gas World, LEB and the DDBT strategies takes place in the 2020s to accommodate onshore wind in Scotland and offshore wind in northwest and northeast of England. No further transmission expansion beyond 2020 is expected in LEB and DDBT, as the electricity demand gradually drops to 353 TWh in LEB and 271 TWh in DDBT by 2050.

4.4.7 Annual gas supply volumes and sources

There is a significant difference in the level of electrification that takes place in the heat sector across the strategies. In addition to this there is also substantial variation of electricity generation via gas-fired power plants. This results in total gas supply varying from roughly 40–140 bcm by 2050 across the strategies (Figure 4.14). The scale of gas supply is very different with almost 55% rise in the MPI-A and GasWorld strategies from 2010 to 2050, but significant reduction in the LEB, DDBT and all EHT strategies, especially where there is limited reliance on gas based CCS power plants.

The volume of annual gas supplies follow very different trends, but all strategies show some common features in terms of sources of gas supplies (Figure 4.15). As current sources of gas decline, first U.K. Continental Shelf (UKCS) and then Norwegian imports, the slack is taken up by imports from Eurasia across the continental interconnectors and increasingly by LNG, which makes up the bulk of gas supplies from 2030 onwards (Figure 4.15).

In the Gas World strategy, the exploitation of shale gas in north England reduces the need for LNG imports compared to MPI-A (Figure 4.15). The total gas supply in 2050 slightly

Table 4.8 Gas import dependency across the strategies in 2050 (in 2010 it was 55%)	
Strategy	Import dependency (%)
MPI-A	94
GasWorld	74
EHT – Nuclear	84
EHT – CCS	91
EHT – Offshore	85
LEB	85
DDBT	82

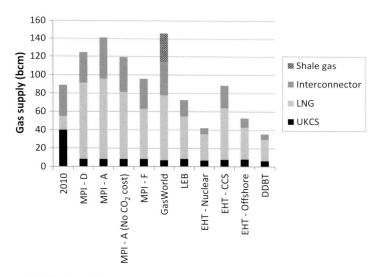

Figure 4.15 Sources of gas supplies 2010 vs. 2050

increases compared to MPI-A due to availability of competitively priced gas resources which increases gas demand for power generation. By 2050, shale gas accounted for 31.6 bcm of supplies and, as a result, the share of LNG imports was reduced significantly (71.4 bcm vs. 87.8 bcm in MPI-A with carbon price floor). Almost 11 bcm reduction in gas imports via the interconnector occurred, due to increased share of shale gas.

Import dependency grows from 55% in 2010 to more than 80% by 2050 for all strategies unless domestic shale gas is exploited (Gas World) (Table 4.8). Due to the decline in gas supply from UKCS and to meet gas demand in 2050 major investment in new LNG capacity was projected in all strategies (see Figure 4.15).

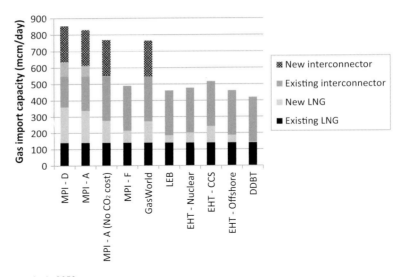

Figure 4.16 Gas import capacity in 2050

4.4.8 Gas import capacity expansion

Considerable investment is required to expand gas import capacity (LNG and interconnector) to meet gas demand including to satisfy 1-in-20 year peak day demand (i.e. the highest level of demand expected in one out of twenty winters (National Grid, 2013)) (Figure 4.16).

In all strategies, except DDBT, LNG terminal expansion is required. Lower levels of gas system expansion, compared to MPI-A, take place in the EHT strategies due to considerably lower gas demand for the heating sector which has been largely electrified. Apart from the MPI-A and Gas World strategies, no new interconnectors are built to boost the capacity of continental gas imports to GB. The MPI-A strategy leads to more imports despite greater gas demand in the Gas World strategy due to availability of indigenous shale gas supplies in the latter. The largest investment in gas supply infrastructure is predicted for MPI-D due to higher peak gas demand (517 mcm/day excluding gas power demand).

4.4.9 Total investment and electricity generation CO$_2$ emission intensity across strategies

For all the strategies, the total discounted cumulative costs (2010–2059, where each planning time step represents ten years) and the CO$_2$ emission intensity from the electricity sector in 2050 are shown in Figure 4.17.

In MPI-A without a carbon price floor, the CO$_2$ emissions intensity is almost constantly above 500 g/kWh throughout the time horizon. In MPI-A with a carbon price floor, the CO$_2$ emission intensity of the electricity system falls by 2030, mainly due to replacement of coal power plants with new CCGT and nuclear plants. After 2030 the emission intensity again starts to increase toward 2050 due to a slight increase in electricity demand and

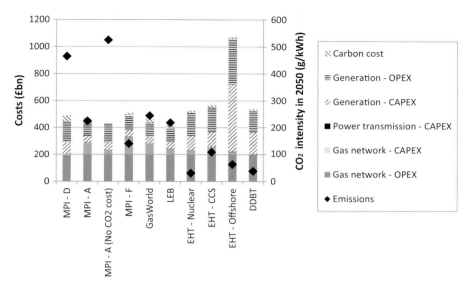

Figure 4.17   Total cumulative discounted costs (to base year 2010) and emission intensity from the electricity sector in 2050

decommissioning of wind farms that were built before 2030. In 2050 the CO_2 emission intensity stands at around 224 g/kWh (Figure 4.17).

The assumption of falling fossil fuel prices in MPI-D makes the carbon price floor insufficient to price coal-fired generation out of the market, resulting in coal power plants satisfying a significant share of electricity demand. Therefore, the CO_2 emission intensity starts to rise gradually after 2020 and reaches 464 g/kWh by 2050. On the other hand, in MPI-F, increasing fuel prices in addition to the carbon price floor make investment in coal power plants economically infeasible and also results in a reduction of power generation from CCGTs. The CO_2 emission intensity achieved in MPI-F is 139 g/kWh by 2050.

Overall the CO_2 emission intensity from the electricity system ranges from high in the MPI-A without a carbon price floor strategy to very low in the EHT-Nuclear (31 g/kWh) and DDBT (38 g/kWh) strategies.

Taking into account the carbon price floor in MPI-A has added to the total cost of the energy system through changing the optimal generation mix and also additional costs incurred for emitted CO_2. Larger gas and electricity demand in MPI-D, in spite of lower fuel prices, leads to higher costs especially for generation expansion and operation. Despite the low energy demand in MPI-F, due to higher fuel prices, the total cost is slightly higher than the other strategies, with costs of gas network operation (including costs of purchasing gas) contributing the largest amount to the total cost.

Results for the EHT strategies indicate that electricity system capacity expansion and operation account for the bulk of costs out to 2050 as the overall energy system is electrified.

Analysis shows that in the EHT-Offshore strategy the development of an additional 12 GW power interconnection (approximate cost £9 billion) between Britain and Western European countries could contribute to the provision of backup for wind farms and results in roughly £15 billion in cost savings.

The costs incurred in the gas and electricity distribution networks (such as distribution network reinforcement and micro-generation installation) and demand side management (such as efficiency improvement and smart meters) are not shown in Figure 4.17.

4.5 Conclusions

4.5.1 Key insights from strategy results

No regret transition options

This analysis helped identify four options as 'no regret', as these have the potential to play a role in all future strategies by helping to meet security and environmental objectives, and because they are relatively inexpensive. The enabling and complementary role of these options mean pursuing their early implementation would help to reduce the uncertainty. The four identified options are as follows:

Energy efficiency (including passive design) measures in buildings, industry and transport – there is evidence of significant cost-effective energy efficiency potential in all end use sectors. Reduced annual consumption contributes to reducing import dependence (and hence budget and trade deficits), end-user energy costs and price volatility and increases energy security. Additionally, energy efficiency can minimise the need for new energy infrastructure (e.g. through reduced peak load) and equipment sizes at end-use level (e.g. such as for boilers, heat pumps, HVAC piping, etc.). It directly contributes to the GHG reduction and enhancement of comfort. For optimal operation of certain key technologies, such as heat pumps, high energy efficiency (of building envelope) is also a pre-requisite.

Strengthening electricity transmission in major corridors across the country and with the EU – as our analysis shows, transmission infrastructure costs are likely to be significantly less than new generation capacity costs. Transmission infrastructure in major corridors would enable exploitation of geographically dispersed resources, for example, in Wales and Scotland. This, along with strengthening connections with other European grids and with a possible future super grid, has the potential to increase flexibility in Britain's electricity system while reducing costs.

Reinforcements at the electricity distribution level – reinforcement of distribution networks will be necessary to accommodate significant volumes of distributed generation, heat and transport electrification and/or distributed demand-side management measures like V2G. These are likely to be common features in sustainable energy futures, although the exact mix is uncertain and will depend on the strategy pursued. According to Pudjianto et al. (2013), cumulative distribution reinforcement investment requirements between 2010 and 2050 is about £35 billion to meet a peak load three times the base year level by 2050 (in our case, EHT strategy peak load is ~2.7 times the base year level). This amount is about 3% of electricity generation capacity and transmission costs in our EHT-Offshore scenario, which is relatively low.

Smart meters and a smarter grid – these are considered pre-requisites for maintaining grid flexibility in a high variable renewable future. They enable effective exploitation of distributed demand-response capabilities (such as from V2G, smart appliances and storage), optimal use of technologies such as heat pumps and can improve the resilience of the grid under increased vulnerabilities from climate change.

Role of fuel switching

The results indicate that improving the energy efficiency of existing systems in buildings, transport and industry alone will not be sufficient to meet ambitious carbon, energy security and affordability targets. Fuel switching, mainly electrification of key heating and transport services, will also be needed. Significant (technical) potential exists in switching from gas for heating in buildings and industrial low temperature processes.

Fuel switching in heating services coupled with energy efficiency (in building envelopes, lighting and appliances) can largely decouple energy consumption and carbon emissions from socio-economic change. With higher GDP and population growth, greater energy efficiency and fuel switching would be required to arrest growth in consumption and import of fossil fuels.

Electricity and gas consumption

Decarbonisation of electricity and gas, which together are responsible for ~80% of ex-transport energy consumption, is key to the deep decarbonisation of the energy system. Alternative strategies may result in significantly different evolutions and final levels of electricity consumption. On the other hand, the level of natural gas consumption by final users by 2050 needs to be significantly lower than the current level to meet climate goals. Trends in gas consumption, however, differ by strategy, impacting planning for tackling disuse and decommissioning and/or conversion of the gas grid.

Long-term electricity demand on the grid will be determined by a combination of factors, namely, the uptake of efficiency measures and onsite generation and electrification of key services. With high uptakes of distributed onsite generation technologies (PV and CHP), net demand on the grid from the residential sector can be insignificant (as seen in our DDBT and LEB strategies). However, this may not translate to reduction in electricity peak demand on the grid. This is an area of further research, especially on the impact of these technologies using finer temporal and spatial scales.

Electricity peak load

For a given annual electricity consumption, electricity peak load on the grid depends on a combination of factors, such as peak hour charging and discharging regimes of the electric vehicles, peak hour use of (non-electric) backup systems for electric heat pumps, peak hour import availability from interconnection and availability of other peak demand response measures. Our analyses show that, aided by a smart grid, a balance of these options can minimise high peak demand. Even in an EHT-like high electrification strategy, peak load can be brought below the MPI strategy level. In a high intermittent renewables future like the EHT-Offshore strategy, timely demand-side management measures can help reduce investments in, and GHG emissions arising from, gas-fired peaking plants.

Implications of deep decarbonisation

The analysis indicates that a number of supply-side strategies exist for deep decarbonisation of Britain's electricity system. However, for decarbonisation of the entire energy system, there is no single technological solution – ambitious transformation will be required in all sectors both in end-use and supply. With high levels of CCS in electricity generation capacity and in industry, deep decarbonisation is technically feasible even with an energy mix similar to that of today. However, even in a high CCS-like transition, transport and heating will have to be largely electrified and/or converted to biofuels. In the latter case, biofuels will have to meet the test of carbon neutrality and minimising concerns about competition with food production – this essentially implies use of second or later generation biofuels in such a transition.

Gas dependency and infrastructure

The analyses showed a significant increase in LNG capacity in virtually all the strategies in response to depletion of UKCS reserves. Electrification of the heat sector and/or energy efficiency measures reduces the demand for gas, and therefore no additional interconnectors are required in the DDBT, LEB and EHT strategies. In all strategies, gas import dependence increases by 2050, mainly due to depletion of Britain's domestic gas. A significant fraction of gas imports is through LNG, as gas from Norway diminishes (due to decline in Norwegian gas reserves), and gas from continental Europe is expensive. In all the strategies except DDBT, the increasing share of LNG imports through Milford Haven and Isle of Grain requires increased gas flow capacity from west and east to central and southern England.

Shale gas exploitation

Despite having the largest gas demand, exploitation of shale gas in the Gas World strategy results in gas import dependence of 74% by 2050, which is the lowest across all the strategies. According to our analysis, if Britain's shale gas resources are technically, environmentally and economically attractive, exploitation could reduce gas imports and increase security of supply.

High electrification with offshore renewables

EHT-Offshore was shown to be the most expensive strategy, mainly due to our assumptions about the capital and fixed costs of offshore generation technologies, and the need for expansion of CCGT capacity to provide backup for variability of wind generation to ensure security of supply. Significant reinforcement of the transmission network is also required to accommodate large scale offshore generation. Given the substantially higher total cost of EHT-Offshore strategy compared to the other strategies, it can be concluded that affordability of electricity supplies would be of great concern. However, long-term costs of offshore wind in comparison to other low carbon options, such as nuclear and CCS are very uncertain.

CCGT plant margins and supply security implications

The economics of CCGTs are currently unattractive, mainly due to the recent drop in the global coal price. This has resulted in many CCGT plant plans being put on hold or even cancelled. The modelling shows CCGT load factors vary from upwards of 50% to under 10% by 2050 across the strategies. A do nothing (MPI) or gas focused (Gas World) strategy leads to higher load factors as opposed to a strategy where the heat and transport sectors are largely electrified. In the EHT strategies, CCGTs perform a different but crucial role in balancing the power system, as backup plants to compensate the variability and/or inflexibility of low carbon generation. This flexibility function is vital in these particular strategies and as such should be rewarded. Without incentives such as capacity payments it is unlikely that these plants would be built.

Implications on the future of the gas network

The gas demand implications of high electrification of residential heating are profound. Large scale replacement of gas-based systems would reduce final demand for gas, resulting in significant over-capacity, or even redundancy in parts of the existing gas network. The EHT strategies show a large overall reduction in annual gas demand which would result in gas network connection and maintenance costs being distributed amongst a smaller number of users. This could lead to a vicious cycle of further loss of users and thus higher costs for the remaining users, who might be vulnerable households or operators of gas-specific processes. If this is to be avoided and if the gas network is seen as potentially useful as an alternative energy infrastructure, then some socialisation of these costs and/or reengineering of the network for other uses (e.g. biogas, hydrogen) may be needed.

Limitations of the set of strategies investigated

The set of strategies investigated here includes the majority of options likely in future energy systems. But the list is not exhaustive and there are inevitable 'unknown unknowns' that are not represented by the models and could not be considered in the analysis. There are a number of options the strategies did not fully address and could benefit in future analyses. These include: (i) the use of bio-energy with carbon capture and storage (BECCS), which provides the possibility of net negative emissions; (ii) strategies for decarbonising the gas grid (and hence the heating systems) using hydrogen (from electrolysis or CCS) or biogas; and (iii) a variety of alternative heating systems for buildings that can reduce the heating peak load problems associated with heat pumps, including gas absorption heat pumps and hybrid gas boiler and electric heat pump systems.

4.5.2 Policy implications and recommendations

We conclude that high levels of decarbonisation are incompatible with continued significant use of either natural gas for heating or petroleum based fuels for transport. Systemic change is therefore required in the energy system if climate mitigation goals are to be met. A number of different strategies can deliver high levels of decarbonisation, but there are concerns and uncertainties about them all, and therefore there is no unambiguously optimal strategy. Any of the transitions will involve changes in supply, demand and the interconnecting elements of the system, such as networks and storage, so policy needs to take a system-wide approach.

Strategies involving high levels of electrification of transport and heat are attractive for the decarbonisation agenda, but could require very large increases in electricity generation capacity to meet peak demand. The extent and cost of this can be mitigated both by energy efficiency and demand response (load shifting in time). More detailed work is required at the level of different users and uses to understand the plausible extent of demand response. High electrification strategies also imply public acceptance of a move away from gas central heating and cooking, which is untested.

We identify some 'no regrets' options – smart grids, energy efficiency and stronger interconnection. These merit more urgent attention as currently, only the timetable for the smart meter roll-out is clear. Shale gas exploitation could help with energy security goals, through reduced imports, but neither the practical U.K. resource nor the economics of its extraction are yet known.

The flexibility offered by gas-fired power generation is important in all strategies, but especially those with high levels of electrification. However, the future of the gas network in decarbonised energy systems needs to be explored in more detail. The cost of operating the low pressure gas grid for residential customers might become prohibitively high. Whether the network can and should be re-engineered for other uses (e.g. hydrogen or biogas) is not explored in our strategies: this needs to be evaluated.

This implies a number of early priorities for policymakers. Policies designed specifically to secure investment in energy efficiency, demand response, smart grids and interconnection need attention alongside those to support low carbon electricity generation. This requires significant changes to electricity market design. Before a firm commitment is made to high levels of electrification of heat and demand, the feasibility and economics of meeting peak demand, the potential for demand response, the social acceptability of the end user technologies and the fate of the gas grid all need more detailed examination. All strategies will be easier to deliver if the costs of relevant technologies fall, and therefore potentially game-changing technologies such as energy storage merit more attention.

References

Andrews, I. J. (2013). *The Carboniferous Bowland Shale gas study: geology and resource estimation*. London, UK, British Geological Survey for Department of Energy and Climate Change.

Baruah, P., N. Eyre, M. Qadrdan, M. Chaudry, S. Blainey, J. W. Hall, N. Jenkins and M. Tran (2014). "Energy system impacts from heat and transport electrification." *Proceedings of Institution of Civil Engineers – Energy* 167(3): 139–151.

BNEF (2013). The economic impact on UK energy policy of shale gas and oil. Bloomberg New Energy Finance, in response to UK House of Lords Select Committee on Economic Affairs Call for Evidence.

BP (2010). *BP statistical review of world energy.* London, UK, British Petroleum.

Chaudry, M., N. Jenkins, M. Qadrdan and J. Wu (2014). "Combined gas and electricity network expansion planning." *Applied Energy* 113: 1171–1187.

Climate Change Act (2008). *Climate Change Act 2008.* London, UK, HM Government.

Committee on Climate Change (2008). *Building a low-carbon economy – the UK's contribution to tackling climate change.* London, UK.

DECC (2010). *2050 Pathways analysis.* London, UK, Department of Energy and Climate Change.

DECC (2013a). *Digest of UK Energy Statistics (DUKES).*London, UK, TSO, Department of Energy and Climate Change.

DECC (2013b). *Fuel poverty report.*London, UK, Department of Energy and Climate Change.

HM Government (2009). *The UK low carbon transition plan: national strategy for climate and energy.* London, UK, TSO.

HM Government (2011). *The carbon plan: delivering our low carbon future.* London, UK, TSO.

HMRC (2014). *Carbon price floor: reform and other technical amendments.* London, UK, HM Revenue & Customs.

IEA (2012). *Energy technology perspectives 2012: Pathways to a clean energy system.* Paris, France, International Energy Agency.

IEA (2013). *World energy outlook 2013.* Paris, France, International Energy Agency.

IEA (2014). International Energy Agency Country Policy and Measures Database.

National Grid (2013). *Gas ten year statement 2013.* Warwick, UK, National Grid.

OECD (2012). OECD Green Growth Studies: Energy. Organisation for Economic Co-operation and Development.

Ofgem (2013). *Electricity capacity assessment report 2013.* London, UK, Ofgem.

Pudjianto, D., P. Djapic, M. Aunedi, C. K. Gan, G. Strbac, S. Huang and D. Infield (2013). "Smart control for minimizing distribution network reinforcement cost due to electrification." *Energy Policy* 52: 76–84.

RAE (2013). *GB electricity capacity margin. A report by the Royal Academy of Engineering for the Council for Science and Technology.* London, UK.

Skea, J., P. Ekins and M. Winskel, Eds. (2010). Making the transition to a secure *low-carbon energy system*, Earthscan.

Taylor, C. and D. Lewis (2013). Getting shale gas working, Institute of Directors, Infrastructure for Business 2013 #6.

Transport systems assessment

SIMON P. BLAINEY, JOHN M. PRESTON

5.1 Introduction

This chapter considers the challenges facing national transport infrastructure systems now and in the future. It introduces the wide range of issues which are currently facing such systems around the world, before considering how future changes in demand for such systems can be modelled, based on a case study of Britain. The development of a new model of Britain's transport system is described, followed by a discussion of the results produced by this model when used to predict the impact of a set of potential future transport strategies under a range of external conditions. Finally, the policy implications of these results are discussed along with the potential impacts of interdependencies with other infrastructure systems. For the purposes of this chapter, national transport infrastructure systems are defined as comprising the road and rail networks along with all major airports and seaports. While the vehicles which operate on this infrastructure are not usually considered to form part of the infrastructure system, they are nonetheless crucial, as changes in factors such as vehicle efficiency and fuels will be key in determining levels of impacts such as carbon emissions, as well as the demands on other sectors, particularly energy.

A significant proportion of transport infrastructure currently experiences congestion for at least part of the day, with the peaks caused by large numbers of people travelling to school and work with short arrival windows placing particular pressure on capacity. A key challenge for transport planners is therefore finding ways to address this congestion and its detrimental effects on the economy and the environment. This is linked to the question of whether it is possible to accommodate existing and predicted traffic on the existing infrastructure system by increasing the efficiency of its operation, or whether the construction of additional infrastructure is required. It seems inevitable that at least some additional construction will be necessary to address key pinch points, but the latent traffic phenomenon means that this will not necessarily reduce congestion (Goodwin, 1996). While growth in the demand for travel by some modes (notably the car) has been levelling off in recent years (see Figure 5.1), the substantial population growth predicted in coming decades means that total travel demand will almost certainly increase even if travel per person is static or declining. Alongside population growth the proportion of older people within the population is also expected to increase, and this will place different demands on the transport network, with, for example, public transport accessibility becoming increasingly important (Mackett, 2015). Land use changes are also likely to affect travel patterns, with, for example, inner city redevelopment and gentrification significantly altering local commuting flows.

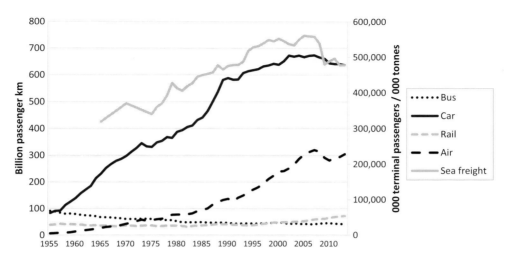

Figure 5.1 Transport trends in Great Britain 1952–2013 (bus, car and rail passenger kilometres, air terminal passengers, sea freight tonnes) (Source: DfT, 2013b)

Another important challenge facing transport planners is environmental, and involves identifying and implementing the interventions required to meet carbon reduction targets. Pursuing policies which disregard the need to meet such targets is not a sensible option, given the implications for societies worldwide, and achieving this goal requires planners to take action in the short term. This will require some tough decisions, and therefore one of the biggest challenges will be overcoming the political tendency towards short-termism, and persuading politicians to follow the advice of scientists rather than corporate lobbyists. This is linked to the wider social challenge of persuading individuals that it is worth accepting a relatively small disbenefit now to avoid a much greater level of disbenefit in the future. Climate change is not the only environmental challenge facing transport planners, and problems of air pollution, noise and land-take will also need to be addressed.

In addition to the challenges described above, transport infrastructure systems also face a number of risks. As long as transport infrastructure has existed, it has been at risk of damage from extreme weather events, but as climate change is expected to lead to an increased frequency of such extreme weather events, the risk of disruption to the transport network should also be expected to increase in the future (Koetse and Rietveld, 2009). This could include an increased risk of flooding, both in coastal areas and from rivers, and an increased likelihood of temperature-related damage such as track buckling and road surface melting during heatwaves, and disruption linked to ice and snow during cold periods. While it is often possible to take actions which prevent or mitigate the effects of such extreme events, such preventative action is likely to become increasingly expensive as the severity of climate extremes increases, and particularly in coastal areas there may eventually be no alternative to moving transport infrastructure further inland if sea level rise continues.

The demand for transport is significantly affected by demographic and economic factors. For example, as population grows, the additional people living in an area will have the same basic requirements for transport to access work, education and healthcare. These interdependencies bring with them risk associated with uncertainty in forecasting future changes in population and the economy. With respect to population, there is both a low-side risk that more infrastructure could be built than was actually needed by the future population, and a high-side risk that the transport infrastructure simply cannot cope with the volume of passengers who want to use it. Transport infrastructure requirements and revenue will also depend to some extent on continued economic growth, and there is a risk that if countries find themselves being outcompeted on the global market they may end up spending money they can ill afford building infrastructure to meet the requirements of economic growth which does not in fact occur. Even in stable economic conditions, transport infrastructure projects still face risks associated with costs, as if actual usage does not meet demand forecasts then the financial sustainability of infrastructure assets may be jeopardised.

There are also major interdependencies with the energy sector, as transport provision is predicated upon the supply of a sufficient quantity of energy at a suitable cost. This means that transport costs are linked closely to energy price changes and therefore to both energy supply policy and changes in energy demand elsewhere. This could place future supply of affordable transport at risk, if, for example, no transition was made away from 'conventional' fuels for road transport and fossil fuel supplies were to dwindle in the latter part of the century while demand increases in other industry sectors and countries. This is of course a two-way relationship, with energy prices influenced by the level of demand from the transport sector, and this could be a particularly key issue if the transport sector is substantially electrified (see Chapter 11).

Finally, perhaps the greatest level of uncertainty associated with transport infrastructure planning relates to the future impact of developments in information and communication technologies (ICT) on transport demand (see Chapter 9). While in the past ICT has not had as great an impact on travel patterns as might have been expected (DfT, 2013a), the scale and speed of development in this field of technology means that the potential future impacts remain huge. This introduces a high degree of risk into infrastructure planning, and makes it harder to 'future proof' transport assets.

5.2 Transport systems in Britain

5.2.1 System description

In 2012, there was estimated to be 394,933 km of roads in Britain, with a total of 487.8 billion vehicle kilometres travelled on these roads (DfT, 2013b). While there has been sustained growth in road traffic since the late nineteenth century, there is some evidence to suggest that this growth may recently have ceased with declines in some sectors (such as young males and company car use) being balanced out by continued growth in others (such as rural areas outside the southeast) (Le Vine and Jones, 2012). The British railway system carried 58.4 billion passenger kilometres in 2012/2013 on a network covering 14,504 route kilometres, with the former figure almost certainly at its highest ever level, alongside 21.5 billion freight tonne kilometres (DfT, 2013b). British seaports handled 257.9 million tonnes of imports in 2012 (almost double the figure in 1980), and 135.1 million tonnes of exports, while British airports were used by 220.6 million terminal passengers, a figure which is still 7% down on the peak in 2007 (DfT, 2013b). Figure 5.1 shows trends in road, rail and air passenger transport and maritime freight over time, and demonstrates the substantial growth in total travel which took place in the second half of the twentieth century. The greatest proportion of the total carbon emissions from transport in Britain (including both domestic and international transport) is generated by cars and taxis (40.3%), with road traffic in total responsible for 67.5% of transport emissions. However, the next biggest contribution is made by air transport, with international and domestic aviation together responsible for 21.6% of transport emissions, a proportion that has increased from 11% in 1990 (DfT 2013b).

5.2.2 Challenges and opportunities

The challenges facing Britain's transport system can be divided into three broad categories, relating to cost, congestion and carbon. In many countries including Britain there is likely to be continued pressure to reduce levels of public subsidy for transport systems, and while capital projects may be being promoted by governments to stimulate economic growth their costs are often subject to intense scrutiny. New transport infrastructure can be very expensive to build, with, for example, the Crossrail project in London expected to cost £14.8 billion (Crossrail Ltd, 2014), and widening of 14 miles of the M1 motorway in Central England costing £316 million (Highways Agency, 2011). Cost overruns can also reduce confidence in the ability of government and contractors to deliver projects on budget and on time, as illustrated by the problems with the Edinburgh tramway scheme (Audit Scotland, 2011). Restrictions on government spending can often lead to the use of Public-Private Partnership (PPP) schemes to finance transport projects, but these can also run into problems as seen with the collapse of the Metronet infrastructure company which had been responsible for part of the London Underground (NAO, 2009). There may also be pressure to address regional imbalances in infrastructure investment, and problems caused by a lack of sufficiently skilled workers.

A wide range of potential policy options are available to address the challenges discussed above. One of the most fundamental choices to be made is whether transport infrastructure planners should aim to constrain transport growth or whether they should build and innovate to accommodate it within the infrastructure network. In the past the latter approach has commonly been adopted, with planners attempting to expand the transport infrastructure network to keep pace with growth in demand. The EU's Transport 2050 Roadmap says that 'curbing mobility is not an option' (European Commission, 2011), perhaps because of the perceived risk of slowing economic growth (Marsden et al., 2014), but it seems question-able whether unconstrained growth in travel can in any way be compatible with meeting emission reduction targets. While there are policy options available which could help to decarbonise the transport sector, this task would be made much easier if travel demand growth could be curbed. While some aspects of transport demand result from people's need to access their workplaces and essential services such as education, a significant proportion of current travel is discretionary. Decarbonisation options include a transition to alternative fuels such as low carbon generated electricity for road and rail vehicles, and technological developments which improve the fuel efficiency of conventional vehicles (Committee on Climate Change, 2013). Intelligent transport systems could also help to increase the efficiency of transport operations, for example, through enhanced vehicle-user-infrastructure communications, and increased investment in research and pilot projects might speed the transition to a technology-driven future.

An alternative to direct involvement in technological development is to use regulation to specify maximum levels of (e.g.) vehicle emissions, and pass legislation which compels industry to find a solution. The success of the European Union road vehicle emission standards shows that such measures can in certain circumstances be highly effective. However, it is important that legislation is related to available technology at scale and backed up by comprehensive enforcement measures, as otherwise regulation may simply be ignored, as is often the case for regulations concerning maximum road speed limits (despite their clear safety and environmental benefits). It should also be noted that a side-effect of such legislation may be an increase in transport prices, as the industry recoups the cost of technological development.

Pricing can itself be used as a policy measure, with, for example, a fuel tax escalator used to encourage mode shift in Britain until political imperatives led to its abolition. More sophisticated pricing policies could include road user charging (Walker, 2011), where drivers are charged a variable (by location or time of day) toll per mile which captures the cost of the congestion they cause, or workplace parking levies to encourage more sustainable travel to work. Sophisticated pricing of this kind could help reduce transport inequality, as driving on uncongested rural roads where no public transport alternative exists would become cheaper. Road user charging (and taxation on other transport modes) might also be extended to include a carbon charge (Fu and Kelly, 2012), potentially linked to an emissions trading scheme. Such trading schemes can be used at a more aggregate scale to offset transport emissions, although it is unclear whether they have a positive environmental effect in the long run.

A major barrier to the introduction of comprehensive road pricing is the potential for widespread public resistance, as shown in Britain when a proposed nationwide scheme was cancelled in 2005. People are more generally often resistant to making changes in their behaviour, and this is linked to the problem of a 'tragedy of the commons' in travel behaviour, where it is in an individual's self-interest to behave in a way which has adverse consequences for society as a whole. Behavioural measures which seek to alter individual behaviour by encouraging people to adopt more sustainable travel patterns may therefore have an important role to play, for example, through the implementation of 'smarter choices' schemes (Cairns et al., 2004). These are often linked to policies aimed at developing a more integrated transport system, something which is often espoused as a policy ambition but seldom fully achieved. This can limit the success of smarter choices schemes, as unless the target population has access to affordable and effective alternatives to car travel no 'smarter choice' can be made.

5.3 Assessment of national transport systems

5.3.1 Modelling approach

A range of long-term transport models have been developed around the world including, for example, the public transport model created for the Rhine/Main Regional Transport Association in Germany (Arnold et al., 2013), the Belgian Federal Planning Bureau's PLANET transport demand model (Gusbin et al., 2010) and New Zealand's National Long-Term Land Transport Demand Model (Stephenson and Zheng, 2013). In Britain there are a number of existing transport models, including the Long Distance Model (URS and Scott Wilson, 2011), the National Transport Model (NTM) (DfT, 2009), the PLANET Long Distance model (HS2 Ltd, 2010), the National Trip End Model (NTEM) (WSP Group, 2011), the Great Britain Freight Model (MDS Transmodal Ltd, 2008), the rail Network Modelling Framework (Steer Davies Gleave and DeltaRail, 2007), the Air Passenger Demand Model (DfT, 2011b) and the National Air Passenger Allocation Model (DfT, 2011a). Each of these models has been developed for particular purposes, and none is fully able to produce multimodal spatially-disaggregated forecasts of transport demand and capacity utilisation based on changes in a range of exogenous and endogenous factors, with short run times to allow a wide range of possible futures to be examined. Our system-of-systems analysis is therefore based on a newly developed spatially disaggregated multi-modal national transport model.

Our transport model forecasts transport demand (and its relationship with transport capacity) by road and rail within and between 144 zones (based on local authorities) covering the whole of Great Britain. Full technical details of the model form and development are provided in Blainey and Preston (2016), with the overall model structure shown in Figure 5.2.

Figure 5.2 shows that the model is made up of six simulation sub-models, covering inter-zonal and intra-zonal road and rail traffic, air passenger traffic and seaborne freight traffic. Inter-zonal traffic is allocated to an infrastructure system made of single aggregated links connecting each pair of adjacent zones, with intra-zonal traffic modelled at the aggregate level. The model differs from most aggregate transport models in that it neither contains nor imputes an origin-destination matrix, as the key point of interest is the volume of traffic on particular links or within individual zones. Seaborne freight traffic is represented via a set of seaport nodes, and air passenger traffic is modelled on the basis of inter-airport links for domestic traffic and airport nodes for international traffic. The model does not include urban congestion disaggregated by road link or detailed consideration of urban public transport systems.

Capacity enhancements are specified in the model inputs prior to the commencement of a model run. A set of strategy files are also included in the model inputs, and these contain data on the values of a range of variables for each model time-step, representing changes in endogenous variables such as tolls and vehicle fuel efficiency. Exogenous changes in population, the economy and fuel costs are taken from a set of scenario files containing

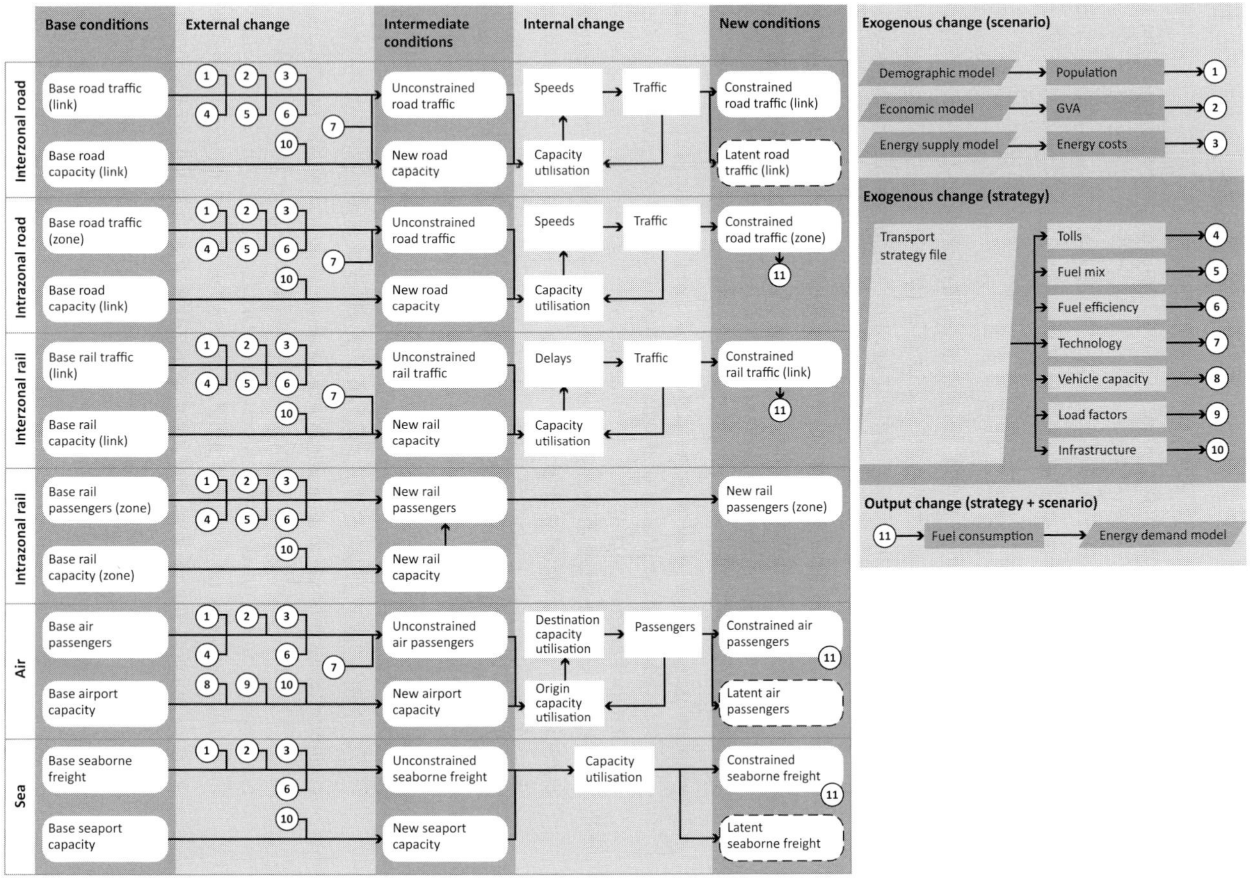

Figure 5.2 Structure of transport system model

forecasts generated externally (see Chapter 3 for scenario generation). The model produces forecasts on a yearly basis for the period 2011–2100, but considers much smaller time intervals during the forecasting process (e.g. to allow a more accurate representation of road congestion). The base data in each year is modified based on changes in explanatory variables and their associated elasticities, which reflect the degree to which a change of a certain magnitude in an explanatory variable leads to a change in the dependent variable (e.g. if the elasticity of road traffic with respect to population was set to 0.9, a 10% increase in population would lead to a 9% increase in traffic). The model generates outputs giving demand, infrastructure capacity utilisation, carbon emissions and fuel consumption for each mode and each time step.

5.3.2 Strategy description

There is a very wide range of strategic options for transport infrastructure provision, ranging right across the policy dimensions shown in Figure 2.2. Both on the demand and supply sides, future policies can be envisaged that range from almost no further intervention to very ambitious measures to manage demand or provide additional infrastructure capacity. Cutting across these strategic options is a range of more or less proven technologies to enable more efficient use of transport infrastructure or to substitute it altogether. Our strategies for the future transport infrastructure system seek to explore viable combinations of infrastructure investments from right across this range of possibilities. Committed transport infrastructure projects where construction work has already started are assumed to be implemented under all the strategies, including the extensive rail electrification projects outlined by government (DfT, 2012a).

TR0 'Decline and Decay' – this strategy describes a future where no replacements are found for fossil fuels, there is no investment in infrastructure beyond that required to maintain the status quo, and no further technological innovation in transport takes place. This is effectively a 'do nothing' strategy, which has been implemented as a baseline for comparative purposes.

TR1 'Predict and Provide' – road policy in Britain during the 1970s and 1980s was driven by an assumption that it was possible to provide sufficient infrastructure to accommodate predicted demand growth. In TR1 we assume that a similar policy was implemented for all modes, and simulate the construction of additional infrastructure when the capacity utilisation level for a particular link or node reaches 90%. Capacity enhancements become effective in the year following that in which capacity first reaches the critical value, representing a situation where demand forecasting has reliably predicted the year in which this would occur and infrastructure construction has then been planned accordingly. Of course, whether such a policy could ever be financially viable is a moot point. Technological innovations for established modes are assumed to continue, and fuel efficiency improvements are therefore modelled based on external evidence for road (Brand, 2012), rail (Schafer et al., 2011), air (Schafer et al., 2011) and sea transport (Hill et al., 2010). Take up of alternative fuels for road vehicles is assumed to be at the lower end of the range obtained from a further review of evidence (Committee on Climate Change, 2010; Hill et al., 2010; McKinsey &

Co., 2009; Page et al., 2004; Skinner et al., 2010). Rail electrification is assumed to occur at a moderately fast rate, with 200 track kilometres per year electrified after 2021. The effects of these changes in fuel consumption and type on travel demand are captured via their effect on the model cost variables.

TR2 'Cost and Constrain' – this strategy models the use of congestion pricing to suppress demand at capacity-constrained nodes and links for all modes. While national road pricing is not yet seen as being politically acceptable in Britain and there continue to be doubts about deliverability (DfT, 2013a) these obstacles may not in fact be as serious as has been claimed (Walker, 2011), and there is a clear economic case for such pricing (Glaister, 2010). Pricing is assumed to vary over time and space and between vehicle types (for road traffic) with maximum charges set to be equivalent to the non-variable costs for particular modes and vehicle types (to avoid correlation with fuel prices). A national workplace parking levy is also introduced. These charges are applied via the model cost variables, acting as a brake on demand via the cost elasticity. A moderate level of take-up of alternative fuels for road vehicles is assumed, with fuel efficiency improvements as in strategy TR1. Rail electrification also occurs at a moderate rate (100 track kilometres per year), and a small amount of additional road, rail and airport infrastructure is constructed.

TR3 'Adapting the Fleet' – this strategy envisages rapid technological development driving a high level of take-up of alternative vehicle fuels, along with major improvements in road vehicle fuel efficiency. Similar improvements are seen in fuel efficiency for other modes, and rail electrification occurs at a rapid rate (300 track kilometres per year). Rail journey times are assumed to be reduced slightly, and maximum aircraft capacities are increased by 20% by 2050 as a result of the use of lighter construction materials. A moderate amount of additional infrastructure is constructed.

TR4 'Promo-Pricing' – in addition to congestion charging (as in TR2), this strategy also sees the introduction of emissions-based pricing. This is accompanied by a high level of take-up of alternative (low emission) vehicle fuels, along with fuel efficiency improvements as in TR1 and TR2. The strategy also includes a moderate quantity of infrastructure construction, along with rapid rail electrification. Aircraft load factors are assumed to increase as a result of the highly disaggregated pricing policy using discounted fares to minimise the number of empty seats, but maximum capacities remain unchanged.

TR5 'Connected Grid' – developments in ICT and its application to transport see maximum capacities increased on both roads and railways through the introduction of advanced control systems and ultimately autonomous vehicles. There is a high level of take-up of alternative road fuels, along with large improvements in vehicle fuel efficiency for all modes. The Gross Value Added (GVA) elasticities in all models (which capture the impact of economic growth on transport demand) are progressively reduced by up to 50% over the study period, to reflect the negative impact of ICT improvements on the propensity to travel. As this strategy focuses on technology, only a small amount of additional infrastructure is constructed.

TR6 'Smarter Choices' – a range of 'soft' interventions are introduced in this strategy, including smarter choices, smart logistics and urban freight innovation schemes. These are all assumed to lead to a 10% reduction in relevant road traffic (similar to the 'high intensity' scenario considered by Cairns et al. (2004)), and are accompanied by a high level of take-up of alternative fuels and moderate levels of rail electrification and vehicle efficiency improvements. Again, only a small amount of additional infrastructure is constructed.

5.4 Strategy analysis and discussion

Our national transport systems model was used to analyse each of the seven strategies under three combinations of the external scenarios, which were: high population growth, high economic growth and low energy costs; medium population growth, central economic growth and central energy costs; low population growth, low economic growth and high energy costs. In interpreting the results we first examine the predicted future performance of the system under the minimum intervention transport strategy, TR0. While such a strategy may not be seen as desirable, given that the government suggests that in their current state, national transport networks will act as a constraint to sustainable economic growth (DfT, 2013a), the results from this strategy form a useful reference case to compare alternative strategy options.

5.4.1 Performance of the reference strategy across modes

Road transport

We begin by validating our reference strategy TR0 against existing forecasts. Figure 5.3 compares our simulations of total road traffic in England up to 2035 with those from the National Transport Model (NTM) (DfT, 2012b). This shows that the traffic forecasts produced by both models are of the same magnitude, although the low and central forecasts from our national transport system model are noticeably lower than the equivalent forecasts

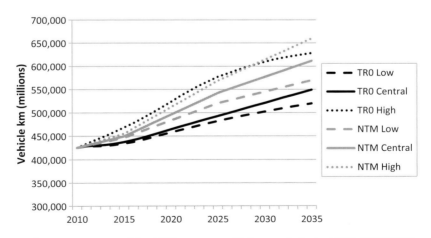

Figure 5.3   Comparison of transport systems model and NTM aggregate traffic forecasts for England 2010–2035 (NTM Source: DfT, 2012b)

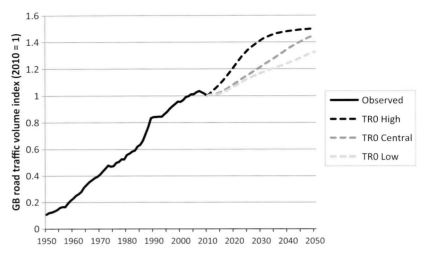

Figure 5.4   Observed and forecast aggregate road traffic 1950–2050 for Britain (Source: DfT, 2013b)

from the NTM. The high growth curves also diverge towards the end of the period, and this is at least in part because the reduction in growth rates predicted by our simulations as a result of traffic congestion is not replicated in the NTM forecasts. Figure 5.4 combines our simulations with observations of traffic volumes for the period 1950–2010, and shows that, when considered as a whole, the high growth forecast is most consistent with observed behaviour over this period. However, the observed growth trend has flattened out in recent years, and if this trend continues then it is possible that even the lower simulations may be too high.

Figure 5.5 now shows the total intra-zonal road traffic across Britain with strategy TR0 using three external scenarios (low, medium and high population/economic growth). There is, as would be expected, considerable variation in the patterns of traffic growth between the scenarios, with traffic growing much faster under the high scenario than under the low scenario. However, under all three scenarios traffic growth eventually levels out as the road network becomes full to capacity, with this point being reached approximately fifty years later under the low growth scenario than under the high growth scenario. Both population and economic growth are assumed to have a direct impact on transport demand, with the elasticity values meaning that the unit impact of an increase in population is greater than the impact of an equivalent increase in GVA.

One of the features of our transport system simulation model is that it produces results which can be spatially disaggregated, and Figure 5.6a shows the predicted spatial variations in road traffic growth under the central scenario with strategy TR0 for the period between 2010 and 2035. This shows that there are big differences in the level of traffic growth between different local authorities, with the highest levels of growth observed in London and in small urban authorities such as Nottingham and Portsmouth. In contrast, relatively little growth in traffic volumes is observed in some more rural areas such as Dorset and Powys. This situation does not merely represent an urban-rural dichotomy, though, as traffic

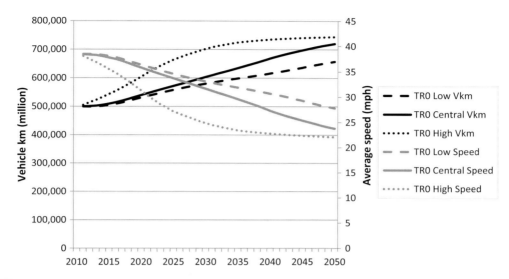

Aggregated road traffic (VKm) and average speeds (mph) under strategy TR0 with low, central, high scenario inputs.

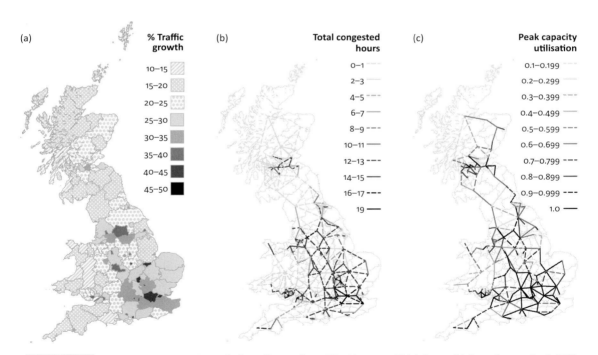

Spatial variations in model results (a: traffic growth to 2035 with strategy TR0; b: hours of daily road congestion in 2100; c: peak inter-zonal rail capacity utilisation in 2100)

is predicted to grow relatively fast in some rural counties such as Norfolk. Comparison with base level data showed that there was no clear correlation between traffic growth and base traffic density, as while in some cases traffic grows rapidly in areas where traffic densities were already high, in other cases it grows fast from a low base. This is related to spatial variations in population and economic growth, and comparison with Figures 3.7 and 3.10 shows that there is a clear correlation between areas of high population and economic growth and high growth in traffic levels.

Similar spatially disaggregated maps can be produced for inter-zonal road traffic forecasts, and Figure 5.6b shows the number of hours during an average day in 2100 when roads on particular inter-zonal links are congested to the extent that they are effectively full (in other words when no additional traffic can be accommodated). This shows that the greatest levels of congestion are found around London, but that most inter-zonal links in central and southern England would experience significant congestion for several hours per day with the TR0 strategy. In contrast, congestion in more rural areas is in general insignificant, although two links involving bridges (from Gwynedd to Anglesey and from Plymouth to Cornwall) are exceptions to this rule.

Rail transport

Under all three scenarios with strategy TR0, our transport systems model predicts that rail traffic will grow throughout the study period, but that in the high and medium growth scenarios this growth will be constrained towards the end of the period by increasing levels of delays. These delays result from increasingly high levels of capacity utilisation on the network, with no additional infrastructure being constructed to alleviate this congestion. As with road traffic, there is a clear link between the level of economic and population growth and the growth in traffic. There is also a high degree of spatial variation in this capacity utilisation, as shown in Figure 5.6c for the central scenario in 2100. Heavy congestion can clearly be seen along the southern end of the West Coast Main Line corridor, and on links radiating from Greater London, as well as in the area around Cardiff and Central Scotland.

Air transport

Under all three scenarios with Strategy TR0 demand for air travel will initially grow very rapidly, but then plateau as large airports become effectively full up in the second quarter of the century. The time taken for capacity to become fully utilised varies widely between airports, as shown in Figure 5.7, with Heathrow, Gatwick and Luton becoming full up within the next decade, but with some other quieter airports such as Teesside and Inverness retaining spare capacity throughout the study period. However, it should be noted that the model treats airports as individual unlinked entities; whereas it is likely that in reality some of the demand which could not be accommodated at 'full' airports would transfer to take up the spare capacity at these quieter airports. The volume of transfer would depend on the proximity of busier and quieter airports and the quality of the transport links between them, and might therefore be expected to be greater at quieter airports such as Teesside which are relatively close to other much busier airports.

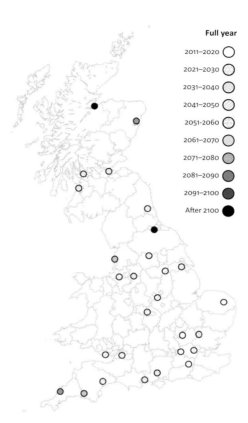

Figure 5.7   Decade in which British airports become full under central scenario with strategy TR0

Maritime freight transport

Strong baseline growth in total maritime freight from British ports is simulated by our model, with this growth being particularly marked for the high population and economic growth scenario. In the model the primary drivers for port usage are population and GDP, but only very limited data on port capacities were available, meaning that it was not possible to apply a capacity constraint at most ports. In reality it is likely that such constraints would limit growth if capacity was not expanded, and given that all scenarios predict sustained economic growth in the long term, if the relationships described in the literature between such growth and growth in shipping are correct then it is likely that substantial investment in port capacity may be required. It should also be noted that in practice the inter-relationships between the economy and freight volumes are much more complicated than the single elasticity linkage used here, although these interactions are not well documented in Britain.

5.4.2 Assessing alternative transport strategies

In the baseline (TR0) analysis, the main message for road transport is that demand growth increases, but that this growth will be constrained by the supply of infrastructure. The

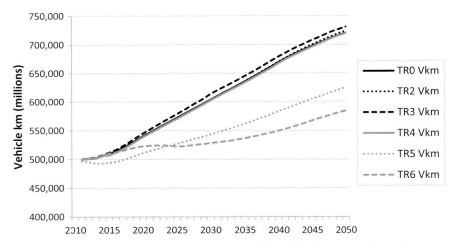

Figure 5.8 Comparison of aggregated road traffic in Britain across strategies (TR0 and TR2-6) using central scenario

transport simulation model has been used to explore the extent to which this pattern might be modified by plausible infrastructure investments and policies. Figure 5.8 shows growth in intra-zonal road traffic under six of the seven transport strategies using the central growth scenario. Strategy TR1 is not shown as this assumes that new infrastructure would be built as soon as capacity utilisation reached 90% of the maximum possible level. Beyond 2050, in the majority of cases road traffic growth plateaus as infrastructure becomes full up.

There are however two exceptions to this general pattern. The first, strategy TR5, models a future where improvements in ICT lead to the substitution of digital interactions for travel for certain trip purposes, such as commuting to work, personal business and shopping. While other trip purposes such as home delivery, commuting to education, business travel and leisure would not be affected, the strategy assumes that trip rates per person would reduce by 1% per year throughout the study period. This reduction in trip rates means that traffic growth initially occurs at a lower rate than under strategies TR0, TR2, TR3 and TR4. However, this strategy also assumes that technological developments such as the introduction of autonomous vehicles allow the effective capacity of existing roads to be significantly improved over time. This means that in the latter part of the century traffic levels exceed those achieved under other strategies, as exogenous growth in population and the economy outpaces the reduction in individual trip rates and technology alleviates the capacity constraints which would otherwise have acted to limit growth. While it may seem counter-intuitive that system-wide efficiency improvements would lead to an increase in traffic, the release of latent traffic following congestion relief means that in fact this is an entirely plausible outcome. It is also not certain that ICT improvements will lead to reductions in trip rates, and the government's National Policy Statement for National Networks suggests that they are not expected to have a significant impact on travel demand (DfT, 2013a).

The other exception to the general pattern of constrained growth is strategy TR6, which models a future where smarter choices schemes are successful in encouraging people to limit their car use, and where innovations in urban freight provision such as freight consolidation centres and drop-off boxes manage to reduce goods vehicle mileage. While traffic still grows for most of the century, albeit at a lower rate than with other strategies, towards the end of the century it begins to plateau at a level which is around 15% below that seen with the majority of strategies. This is backed up by findings from other research projects, which show that smarter choices schemes can reduce urban travel by up to 11% (Cairns et al., 2004), although the impact on latent traffic is less well understood.

While they have not been modelled under the current set of strategies, it is possible that similar trip rate reductions to those modelled in strategy TR5 may occur in the future as a result of demographic and economic phenomena, such as younger adults moving towards cities and older adults moving out, or saturation of growth in female driving license holding and car ownership. These factors along with other trends such as the relationship between vehicle speeds and constant travel time budgets, and the declining marginal utility of additional destinations, may be contributing towards an apparent 'peak' in road vehicle traffic (Millard-Ball and Schipper, 2011). In addition to changes in underlying trip rates, this phenomena could also mean that some of the model elasticities (e.g. with respect to economic growth) require revision over time, but further research would be needed to establish what level of revision was most appropriate, particularly given that not all researchers agree that this 'peak' has actually occurred (DfT, 2013a). It should also be noted that even in a scenario where individual travel has peaked, demographic changes (particularly population growth) could still lead to growth in aggregate road traffic, although this is not a foregone conclusion, and will be strongly influenced by the response of land use planning to demographic growth (Metz, 2012). For instance, housing developments, new employment opportunities and the development of other large infrastructure projects could have significant impacts on the use of transport networks (DfT, 2013a).

A similar comparison between strategies for inter-zonal rail traffic found that all strategies including infrastructure interventions generated a greater level of traffic by 2100 than the 'do nothing' strategy TR0. Once again, strategy TR5 permits the highest level of growth as a result of technological developments increasing maximum capacities (through the introduction of the European Rail Traffic Management System). The impact of the introduction of congestion- and carbon-based pricing under strategies TR2 and TR4 could also clearly be seen, although the resultant reduction in the number of trains operating is only temporary, as it is soon cancelled out by the impacts of exogenous growth.

As stated above, infrastructure capacity constraints can be a major brake on traffic growth, and (assuming that unconstrained transport is seen as a desirable goal) an obvious way to address this would be to build additional infrastructure. However, while this may provide a temporary solution at key pinch-points, underlying growth in demand coupled with the release of latent traffic previously deterred by the capacity constraint mean that congestion relief is likely to be short-lived. This is illustrated by Figure 5.9, which shows both unconstrained (from strategy TR1) and constrained (from strategy TR3) road traffic on an individual inter-zonal link (note that PCU stands for 'Passenger Car Unit' and is a measure of traffic volumes which takes account of the different physical size of different vehicle types). New infrastructure (in the form of additional road lanes) is constructed on each of the three road types during the study period, but while in all cases this permits

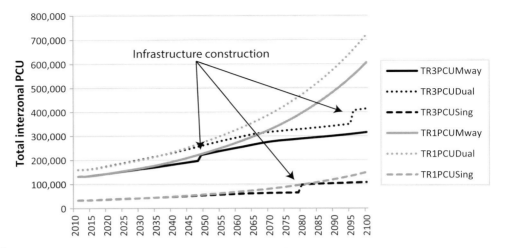

Figure 5.9   Inter-zonal passenger car units for flow 179 with strategies TR1 and TR3

a brief period of rapid traffic growth, traffic soon plateaus again at levels well below the unconstrained demand curve. This strongly suggests that a policy of 'predict and provide' is not a sustainable option for dealing with the issue of transport demand growth. It should be noted that while the traffic growth might in reality have more of a time lag than is predicted by the model, the general pattern of capacity release followed by a renewed constraint at a higher level of traffic remains valid.

Making use of an example from an individual link serves as a reminder that trends in traffic growth are not the same everywhere. Figures 5.6 and 5.7 show that there is a high level of spatial variation in the patterns of growth in road, rail and air transport, linked both to local capacity constraints and to variations in the underlying demographic and economic trends. However, it should also be noted that a lot of these variations are in 'when' rather than 'if' transport infrastructure reaches full capacity. One conclusion which could be drawn from this is therefore that supply constraints are one of the most effective way of constraining growth in travel volumes and that, if continued growth in traffic is seen as being undesirable, a policy of not constructing additional infrastructure would be a sensible one to adopt. However, transport congestion has been shown to be sub-optimal with regard to both the environment and the economy. The British government has estimated that in 2010 the direct costs of congestion on the strategic road network were £2 billion per annum (DfT, 2013a).

Indeed, published transport planning goals mean that in some cases the constraints imposed by infrastructure will make capital investment essential if these targets are going to be met. For example, the EU's 2011 Transport White Paper sets out requirements that by 2050 the majority of medium-distance passenger transport should go by rail (European Commission, 2011), which could mean accommodating higher levels of growth than are modelled under any of the strategies considered here (because the model does not explicitly forecast mode shift), given rail's relatively small base market share. As Figures 5.6b and 5.6c show, many of the most congested road links where mode shift is most desirable correspond to similarly congested rail links, and it therefore seems unlikely that the current

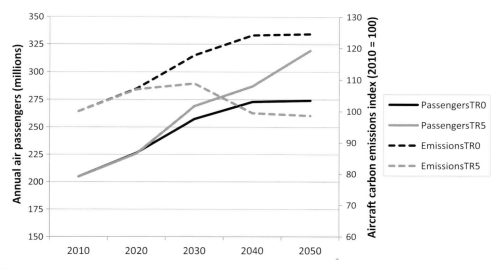

Figure 5.10   Air passengers and emissions in Britain with strategies TR0 and TR5 under central scenario

rail infrastructure will be able to accommodate the volume of mode shift required by the White Paper. This finding is backed up by industry reports which suggest that the West Coast Main Line will be effectively full up within the next twenty years (Network Rail and Passenger Focus, 2012). Figures 5.6b and 5.6c also emphasise the point that with the current configuration of the transport network London is both the locus of most transport congestion and also therefore a potential barrier to national mobility. There may therefore need to be a focus on identifying measures which can increase the capacity of the existing rail infrastructure, such as in-cab signalling to reduce headways.

Equally, though, other targets may mean that unconstrained growth in some forms of transport is highly undesirable. Emerging thinking from the U.K. Airports Commission suggests that additional runway capacity will be essential to resolve the capacity/demand imbalance (Davies, 2013), but this seems to be perpetuating a policy of 'predict and provide' in the aviation sector which has long been discredited in the road sector as a result of the latent traffic phenomenon. Figure 5.10 shows that in a 'do nothing' scenario, emissions would be expected to increase by approximately 25% before airport capacity becomes full up (meaning a similar percentage increase in passenger numbers), whereas under strategy TR5 a combination of limited airport expansion and significant improvements in fuel efficiency and load factors would mean that by 2050 emissions would have remained broadly constant while allowing passenger numbers to increase by 56% compared to base levels. In contrast, a policy of unconstrained growth coupled with limited efficiency improvements (strategy TR1) could lead to a quadrupling of emissions from air transport by 2050.

5.4.3 Decarbonising the transport system

Decarbonising the transport system is a major policy objective of Britain and other governments around the world. Electrification of road transport and the rail network can potentially

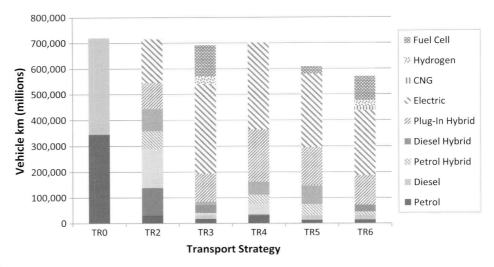

Figure 5.11 Aggregate annual vehicle kilometre disaggregated by vehicle fuel

make a very important contribution to reducing carbon emissions from the transport sector. Several of the strategies that we have analysed assume that electrification of the road passenger vehicle fleet will take place to at least some degree over the first half of the century, as illustrated by Figure 5.11. While the exact pathway to electrification is still uncertain, we have assumed that this will include both plug-in hybrid electric vehicles and 'full electric' battery powered vehicles. We have assumed that the switch to electric vehicles will have no direct impact on the volume of car travel undertaken (other than via differential cost signals). In reality the differing characteristics of electric and conventionally fuelled vehicles (particularly with regard to range) may lead people to alter their travel patterns, but the nature of this change will depend on the future path of technological development. While the switch to electric vehicles will have a more significant impact on emissions this will depend on the future fuel mix for electricity generation. This is illustrated by Figure 5.12, which shows road sector emissions for Strategy TR0 with the minimum policy intervention (MPI) energy strategy, and for Strategy TR3 with the three electrification of heat and transport energy strategies (EHT) (see Chapter 4 for energy strategy descriptions). This shows that the electrification of road transport combined with significant improvements in vehicle fuel efficiency leads to a significant reduction in emissions. It should also be noted that under strategy TR0 growth in traffic leads to an increase in road vehicle emissions of over 50%. While this growth might in reality be mitigated by improvements in vehicle technology and fuel efficiency (even without a shift towards electric vehicles), it suggests that continued growth in road traffic is unlikely to be compatible with meeting emission reduction targets.

Under current pricing and tax regimes, it might be expected that electrification would lead to reductions in out of pocket costs for motorists. However, it is not clear that this would be the case in practice as, for example, at current energy consumption levels, electrifying the road fleet might require a doubling of electricity generating capacity (see Section 4.4.2)

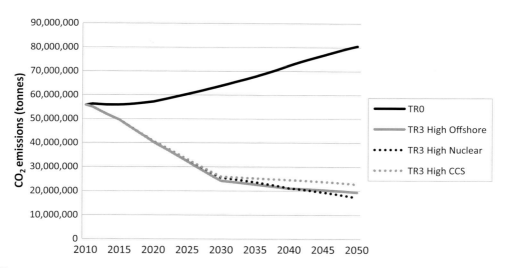

Figure 5.12 Road vehicle carbon emissions under strategies TR0 and TR3 with central scenario

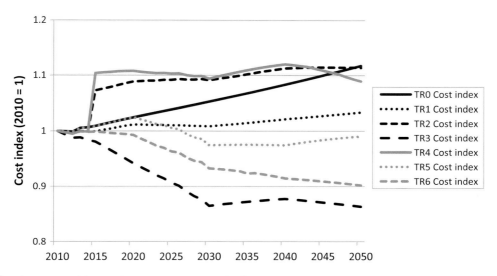

Figure 5.13 Relative cost per kilometre for cars on motorways under all strategies

and substantial energy price rises might therefore be expected. The cost impact of a move to electric vehicles could also be limited by an increase in vehicle purchase costs compared to conventional vehicles. Our modelling only considers vehicle operating costs, and Figure 5.13 suggests that these are noticeably lower for strategy TR3 (high take-up of electric vehicles and no additional charges) than for strategy TR1 (low take up of electric vehicles and no additional charges). Changes in vehicle efficiency also play a part in the cost reduction under strategy TR3, however, and congestion and carbon charges lead to a

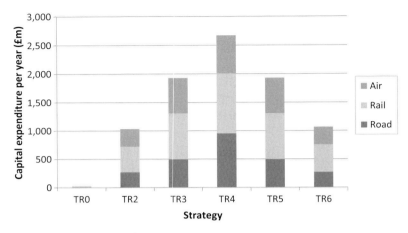

Average annual capital investment costs for transport strategies 2011–2050

noticeable increase in costs under strategies TR2 and TR4. It should though be noted that the low elasticity of motoring demand with respect to cost means that road traffic volumes are relatively insensitive to changes in cost, in comparison with their sensitivity to demographic and economic fluctuations.

Plans for rail electrification are rather more certain, and all the strategies assume that the majority of the network will be electrified during the twenty-first century, with the rate of electrification largely determined by the level of investment available. There is also scope for some electrification of urban public transport through the deployment of tram and trolleybus systems, which would also assist in decarbonising the transport system.

5.4.4 Investment costs for infrastructure strategies

Figure 5.14 shows estimated average annual capital investment costs for six of the seven transport strategies. Strategy TR1 is omitted because the respective costs are an order of magnitude greater than for any other strategy, at over £85 billion per year. It also differs from other strategies in that the majority of capital investment goes towards road and airport infrastructure, with rail accounting for less than 10% of capital investment. In contrast, the other six strategies show a relatively even balance between the sectors. Investment in transport infrastructure over the period 2015–2021 is expected to be £12.1 billion per year on average (HM Treasury, 2013). However, this includes items such as local authority transport maintenance and the whole of Network Rail's direct grant which are not necessarily used for capital projects, and this figure is not therefore directly comparable with those shown in Figure 5.14.

5.5 Conclusions

5.5.1 Cross sector impacts

The transport sector interacts strongly with the energy sector. As Figure 5.11 shows, some transport strategies are predicated upon significant expansion in the electric vehicle fleet, and this is assumed to be accompanied by widespread rail electrification. However, this would make transport increasingly dependent upon electricity generation and distribution infrastructure (while making it less dependent on oil). Our energy and transport system simulations together indicate that this could see a substantial increase in peak electricity demand, requiring a significant increase in generating capacity, and the extent to which this increase is met by sustainable power sources will dictate the extent to which electrification of transport is able to contribute to meeting carbon reduction targets. Electrification of the road vehicle fleet also has implications for electricity distribution networks, as the relatively limited range of electric vehicles (currently around 100 miles maximum (IMechE, 2011)) means that both the density of charging points and the speed of charging will be important in influencing take-up. 'Slow' charging can take place via existing power supplies at homes and workplaces, with the British government stating that it wants to see the majority of charging taking place at home at night, during off-peak hours for electricity demand (OLEV, 2011). Such charging would therefore require relatively little additional infrastructure (in areas where safe secure access to charging points can be easily arranged), but as it can take seven to eight hours to achieve a full charge slow charging is only suitable for vehicles which make relatively short trips that are widely spaced in time. Even though the vast majority of car trips in Britain do fall into this category and would therefore be suitable for electric vehicles, consumers' vehicle purchasing decisions are influenced by the potential to travel further (OLEV, 2011). It has been suggested that while widespread slow charging would do little to increase the attractiveness of electric vehicles for 'all-purpose driving' (as people are unwilling to interrupt their journeys for long periods), fast charging points could be very effective in encouraging drivers to exploit the full potential of electric vehicles (IMechE, 2011). However, even 'rapid' charging is still reported to take at least thirty minutes and therefore for drivers undertaking long journeys battery swap stations may be a more effective solution. Such a system has significant cost implications, though, as both construction of these stations and provision of a stock of spare batteries could prove expensive, and the use of plug-in hybrid vehicles may provide a more pragmatic solution at least in the medium term as reflected in some of the transport strategies.

It has also been suggested that both battery electric vehicles and plug-in hybrids could potentially form important sources of storage in future electricity networks, with remaining charge in vehicle batteries being used to supply domestic demand when the vehicle is plugged in during peak periods, before fully recharging the vehicle overnight (National Grid & Ricardo plc, 2011). While it has been suggested that increased battery cycling (charging and discharging) could significantly increase the life cycle cost of vehicle batteries, recent research indicates that actually the detrimental effect on batteries may be minimal, with fast charging potentially more of a problem (Lacey et al., 2013).

Developments in energy supply may also have an impact on transport fuels; for example, the current 'shale gas revolution' has the potential to drive an increase in the number of gas-powered road vehicles. In the USA the low price of LNG relative to diesel fuel has led many logistics companies to switch to vehicles powered by natural gas, and a network of fuelling stations is being established as part of the National Gas Highway initiative (PwC, 2013). Lower gas prices could potentially have a similar impact in Britain. Changes in energy mix linked either to shale gas or to the expansion of renewable energy production could also lead to changes in freight transport patterns, with, for example, a potential release of rail capacity on some routes if coal traffic declined.

Increased demand for transport, as predicted under all the strategies considered here, brings with it an increased demand for energy, and it seems likely that this will lead to an impact on energy prices (particularly for electricity). Of course, these increased energy prices may then themselves feedback to reduce the demand for transport as a result of negative cost elasticities.

There are also potential interactions between transport and the other infrastructure sectors. If water scarcity becomes an increasing issue over time then there might potentially be some use of the transportation network to move water during extreme events, and the high priority of such water transport would then impact on the ability of the network to accommodate underlying transport demand. Changes in waste disposal patterns will also have an impact on transport infrastructure capacity utilisation, but as waste transport only forms a small proportion of total freight traffic these impacts are unlikely to be significant at a national scale.

5.5.2 Policy insights

The analysis and simulation modelling presented in this chapter has a number of implications for future transport policy making. Perhaps the most obvious is that a policy of 'predict and provide' for transport infrastructure is not sustainable in terms of cost and carbon emissions and that it may therefore be necessary to increase the threshold benefit cost ratio criteria for investment which aims to provide capacity relief. The results also suggest that if the cost elasticities used in the model (and taken from the best available existing evidence) are correct then it may be difficult to use pricing as a tool to constrain growth in road traffic. However, it may be more useful as a means of redistributing traffic at a local level, as shown by the success of the congestion charging schemes in London and Durham.

Both the electrification of road transport and improved vehicle fuel efficiencies have the potential to play a significant role in decarbonising land-based transport. This is dependent on the necessary technological advances being adopted on a large scale. We have demonstrated that significant interdependencies between the transport and energy sectors, in particular, mean it is important to ensure that sufficient and suitable low-carbon electricity generating capacity is developed to gain the full potential carbon reduction benefits of electric vehicles. We will discuss these interdependencies further in Chapter 11.

Our analysis has explored the potential of behavioural measures (such as 'smarter choices' schemes) to encourage more sustainable travel and therefore tackle congestion and carbon emissions. However, the costs and wider economic implications of such measures are not fully understood, and so future work should more comprehensively compare them with the range of other infrastructure interventions that have been analysed in this chapter. The use of technology to increase the efficiency of transport could also be accompanied by policies encouraging the use of alternatives to travel, such as videoconferencing and teleworking. While the transport impacts of ICT developments are often difficult to predict, it is possible that there will be important linkages between policies enabling transport and communication technologies, a topic to which we will return in Chapter 9.

References

Arnold, M., B. Hajos and T. Busch (2013). Long term demand and financial forecast methodology for a large scale regional public transport network in Germany. *European Transport Conference*. Frankfurt, Germany.

Audit Scotland (2011). Edinburgh trams: interim report. *Audit* Scotland. Edinburgh, Scotland.

Blainey, S. P. and J. M. Preston (2016). "Determining future infrastructure pathways: a long term model of transport demand and capacity." *Transportation*. Submitted.

Brand, C., M. Tran and J. Anable (2012). "The UK transport carbon model: An integrated life cycle approach to explore low carbon futures." *Energy Policy* 41: 107–124.

Cairns, S., L. Sloman, C. Newson, J. Anable, A. Kirkbride and P. Goodwin (2004). *Smarter choices: changing the way we travel*. London, UK, Department for Transport.

Committee on Climate Change (2010). *The fourth carbon budget – reducing emissions through the 2020s*. London, UK.

Committee on Climate Change (2013). *The fourth carbon budget review – technical report: Sectoral analysis of the cost-effective path to the 2050 target*. London, UK, Committee on Climate Change.

Crossrail Ltd (2014). "Funding." Retrieved from www.crossrail.co.uk/about-us/funding.

Davies, H. (2013) Aviation capacity in the UK: emerging thinking.

DfT (2009). *National Transport Model: high level overview*. London, UK, Department for Transport.

DfT (2011a). *Response to the Peer Review of NAPALM*. London, UK, Department for Transport.

DfT (2011b). *UK aviation forecasts*. London, UK, Department for Transport.

DfT (2012a). *Railways Act 2005 Statement for Control Period 5*. London, UK, Department for Transport.

DfT (2012b). *Road transport forecasts 2011: Results from the Department for Transport's National Transport Model*. London, UK, Department for Transport.

DfT (2013a). *Draft National Policy Statement for national networks*. London, UK, Department for Transport.

DfT (2013b). *Transport Statistics for Great Britain*. London, UK, Department for Transport.

European Commission (2011). *Roadmap to a single European transport area – towards a competitive and resource efficient transport system*. White paper. Brussels, Belgium.

Fu, M. and A. J. Kelly (2012). "Carbon related taxation policies for road transport: Efficacy of ownership and usage taxes, and the role of public transport and motorist cost perception on policy outcomes." *Transport Policy* 22: 57–69.

Glaister, S. (2010). *Governing and paying for England's roads*. London, UK, R. Foundation.

Goodwin, P. (1996). "Empirical evidence on induced traffic." *Transportation* 23(1): 35–54.

Gusbin, D., B. Hoornaert, I. Mayeres and M. Nautet (2010). The PLANET model methodological report: modelling of short sea shipping and bus-tram-metro. Brussels, Belgium, Federal Planning Bureau.

Highways Agency (2014). M1 Junctions 25–28 Widening Scheme. Retrieved from www.highways.gov.uk/roads/road-projects/m1-junctions-25-28-widening-scheme/.

Hill, N., M. Morris and I. Skinner (2010). SULTAN: Development of an illustrative scenarios tool for assessing potential impacts of measures on EU transport GHG. EU Transport GHG: Routes to 2050? Task 9 Report VII produced by EC Directorate-General Environment and AEA Technology plc.

HM Treasury (2013). *Investing in Britain's future*. London, UK.

HS2 Ltd (2010). *High Speed Rail – London to the West Midlands and beyond: HS2 demand model analysis*. London, UK, HS2 Ltd.

IMechE (2011). *Electric Vehicle Recharging Infrastructure*. London, UK, Institution of Mechanical Engineers.

Koetse, M. J. and P. Rietveld (2009). "The impact of climate change and weather on transport: An overview of empirical findings." *Transportation Research Part D: Transport and Environment* 14(3): 205–221.

Lacey, G., G. Putrus, T. Jiang and R. Kotter (2013). The effect of cycling on the state of health of the electric vehicle battery. *48th Universities' Power Engineering Conference*. Dublin, Ireland.

Le Vine, S. and P. Jones (2012). *On the move: making sense of car and train travel trends in Britain*. London, UK, RAC Foundation.

Mackett, R. (2015). "Improving accessibility for older people – Investing in a valuable asset." *Journal of Transport & Health* 2(1): 5–13.

Marsden, G., C. Mullen, I. Bache, I. Bartle and M. Flinders (2014). "Carbon reduction and travel behaviour: Discourses, disputes and contradictions in governance." *Transport Policy* 35: 71–78.

McKinsey & Co (2009). *Roads toward a low-carbon future: Reducing CO_2 emissions from passenger vehicles in the global road transportation system*. New York, USA, McKinsey & Company.

MDS Transmodal Ltd (2008). GBFM Version 5.0 report. MDS Transmodal Ltd.

Metz, D. (2012). "Demographic determinants of daily travel demand." *Transport Policy* 21: 20–25.

Millard-Ball, A. and L. Schipper (2010). "Are we reaching peak travel? Trends in passenger transport in eight industrialized countries." *Transport Reviews* 31(3): 357–378.

NAO (2009). *Department for Transport: the failure of Metronet*. London, UK, National Audit Office.

National Grid and Ricardo plc (2011). Bucks for balancing: can plug-in vehicles of the future extract cash – and carbon – from the power grid? London, UK, *Ricardo plc*.

Network Rail and Passenger Focus (2012). *Future priorities for the West Coast Main Line: released capacity from a potential high speed line*. London, UK, Network Rail.

OLEV (2011). *Making the connection: the plug-in vehicle infrastructure strategy*. London, UK, Office for Low Emission Vehicles, DfT.

Page, M., C. Kelly and A. Bristow (2004). Exploring scenarios to 2050 for hydrogen use in transport in the UK. *European Transport Conference*, Strasbourg.

PwC (2013). Transportation & logistics companies reap the benefits of US shale gas boom. PwC Press release.

Schafer, A., L. Dray, E. Andersson, M. E. Ben-Akiva, M. Berg, K. Boulouchos, P. Dietrich, O. Froidh, W. Graham, R. Kok, S. Majer, B. Nelldal, F. Noembrini, A. Odoni, I. Pagoni, A. Perimenis, V. Psaraki, A. Rahman, S. Safarinova and M. Vera-Morales (2011). TOSCA Project Final Report: Description of the main S&T results/foregrounds. EC FP7 Project Report.

Skinner, I., H. van Essen, R. Smokers and N. Hill (2010). Towards the decarbonisation of the EU's transport sector by 2050. EU Transport GHG: Routes to 2050? Final Report produced by EC Directorate-General Environment and AEA Technology plc.

Steer Davies Gleave and DeltaRail (2007). *Network Modelling Framework – background documentation: overview document*. Scotland. London, UK, DfT, ORR, Transport.

Stephenson, J. and L. Zheng (2013). *National long-term land transport demand model*. Wellington, New Zealand, NZ Transport Agency.

URS and Scott Wilson (2011). Modelling longer distance demand for travel phase 3: Final Report – Volume 1 Main Report. URS/Scott Wilson. Basingstoke, UK.

Walker, J. (2011). *The acceptability of road pricing*. London, UK, R. Foundation.

WSP Group (2011). *NTEM Planning Data version 6.2: Guidance Note*. London, UK, Department for Transport.

6 Water supply systems assessment

MIKE SIMPSON, MATTHEW C. IVES, JIM W. HALL, CHRIS G. KILSBY

6.1 Introduction

Improved drinking water supplies now reach 89% of the world's population, an increase of 2.3 billion people since 1990. In the developed world drinking water provision is effectively universal (World Health Organisation and UNICEF, 2014). However, maintaining this level of supply in the context of increasing demand, deteriorating infrastructure and a changing climate will require sustained investment. Natural systems that provide water supplies can be augmented through engineered infrastructure, in order to cope with heterogeneity of availability in space and time. The current global challenge is to manage the resulting trade-offs between requirements of the users of water and sustainability of the natural environment.

Water infrastructure can be separated into raw water, treated water and wastewater systems. Raw water infrastructure abstracts, transports and stores water prior to treatment. Usually, raw water is acquired from a lake or reservoir, directly from a river or from groundwater. The treated water system begins at the treatment works and distributes water in pipes from the treatment works to the various users of water, via local treated water storage units. The system for collection, treatment and discharge of wastewater is dealt with in Chapter 7.

As with any infrastructure, water supply infrastructure deteriorates over time, so resources are required to maintain systems. Climate change is anticipated to influence extreme and average temperatures and precipitation, with consequent impacts on the patterns of water availability across the country. With rising population there is increasing awareness of the impact of abstraction on the natural environment (Acreman, 2001) and pressure to address projected water supply challenges through the most efficient and sustainable solutions.

The challenge for water supply in the twenty-first century lies in the implementation of appropriate solutions that can meet changing requirements without compromising the environment and other users of water, or placing excessive financial burden on citizens. As such, traditional storage and transfer technologies are required in parallel with new ideas for water supply and efficiency of water use. Establishing the appropriate combination of these options is a complex engineering and economic systems problem. With growing pressure on water supplies and major investment decisions ahead, a more strategic national approach is required, which we describe here as part of the system-of-systems methodology presented in Chapter 2.

6.2 Water supply systems in Britain

6.2.1 System description

The water resource system for England and Wales currently provides around 13.7 billion cubic metres of water per year (Defra, 2013). Much of this is from surface water, which accounts for 85% of provision in England and Wales (EA et al., 2012). The largely impermeable geology and high annual rainfall of the north and west of the country give rise to dense networks of small stream systems in which almost all water is carried overland, resulting in very quick responses to precipitation events and low catchment storage. The upland nature of these areas leaves them naturally suited to reservoir construction. By contrast, the southeast of the country is set on permeable expanses of sandstone and limestone into which the surface hydrology can drain. This reduces the surface drainage density, generating a sparse network of streams which are notably less responsive to rainfall due to the attenuating influence of slow flows underground. Lowland landscapes are less suitable for large-scale reservoirs, therefore the availability of groundwater as a resource is invaluable in these areas. Groundwater contribution gives lowland rivers much greater resilience to drought, at least in the short- and medium-term, than upland rivers. Groundwater represents 100% of the water resource in many lowland water resource zones (e.g. Anglian Water, 2014).

Demand for water can be divided into domestic and non-domestic uses. For ease of calculation a Per Capita Demand (PCD) figure is used to estimate and monitor average water use. This figure, multiplied by population, can be used to determine the domestic component of demand. Non-domestic demand for water comes from industry, power generation, primary production and commerce (EA, 2012). Along with the demands for water by end users, there is water use in the processes of treatment and distribution by water companies. Of these, leakage is the most significant, making up 23% of the total distribution input (Tooms et al., 2011). Each of these components varies regionally. They have each shown significant change over time, and forecasts of these components contribute to the uncertainty in future water resource planning. Disaggregating the total demand allows for detailed comparison with similar systems elsewhere. For example, the mean UK PCD of 150 litres per person per day (L/p/d) is broadly in line with other European countries, although Germany and the Netherlands achieve PCDs of less than 130 L/p/d and Slovakia, Belgium, the Czech Republic and Estonia all manage less than 110 L/p/d (Defra, 2008), suggesting that an increase in domestic efficiency is feasible.

6.2.2 Challenges and opportunities

Britain is a temperate country and average rainfalls are high compared with many parts of the world. However, population densities in the south and east of England result in per capita rainfall comparable with arid regions of the world (Defra, 2008). Uncertainties in security of supply and projected demand suggest that provision of water infrastructure will become increasingly challenging for the foreseeable future.

Climate change is manifested as changed patterns of average and extreme rainfall and evaporation, meaning that future water resource availability cannot be assumed to operate to steady-state conditions (Bates et al., 2008). The translation of global climate models to the point scale has received significant attention (Fowler et al., 2007), so for Britain, models exist for projection of hydrological variables which represent future changes in space and time (Prudhomme et al., 2012a). The anticipated changes broadly show increase of the highest flows and decrease of the lowest flows, and are expected to increase the risk of drought in Britain. This perturbed climate is also expected to change the water cycle, leading to increased drought throughout northwest Europe (Rahiz and New, 2013).

However, climate projections of precipitation are highly uncertain: climate models show biases when compared to observations and different models have conflicting future projections, showing the potential for both positive and negative change. The transformation of climate variable time series into water resource yields is not linear. Storage allows an increase in yield as a consequence of the capture of higher flood flows, even though overall flow may be reduced. Conversely, areas that have low storage capacity (either as groundwater, lakes or reservoirs) may be dependent on high mean flows to sustain supply throughout dry periods, resulting in reduced yields even when low flows increase. The impact of climate change on demand is expected to be an increase of around 2% for domestic and most non-domestic use, but potentially around 20% for agricultural demand (Downing et al., 2003). The identification of a representative model for water resource planning is therefore important, as is an understanding of the full range of possible consequences of climate change for water resource yield and demand.

As well as these physical changes, there is increasing social and legal pressure to manage water resources sustainably (Acreman, 2001). Historic rights to water were tied to land ownership with little regard for the consequences on the wider environment. The relationship between ecological damage and high levels of local abstraction was formalised in the 1960s (Hannah et al., 2004), with legal abstraction limits determined at that time still in force today (Defra, 2011). Recent regulatory pressure to improve environmental standards has come from the Water Framework Directive (WFD) (European Parliament and Council of the European Union, 2000), which introduces a set of legal standards for the water environment in terms of physicochemical and biological conditions to be implemented by 2015 (Defra et al., 2006). Approval for new abstraction licences is now considered in a catchment context, with Catchment Abstraction Management Strategies (CAMS) implemented for all major catchments in England and Wales since 2001 (EA, 2013). In practical terms, new licences for abstraction are typically constrained by hands-off flows (the minimum flow permitted in the river) and/or maximum volumes and are restricted to areas deemed to have available water. However, existing licence holders may retain the legal power to operate without a hands-off flow restriction, with potential consequences for ecosystems that are dependent on river flow. These issues are being addressed through a process of reform of the abstraction regime in England (Defra, 2011).

Despite these challenges, solutions for water provision are available both in terms of the implementation of existing solutions at new locations or scales and as novel technologies. Planning of water supply infrastructure takes place at the scale of water resource zones. As pressure on supplies increases, co-ordinated planning between water resource zones and water companies becomes more common. This approach is most prevalent in the southeast of England, where water availability and demand are closely balanced (Critchley and Marshallay, 2013). However, the southeast of England is not the only place where major infrastructure investments are anticipated. Across Great Britain, strategic choices need to be made to balance measures to reduce demand, infrastructure investments (including new supply sources and leakage reduction) and constraints upon the availability of water to abstract from the environment. Infrastructure investments continue to be controversial because of the environmental, landscape and other impacts and because of the cost which is passed on to water customers. Leakage reduction has already reached a point where many water companies argue that because of the high cost of fixing leaks, it is not economical to reduce leakage much further. Water metering and associated steps to manage demand are also controversial because of fears about curbing personal freedoms, excessive regulation and because price incentives may disproportionately affect some customers. Nonetheless, there is likely to be the need for large scale infrastructure investments to provide additional storage (in aquifers as well as reservoirs), to provide flexibility to transfer resources between different locations, and to expand new (but more energy-intensive) supplies such as desalination and direct reuse of water by treating effluent from wastewater treatment works. The following summarises important strategic options.

Storage

The most conspicuous water infrastructure assets in Britain are reservoirs. Reservoirs buffer the natural variability in water availability and so provide storage for water supply, flood control and hydropower generation. Reservoirs are often controlled as infrastructure systems, with sequences of reservoirs down a river valley regulating river flows. The size of reservoirs increased throughout the nineteenth and twentieth centuries (Charles et al., 2011), culminating in the construction of Kielder Water, Britain's largest reservoir, in 1981. Rutland Water, the next largest reservoir, was just 60% of this size. Kielder was both a demonstration that this scale of reservoir construction is feasible and a case study for the limitations of large water infrastructure projects. Kielder's size was criticised as excessive and uneconomical (McCulloch, 2006), and high projections of local demand from population growth and water-intensive industry were not realised as the regional economy underwent industrial decline. Partly as a consequence of Kielder's perceived failure no major dams have been constructed in the U.K. in the last twenty-five years.

However, even Kielder Reservoir is small compared to the large natural lakes in the upland of Britain. At 7,500 GL, Loch Ness is more than 35 times larger than Kielder, with Scotland's ten largest lochs totalling 22,000 GL. This capacity is not necessarily available for abstraction due to environmental concerns, but underlines the geographical heterogeneity of water availability in Britain.

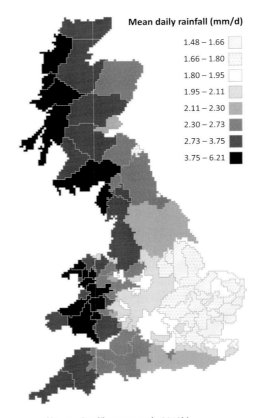

Mean daily rainfall (mm/d)

1.48 – 1.66	
1.66 – 1.80	
1.80 – 1.95	
1.95 – 2.11	
2.11 – 2.30	
2.30 – 2.73	
2.73 – 3.75	
3.75 – 6.21	

Figure 6.1   Mean rainfall by water resource zone (Source: Prud'homme et al., 2012b)

Transfers

The ratio of water availability to demand varies significantly across Britain. While an east–west gradient of total annual rainfall and a north–south gradient of annual evaporation means that the southeast is naturally drier than the northwest (Figure 6.1), the difference in population density means that regional demand varies by several orders of magnitude. As these factors are inversely correlated, water shortage in the southeast of Britain is a significant challenge, whereas water shortage in northwest Scotland as a consequence of natural regional scarcity is inconceivable (although historic underinvestment and environmental regulation results in short-term shortages, see Marsh & Anderson (2002)). An intuitive response to this situation that has been applied successfully in other countries (Jenkins et al., 2004) is the long-distance transfer of water.

The cost to transfer 1,100 megalitres per day (ML/d) of water from the north of England to the southeast via pipeline has been estimated at £9–12 billion making such schemes more expensive than programmes for improved local supply in the southeast (EA, 2006). Moreover, environmental consequences such as altered flow regimes, differing water chemistry and the potential for transfer of non-native species are factors which have been cited in opposition to open water transfers, although these have also been challenged (Johnson, 2011).

In contrast to these contested long-distance transfers, transferring water between neighbouring water resource zones is commonplace. When sequences of neighbouring catchments are interconnected, as they are in the northeast of England, an effective transfer system is already in place. An alternative form of national water grid is where connections between companies allow a displacement of temporary shortage through sharing of resources (ICE, 2012). This type of system is emerging, as agreements for the movement of large quantities of water from one region to another are already in place (e.g. Welsh Water (2014)).

Emerging supply technologies

Three important emerging technologies for Britain are desalination, effluent reuse and aquifer storage and recovery. Desalination is the process of turning brackish or saltwater to freshwater, an intuitive solution to water shortage in coastal regions. High energy use and challenges in disposing of the large quantities of saline effluent produced mean that desalination is currently an expensive solution and not without environmental implications. Developments in membrane technology and the use of forward osmosis are anticipated to improve the efficiency of desalination towards the thermodynamic limit. The ability to produce water anywhere with a coastline using a small amount of land makes desalination a unique solution, especially as a means to supplement conventional sources during times of shortage (IChemE, 2013). Britain's first operational desalination plant for public use was constructed at Beckton; operating during times of shortage it can supply up to 150 ML/d.

Effluent reuse is the direct recycling of water from wastewater treatment plants into the water distribution system. It suffers from low public acceptability (Aitken et al., 2014). In catchments in the south of England (notably the Thames), where treated wastewater discharges are sited upstream of water abstractions, effluent is effectively being reused, but only after a period of mixing with river water before being treated to drinking water standards. Direct reuse avoids this intermediate stage in the river. Water recycling uses around a third of the energy used for seawater desalination (Stokes and Horvath, 2006), making it an attractive option. There are currently no effluent reuse schemes in operation in Britain (EA, 2011), although there is significant interest in such schemes in the current round of water resource management plans. Additionally, recycled water can be used specifically for non-domestic purposes, such as industrial use, reducing the demand for non-recycled water.

Aquifer Storage and Recovery (ASR) makes use of surplus groundwater storage by pumping water back to groundwater in times of high water availability. This technique is only appropriate in areas with existing groundwater storage, although it is possible to implement in areas with groundwater which is saline or otherwise non-potable. While ASR is a strong candidate due to its sustainability in terms of low energy use, potential exists for damage to nearby ecosystems even at a range of several kilometres due to unforeseen alteration of water levels in rivers or wetlands. Earlier reports suggest that recovery is compromised by limited understanding of hydrogeology and unforeseen contamination of recovered water by soluble elements in the rock (Gale et al., 2002). Poor rates of recovery at several trial sites led to a perception of ASR as a high risk option (Eastwood and Stanfield, 2001). Nevertheless, ASR schemes are present in several current water resource strategies.

Leakage reduction

Leakage currently accounts for 23% of the water put into distribution (Tooms et al., 2011). This figure has been steadily reducing since privatisation of the water industry in England and Wales in the 1990s, although it has stabilised in recent years. Fixing leaks and replacing pipes is costly, particularly in built up areas. Depending on the economic value of water, a point is reached when it is no longer efficient to invest in reducing leakage. The quantity of water lost at the economically optimal point between allowing and repairing leaks is known as the 'economic level of leakage'. However, if the cost of new sources of supply is going up, the value of water saved through leakage reduction will also increase.

Current strategies for reducing leakage include the detection and repair of leaks, management of water pressure in the distribution system, implementation of customer metering to allow improved leakage detection and rolling programmes of pipe replacement. A fully implemented leakage reduction schedule would reduce losses by two thirds, with anticipated forthcoming technologies potentially reducing this further to 17% of current leakage, or less than 4% of supply. The cost of this would be around £2,500/property, or around £60 billion for Britain in total. The majority of this cost is in mains replacement, but even excluding that component would still result in anticipated leakage of under 8% of supply for £7 billion spend over twenty-five years, saving 2,600 ML/d nationally. While a complete system refit is clearly expensive, the lower cost options have substantial benefit, making leakage reduction a viable alternative for water companies (Tooms et al., 2011).

Demand reduction

Reduction in PCD can be achieved using several approaches, categorised by Herrington (2005) as: economic (pricing structures and taxes); educational (audits, formal education and campaigns); regulatory (building and plumbing codes); and restrictive (rationing). The pressure on low-income households caused by rising water bills (Defra, 2011) coupled with the unacceptability of disconnecting households from the water supply even if they cannot or will not pay bills (Walker, 2009) make increasing the cost of water increasingly impracticable. Domestic metering, used by a third of domestic households at the time of writing, typically reduces PCD by 15 L/p/d (Walker, 2009) and is currently used without further incentivisation for water use reduction. Whilst mandatory national rollout of metering is not currently planned, the number of metered households increases by around 2% of all households per year in England and Wales (Walker, 2009). Water companies are mandated to conduct rolling efficiency campaigns, which currently includes the supply of

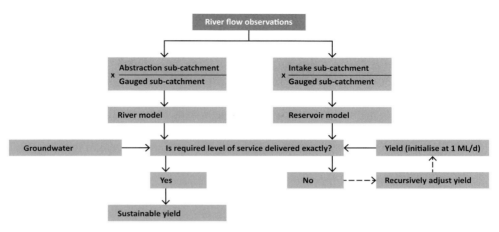

Figure 6.2   Methodology used to generate the water yield estimates for each WRZ

water-saving equipment such as cistern devices and conducting water efficiency audits for
households and businesses.

6.3 Assessment of national water supply infrastructure systems

6.3.1 Modelling approach

There have been a number of analyses of water infrastructure modelling in Britain published
in the last two decades, although the majority of such efforts are focused on the city,
watershed or regional scale (Bekele and Knapp, 2010; Urich et al., 2011; Ribeiro Neto
et al., 2014). Our approach focuses on national-scale water infrastructure systems, using a
new model built specifically for Britain. As such it has similarities with previous models
but with some simplifications necessary for model performance and parsimony.

The water supply systems model is based on existing regional water resource management
arrangements in Britain. Every area in the model falls within a water resource zone (WRZ).
Due to the size of the territory managed by Scottish Water the company's 'megazones' are
treated as WRZs.

Sub-models capturing the hydrology and the water availability of each WRZ are implemented to determine values for potential water yield (deployable output) as shown in Figure
6.2. The sub-model components include river intakes, reservoir intakes and groundwater,
with each component present or absent depending on whether the component is reported
by the water company for that zone. A single reservoir and river intake are included in each
WRZ and represent the total available reservoir capacity and river resource. All increases
in reservoir capacity are represented as an increase in the capacity of the existing reservoir.

Where river intakes and/or reservoirs are present, a relevant catchment in which river flows are gauged is identified from the National River Flow Archive (Centre for Ecology and Hydrology, 2014), with recorded flow proportionally adjusted to the representative river intake flow or reservoir watershed on the basis of sub-catchment area. Where a river and reservoir both exist, the operating rule is to remove as much as possible from the river within licenced conditions and, where this is not greater than yield, to remove deficit water from the reservoir. Groundwater is taken as a steady state input which is subtracted from yield. The maximum sustainable yield is determined as the maximum abstraction which can be removed from the system without causing a breach of the lowest return period level of service for the WRZ. Yield is initialised at 1 ML/d and increased until the reported frequency of shortages in the WRZ is reproduced by the model.

Investment in increased capacity of a reservoir implies an increase in capital expenditure (CapEx) and ongoing operating expenditure (OpEx). Alternative water supply options are available for exploitation in the future depending on the infrastructure strategy being employed. These options include desalination, aquifer recharge, effluent reuse, inter-company transfers, and new groundwater supplies. Investment and operating costs for each of these options are based on a series of regression analyses of estimated costs against yields taken from examples of water infrastructure projects detailed in the water resource management plans published by water companies (e.g. Anglian Water, 2014).

Demand for each WRZ is the sum of domestic and non-domestic demand. Domestic demand for a WRZ is determined by multiplying the projected population of the WRZ in the given scenario (see Chapter 3) by the average per capita water demand for the WRZ. A range of possibilities for PCD are assumed, which depend upon the scale of demand management efforts. Non-domestic demand is calculated as a percentage of domestic demand for a given WRZ based on the regional break-up of non-domestic demand (EA, 2012) and the industrial and agricultural activity in the WRZ. In most WRZs in Britain agriculture is primarily rain-fed and agricultural abstractions are a very small proportion of the total. Potential trade-offs, with abstractions for cooling of thermoelectric power plants, are dealt with in Chapter 11.

Investment in leakage reduction is limited only by the amount of leakage estimated for a WRZ and incurs a CapEx and OpEx cost as per other investment options. CO_2 emissions for leakage reduction are based on energy use and the effects on traffic in the case of leakage reduction efforts (Tooms et al., 2011).

Existing arrangements for distributing water within each WRZ are assumed to remain. A set of existing and possible future inter-company transfers between specific WRZs is explored in relevant strategies, with a cost determined by geographical location. The cost to the receiving company is estimated based on the length of pipe required between the central points in the areas managed by the associated companies.

Infrastructure investments are determined every ten years and are based on a twenty-year planning horizon. Investment options are chosen based on the management strategy and involve decision rules for minimising long-term CapEx and OpEx costs. Each strategy has a fixed level of demand and leakage reduction with alternative options chosen from those available under the modelled management strategy (listed below).

6.3.2 Assessment of national water resources strategies

Scenarios

Three socio-economic scenarios (high, central and low growth, as described in Chapter 3) and three climate scenarios are used to assess the relative performance of each water supply strategy. The possible impacts of climate change on water availability is assessed through the use of the Future Flows hydrological scenarios (Prudhomme et al., 2013). These scenarios were used to identify yields under each of the eleven Future Flow scenarios at 2025 and 2075. Each of these data sets is interpolated to give a set of transient projections for each WRZ. Of the eleven Future Flows scenarios, examples of increasing, median and decreasing water availability are identified (scenarios 'afixh', 'afixc' and 'afixk', respectively).

Strategies

The strategies developed for investment in water supply infrastructure are intended to span the range of possible options for achieving water supply/demand balance. Each strategy represents different approaches to policy intervention, demand management, choices of technologies to increase yield or decrease demand and policy targets with respect to reliability, cost and carbon emissions. The precise investments are not pre-specified – they are triggered by rules within the model, with the portfolio of possible options dependent on the strategy specification. The main alternative management strategies are summarised as follows:

No build (NB): a strategy involving no new investment in any future year. This strategy was employed as a model validation tool and as a means of analysing the impact of socio-economic and climate change without the mitigating influence of new investment.

Minimal intervention (MI): a low investment strategy that assumes minimal intervention at a national scale with only modest reservoir enhancement and expansion of groundwater resources. No new inter-company transfers are incorporated. There is limited leakage reduction and modest demand action. A limit is imposed upon the capacity of desalination plants in each WRZ and no water reuse is implemented.

Capital expenditure (CE): a strategy that focuses on building large-scale, long-term investments in supply infrastructure to meet increasing demand. This approach involves no demand reduction and medium leakage reduction with an emphasis on reservoir enhancement, new groundwater and with no limit on the number of allowed desalination plants. Water reuse is not entertained as an option but inter-company transfers are allowed between all nearby water companies.

System efficiency (SE): a strategy that focuses on increased efficiency of water use, targeting measures to reduce the need for additional infrastructure. This strategy therefore embraces high demand management action with a high level of leakage reduction, while placing a limit on the capacity of desalination plants and permits no new water reuse facilities or inter-company transfers.

System restructuring (SR): focusing on rethinking the system through innovation and design, with a combination of new delivery options and efficiency schemes. This strategy involves high demand action, leakage reduction, inter-company transfers and water reuse.

Performance indicators

The performance of the system in the context of the scenarios and strategies described above is assessed with the following performance indicators:

Capacity margin: the percentage margin between total water potential yield of the water supply system and the projected water use. The capacity margin for the whole of Britain, presented in the figures below, is an average of all water companies' capacity margins, weighted by the population serviced by the company. It is calculated as follows:

Capacity margin $= (($per capita supply $-$ per capita demand$)/$per capita demand$) \times 100\%$,

where per capita demand $=$ (domestic demand $+$ non-domestic demand $+$ leakage)/ population

$$\text{per capita supply} = \left(\text{yield} + \text{net transfers}^{\dagger}\right)/\text{population}$$

†Note that net transfers will equal zero for the whole of Great Britain.

Total water delivered: is the total water used, including end-use demand and leakage. When this total exceeds the yield of the system, then total demand will not be met, and the yield is recorded as the value of the total water delivered.

Cumulative investment: the total capital and operating expenditure up until the year presented. 2010 is the base year.

Carbon emissions: *an estimate of CO_2-equivalent emissions produced by the operation of water supply assets. This value does not include 'embedded' carbon produced during the construction of the infrastructure assets.*

6.4 Strategy analysis and discussion

6.4.1 Performance of alternative infrastructure strategies

The results from our analysis of the water supply system have been validated against the future water requirements presented in the water companies' 2014 Water Resource Management Plans (WRMPs). At present, demand is being met, but without future investment in managing water supplies (No build (NB) strategy), seven of the twenty-three modelled companies may be unable to meet demand in the central growth scenario, by 2050. With the population growth represented in the high growth scenario an additional five companies (i.e. twelve in total) will not meet demand by 2050. As there is no capital expenditure in the NB strategy, associated cost is exclusively operational expenditure, which is mostly energy costs.

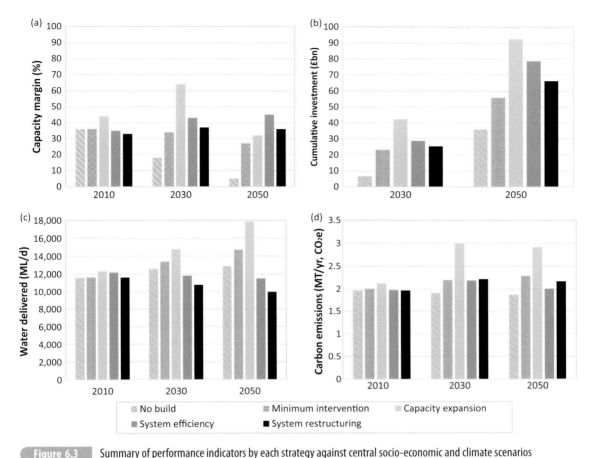

Figure 6.3 Summary of performance indicators by each strategy against central socio-economic and climate scenarios

The results from running the alternative strategies against the central socio-economic and climate scenarios are presented in Figure 6.3. MI is the lowest cost solution but results in the lowest capacity margins by 2050 and medium water delivery and carbon emission levels. CE anticipates the long lead-time on new capacity in order to enable inter-basin transfers to be built to meet future capacity shortfalls in companies with limited options. Due to this heavy investment in infrastructure this strategy results in the highest costs, water delivered and carbon emissions. SE achieves good capacity margins (security of supply) in 2050 with the lowest carbon emissions. SR achieves results that lie somewhere between the capacity expansion and system efficiency approaches which reflect the combination of supply and demand interventions.

Although each strategy achieves a positive overall capacity margin, at the company level the use of standard supply-side approaches in the MI strategy, such as the construction of reservoirs, new groundwater sources and even desalination plants, will not necessarily be able to solve all water resource shortfalls even given modest socio-economic and climatic

Table 6.1 Summary of the 2050 yield in ML/d from new investments that are included in each of the build strategies presented in Figure 6.3 (central socio-economic and median climate scenarios)

Investment option	Minimum intervention	Capacity expansion	System efficiency	System restructuring
New reservoirs	4,380	10,300	1,250	2,620
Desalination plants	0	40	0	0
Aquifer recharge	70	20	30	80
New groundwater	60	200	0	50
Effluent reuse	0	0	0	60
Leakage reduction	1,420	1,510	3,670	4,140
Demand management	4,550	0	11,620	7,780
Inter-company transfers[†]	380	2,670	0	0

[†] Note that net transfers will equal zero for the whole of Great Britain.

changes. Table 6.1 lists the new investments included by 2050 in each of the build strategies presented in Figure 6.3. Without the high demand management and leakage reductions of the SE and SR strategies, significant intercompany transfers are required to meet even the moderate demands of the central growth scenario. Added to this challenge is the fact that all regional shortfalls can only be met if companies with capacity for expansion build extra capacity solely for the purpose of supplying their excess capacity to other companies. The amounts that need to be transferred between companies by 2050 are shown at the bottom of Table 6.1. The costs of these investments are high, but comparable with the costs of the very different SR strategy, which focuses on leakage reduction and demand management.

6.4.2 Strategy performance in the context of climate change

Figures 6.4 and 6.5 present the simulated performance indicators for three alternative runs of the SE strategy against three climate scenarios that show high, central and low overall levels of water flow. The modelled water flow levels change more dramatically for some areas than others between the scenarios, and water availability does not necessarily show uniform positive or negative trends across all water resource zones and climate scenarios. Once these effects have been averaged out, the overall impact of climate change on the capacity margin is great.

In contrast, the impact of alternative socio-economic growth scenarios on water supply under the system efficiency strategy shows a marked decrease in overall capacity margin between the high and low growth scenarios (Figure 6.5). Under the NB strategy the alternative climate scenarios alter the overall capacity margin in 2050 for Great Britain by around 11%. In contrast the alternate socio-economic scenarios alter the average capacity margin by 38% in 2050 without any new infrastructure. All companies experience population

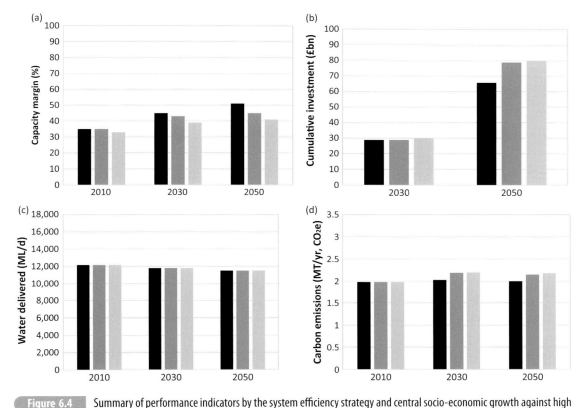

Figure 6.4 Summary of performance indicators by the system efficiency strategy and central socio-economic growth against high (black), median (dark grey) and low (light grey) water flow climate scenarios

increases between the low, central and high growth scenarios with the exception of Thames which has a population that is similar for the central and high scenarios. This results in the cumulative impact on the overall capacity margin being much more pronounced.

The impact on water supplies of the alternate future flows projections varies between regions. As shown in Figure 6.6, companies such as Thames Water, United Utilities, Welsh Water and Scottish Water experience differences in their 2050 water yield of more than 200 ML/d across the alternative future flow scenarios (given no changes in water supply infrastructure). As the model did not evaluate the impact of climate change on groundwater yields, companies that rely entirely on groundwater experience no supply effects from climate change. Despite this artificial advantage, such companies do not generally fare well against socio-economic growth as they have fewer options to increase their yields significantly.

With the exception of the MI strategy, all regional companies are able to meet their water demands for the central growth scenario. Figure 6.7 shows that companies such as Northumbria and Yorkshire Water may be better off with supply-side management (MI and CE strategies), while for companies such as Thames and Bristol, demand management (SE and SR strategies) may be the most cost-effective solution. It should be noted that the

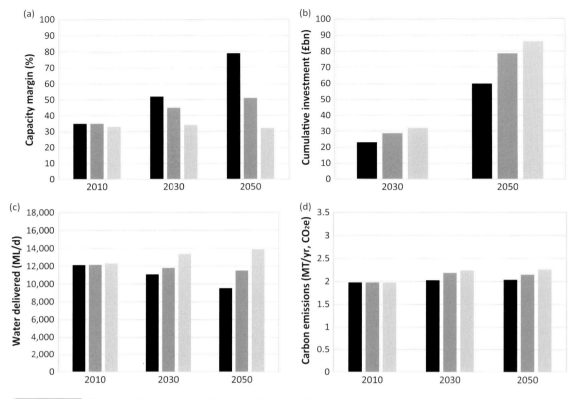

Figure 6.5 Summary of performance indicators by the system efficiency strategy with the median climate scenario against low (black), central (dark grey) and high (light grey) socio-economic growth scenarios

costs shown in Figure 6.4 exclude any revenue derived from inter-company trades, which explains why Scotland and Wales (which become net suppliers of water) show relatively high costs despite having large water resources and small populations.

6.5 Conclusions

Britain's existing water supply infrastructure has provided excellent service up until now. The challenge for water companies in the twenty-first century is to continue to provide similar levels of service under the ever increasing pressures of socio-economic growth and changing climatic conditions. Capital expenditure on infrastructure to maintain supply-demand balance for England and Wales for 2010–2015 is £1.4 billion (Walker, 2009). A major shift has involved the move away from capital expenditure regulation towards total expenditure (TotEx) which has prompted an emphasis within water companies on investment in newer approaches to water management. The modelling efforts presented

Central growth
High flow climate scenario

Central growth
Central flow climate scenario

Central growth
Low flow climate scenario

Balance (ML/d)

−1,700 : −1,400	
−1,400 : −1,100	
−1,100 : −800	
−800 : −500	
−500 : −200	
−200 : 0	
0 : 200	
200 : 500	
500 : 800	

Low growth
Central flow climate scenario

Central growth
Central flow climate scenario

High growth
Central flow climate scenario

Figure 6.6   Company water balances in 2050 under different combinations of modelled socio-economic and climate change under the no build strategy

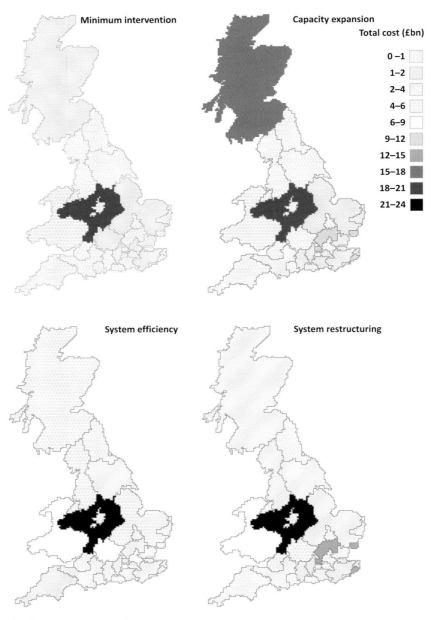

Figure 6.7 Geographical representation of total infrastructure costs (capital and operating costs) to 2050 by water company for each of the four national strategies (under central growth and median climate scenario). Note that costs are given in 2010 terms

here support this approach by proposing that Britain's future water challenges can only be overcome if traditional supply-side infrastructure enhancements are coupled with more innovative efforts. These include demand management, leakage reduction, efficiency measures and wastewater reuse. In the more extreme socio-economic and climatic scenarios large-scale water transfers become cost-effective to alleviate the disparity between regions with water scarcity and those with water abundance.

Though high-level in its approach, the national water infrastructure model provides key insights for national-scale water planning. Unconstrained socio-economic growth without demand management would result in significant constraints being placed on the capacity of new and existing infrastructure to meet future demand for water. Climate change is not expected to impact the total per capita water supply in the same way as socio-economic growth. However, it creates uncertainty around how each region will be affected, with alternative climate model runs resulting in very different water balances for certain regions. The biggest challenge for water supplies presented by climate change may therefore lie in devising management strategies that are robust to uncertain future conditions.

References

Acreman, M. (2001). "Ethical aspects of water and ecosystems." *Water Policy* 3(3): 257–265.

Aitken, V., S. Bell, S. Hills and L. Rees (2014). "Public acceptability of indirect potable water reuse in the south-east of England." *Water Science & Technology: Water Supply* 14(5): 875–885.

Anglian Water (2014). *Water resources management plan*. Huntingdon, UK, Anglian Water.

Bates, B., Z. Kundzewicz, S. Wu and J. Palutikof, Eds. (2008). Climate change and water. *Technical Papers of the Intergovernmental Panel on Climate Change*. Geneva, Switzerland, IPCC Secretariat.

Bekele, E. and H. V. Knapp (2010). "Watershed modeling to assessing impacts of potential climate change on water supply availability." *Water Resources Management* 24(13): 3299–3320.

Centre for Ecology and Hydrology (2014). "National River Flow Archive." Retrieved from www.ceh.ac.uk/data/nrfa/.

Charles, J., P. Tedd and A. Warren (2011). *Lessons from historical dam incidents. Delivering benefits through evidence*. Bristol, UK, Environment Agency.

Critchley, R. and D. Marshallay (2013). Progress towards a shared water resources strategy in the South East of England. WRSE.

Defra (2008). Future water – the Government's water strategy for England. Department for Environment, Food and Rural Affairs. London, UK, TSO.

Defra (2011). *Water for life*. London, UK, Department for Environment, Food and Rural Affairs.

Defra (2013). *Water abstraction from non-tidal surface water and groundwater in England and Wales, 2000 to 2012. Defra official statistics release*. London, UK, Department for Environment, Food and Rural Affairs.

Defra, Department of the Environment Northern Ireland, Scottish Executive and Welsh Government (2006). Water Framework Directive (WFD): Note from the UK administrations on the development of environmental standards and conditions.

Downing, T., R. Butterfield, B. Edmonds, J. Knox, S. Moss, B. Piper and E. Weatherhead (2003). *Climate change and demand for water*. Oxford, UK, Stockholm Environment Institute.

EA (2006). *Do we need large scale water transfers for south east England?* Bristol, UK, Environment Agency.

EA (2011). *Effluent re-use for potable water supply*. Bristol, UK, Environment Agency.

EA (2012). *The case for change – current and future water availability*. Bristol, UK, Environment Agency.

EA (2013). *Managing water abstraction*. Bristol, UK, Environment Agency.

EA, Ofwat, Defra and Welsh Government (2012). *Water resources planning guideline*. Bristol, UK.

Eastwood, J. C. and P. J. Stanfield (2001). "Key success factors in an ASR scheme." *Quarterly Journal of Engineering Geology and Hydrogeology* 34(4): 399–409.

European Parliament and Council of the European Union (2000). Council Directive 2000/60/EC on the Water Framework Directive. *Official Journal of the European Communities*. Brussels, Belgium.

Fowler, H. J., S. Blenkinsop and C. Tebaldi (2007). "Linking climate change modelling to impacts studies: recent advances in downscaling techniques for hydrological modelling." *International Journal of Climatology* 27(12): 1547–1578.

Gale, I., A. Williams, I. Gaus and H. Jones (2002). ASR – UK: elucidating the hydrogeological issues associated with aquifer storage and recovery in the UK. London, UK, UKWIR.

Hannah, D. M., P. J. Wood and J. P. Sadler (2004). "Ecohydrology and hydroecology: a 'new paradigm'?" *Hydrological Processes* 18(17): 3439–3445.

Herrington, P. (2005). The economics of water demand management. *Water demand management*. D. Butler and F. A. Memon (Eds.). London, UK, IWA Publishing.

ICE (2012). *State of the Nation: water*. London, UK, Institution of Civil Engineers.

IChemE (2013). "Water challenges make UK desalination plants more likely." Retrieved from www.icheme.org/media_centre/news/2013/water-challenges-make-uk-desalination-plants-more-likely.aspx.

Jenkins, M., J. Lund, E. Howitt, A. Draper, S. Msangi, S. Tanaka, R. Ritzema and G. Marques (2004). "Optimization of California's water supply system: results and insights." *Journal of Water Resources Planning and Management* 130(4): 271–280.

Johnson, B. (2011). Ignore this rain, it's the drought that we need to think about. *The Telegraph*. London, UK, Telegraph Media Group Limited.

Marsh, T. J. and J. L. Anderson (2002). "Assessing the water resources of Scotland – perspectives, progress and problems." *Science of The Total Environment* 294(1–3): 13–27.

McCulloch, C. (2006). The Kielder Water Scheme: the last of its kind? *Improvements in Reservoir Construction, Operation and Maintenance*. H. Hewlett (ed.). London, UK, Thomas Telford.

Prudhomme, C., S. Crooks, C. Jackson, J. Kelvin and A. Young (2012a). *Future flows and groundwater levels – final technical report*. Wallingford, UK, Centre for Ecology and Hydrology.

Prudhomme, C., S. Dadson, D. Morris, J. Williamson, G. Goodsell, S. Crooks, H. Boolee, G. Buys and T. Lafon (2012b). Future flows climate data. NERC Environmental Information Data Centre.

Prudhomme, C., T. Haxton, S. Crooks, C. Jackson, A. Barkwith, J. Williamson, J. Kelvin, J. Mackay, L. Wang, A. Young and G. Watts (2013). "Future Flows Hydrology: an ensemble of daily river flow and monthly groundwater levels for use for climate change impact assessment across Great Britain." *Earth System Science Data* 5(1): 101–107.

Rahiz, M. and M. New (2013). "21st century drought scenarios for the UK." *Water Resources Management* 27(4): 1039–1061.

Ribeiro Neto, A., C. A. Scott, E. A. Lima, S. M. G. L. Montenegro and J. A. Cirilo (2014). "Infrastructure sufficiency in meeting water demand under climate-induced socio-hydrological transition in the urbanizing Capibaribe River basin – Brazil." *Hydrology and Earth System Sciences* 18(9): 3449–3459.

Stokes, J. and A. Horvath (2006). "Life cycle energy assessment of alternative water supply systems." *The International Journal of Life Cycle Assessment* 11(5): 335–343.

Tooms, S., S. Trow and H. Walker (2011). *Long term leakage goals*. London, UK, UK Water Industry Research Limited.

Urich, C., P. Bach, C. Hellbach, S. Robert, M. Kleidorfer, D. McCarthy, A. Deletic and W. Rauch (2011). Dynamics of cities and water infrastructure in the DAnCE4Water model. *Proceedings of 12th International Conference on Urban Drainage*. Porto Alegre.

Walker, A. (2009). *The independent review of charging for household water and sewerage services*. London, UK, Department for Environment, Food and Rural Affairs.

Welsh Water (2014). *Water Resources Management Plan*. Cardiff, UK.

World Health Organisation and UNICEF (2014). *Progress on drinking water and sanitation – 2014 update*. Geneva, Switzerland, World Health Organisation/UNICEF.

Wastewater systems assessment

LUCY J. MANNING, DAVID W. GRAHAM, JIM W. HALL

7.1 Introduction

Effective wastewater collection, treatment and disposal are essential components of civilised life in any society. The manifest impact of improved waste management in reducing epidemic infectious disease in the latter nineteenth century is a testament to its importance. Therefore, how we handle our wastes is central to the well-being of all communities (Lofrano and Brown, 2010). In the developed world, wastewater management systems broadly consist of a sewer network to collect the wastewater, and treatment works to transform the wastes to a state where they can be returned to the natural environment.

In many parts of the world, wastewater management systems are rudimentary, or do not exist at all. Only 64% of the world's population have access to improved sanitation facilities with conditions in many parts of world getting worse. In 2014, 2.5 billion people still lack access to common sanitation despite improved sanitation being a central Millennium Development Goal (World Health Organisation, 2014). Population growth rates in Africa continue to be greater than rates of sanitation provision. Therefore, there is a global need to continue to innovate in sanitation infrastructure, possibly exploiting less capital- or water-intensive technologies (Marlow et al., 2013). Even in advanced economies, wastewater management and treatment systems face multiple challenges, including the accommodation of urban population growth, stricter environmental standards, rising energy costs and ageing infrastructure.

A program to improve waste treatment has been in place in Europe since the early 1990s, driven by legislation to improve environmental quality in rivers and coastal waters (Figure 7.1). The result in Britain has been the minimisation of untreated and primary treated sewage that is discharged into the environment. The pursuit of ever-stricter wastewater treatment standards has led to significant investment in new infrastructure and increased energy consumption. This has resulted in higher tariffs for customers because operating costs are sensitive to energy costs, which may increase in the future. The energy intensity of wastewater treatment infrastructure also makes it a target for policies to reduce greenhouse gas emissions, such as the U.K.'s Carbon Reduction Commitment Energy Efficiency Scheme.

As noted, both sewer networks and wastewater treatment plants are capital intensive. The design life varies between the mechanical plant and the longer-lived fixed 'civil' infrastructure. In Britain much of the civil infrastructure is many decades old, some dating back to the nineteenth century, so the need for maintenance, rehabilitation or replacement is

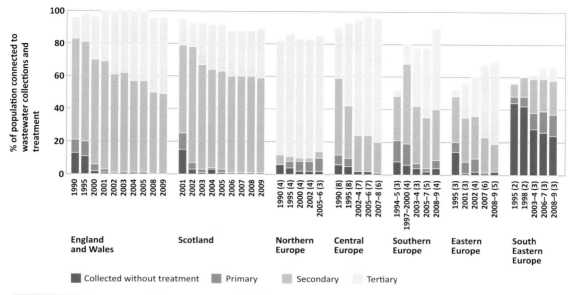

Figure 7.1 Improvements in the maximum standard of European wastewater treatment since 1990 (Source: EEA, 2012) (Numbers in brackets show the number of countries represented)

increasing. This represents a major challenge in the future as human population increases, energy costs rise and environmental targets become stricter. There is, therefore, intensifying pressure to innovate approaches to waste treatment because current methods will almost certainly not be sustainable. Such innovations may take place in the individual technologies that make up the wastewater collection, treatment and disposal system and/or in the system as a whole.

The purpose of this chapter is to examine the current state of Britain's wastewater systems, and then explore challenges and opportunities for the future. Specifically, we evaluate strategic options for wastewater treatment technologies and systems using future scenarios of population, economy and energy prices discussed in Chapter 2 as part of our system-of-systems modelling approach.

7.2 Wastewater systems in Britain

7.2.1 System description

Wastewater management networks perform the key functions of collecting and treating wastewater, discharging treated effluent to the environment and disposing biosolid sludge (Figure 7.2).

In Britain, the infrastructure system consists of 624,000 km of sewer pipe and over 9,000 treatment works, which collectively process wastewater from about 96% of Britain's

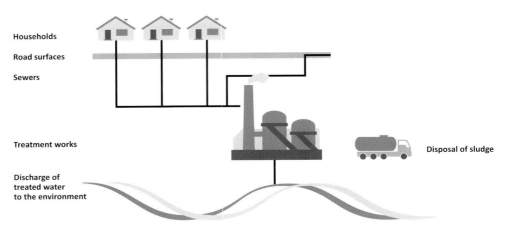

Figure 7.2   Schematic of wastewater systems

population (Defra, 2012b). Treated effluent is discharged to waterways and coastal waters, and sludge is either digested, spread on agricultural land, incinerated to reduce the volume, or disposed in landfills. The volume of wastewater treated reflects per capita daily water use, currently about 150 L/day in Britain, most of which is returned to the sewerage system. However, as well as domestic wastewater, the sewerage system also accepts effluent from industrial users, and older systems (approximately 40% of the current network (Defra, 2012a)) also carry urban surface runoff. Each of these additional sources bring difficulties; industrial users may discharge pollutants that are difficult to remove from the wastewater, whereas urban runoff increases liquid volumes that must pass through the sewers, which is especially problematic at times of severe storms and flooding.

Sewage treatment in Britain involves significant consumption of energy, which results from the active aeration treatment processes frequently used to accelerate natural breakdown of organic matter. The water industry consumes approximately 1% (CIWEM, 2013) of the national energy budget; of this, 40% (Ofwat, 2011) is used in wastewater treatment, the most energy-expensive component of the water industry, a proportion which will continue to increase unless treatment technologies are changed. Currently, anaerobic digestion (AD) offsets part of this energy expense by biologically converting residual sludge to methane, which can be burned. This produced a carbon credit in 2011 of 0.71 $MtCO_2e$ (Ofwat, 2011; Carbon Trust, 2013) and represents approximately 30% of the total energy cost of the wastewater service in England and Wales (Ofwat, 2011). Solid residues from AD systems still must be dispersed by land spreading, landfilling and incineration. In 2010, Britain produced 1.4 million tonnes of sludge dry solids, of which approximately 80% was spread on agricultural land, 18% disposed of to landfill and 1% was incinerated (Defra, 2012b).

In Britain, as in other mature economies, wastewater treatment has evolved during the twentieth century as a highly centralised function, resulting in substantial investment in capital assets. This centralisation results from economies of scale and the ease of maintaining reliable treatment quality at large centralised plants, but leaves the industry relatively inflexible to changes in demand, environmental regulation and treatment technology, which

Table 7.1 Replacement and annual operational costs (£m) of different elements of wastewater treatment in England and Wales (Source: Ofwat, 2011)

	Sewage collection	Sewage treatment	Sludge treatment	Sludge disposal	Total
Capital asset value	180,000	15,000	2,500	700	212,000
Annual operational cost	230	470	185	85	970
Annual energy cost	45	160	10	0	215

can occur over the lifetime of existing infrastructure. While the lowest cost treatment option will always be chosen within these constraints, inflexibility of previously invested assets, together with risk aversion, has led to a conservative approach in asset management.

Britain's sewer network serves a population of approximately 60 million people, and has a replacement value in England and Wales of about £180 billion (Ofwat, 2011, updated for inflation). In addition to the capital value of the sewer network, its operation involves 24% of costs of the wastewater industry, largely in pumping costs (Ofwat, 2011).

Analysis of wastewater industry expenditure in England and Wales shows that wastewater treatment accounts for 48% of operating costs and 72% of energy expenditure in the wastewater systems (Table 7.1). Most of the remaining energy use is in pumping in the sewer network. On the other hand, the vast majority of the capital assets in the system are in the sewer network, a proportion of which dates back to the advent of modern sanitation. Figure 7.3 shows Britain's treatment capacity in terms of population served, disaggregated by size of plant serving different regions.

7.2.2 Challenges and opportunities

A number of external drivers for change must be considered in future wastewater infrastructure systems, including population changes, the emergence of 'new' pollutants, legally imposed environmental standards and energy costs. Climate change may also affect capacity through changes in the hydrological cycle, such as intensified rainfall and runoff, although this is not considered here. Additionally, there is a constant search for improved economic performance. The following summarises important challenges and opportunities for strategic investments.

Population

Population increase will require more sewerage and treatment capacity, impacting both network size and capacity requirements, with the potential need for expanded trunk sewer capacity to accommodate increased wastewater volumes. Continuing current trends, higher population density will affect choices of treatment options, possibly necessitating facilities that take up less space in dense urban areas. However, the increased requirement for treatment capacity in areas of sharp population growth also provides an opportunity for the

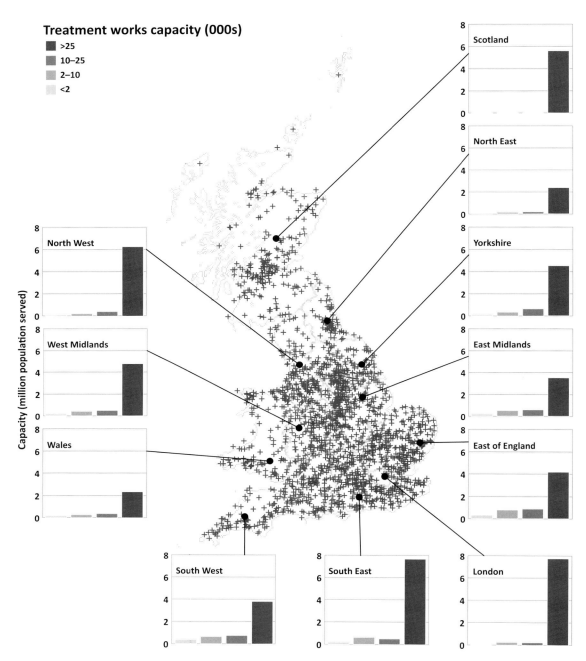

Figure 7.3 Map of GB treatment capacity in terms of population served, disaggregated by size of plant serving the given populations (Source: Ofwat, 2011)

modernisation of treatment technologies, possibly more rapidly than would occur through replacement of infrastructure at the end of its life.

For a given effluent standard, there is a limit to the capacity of receiving waters in the environment to assimilate the waste. This means that even for effluents treated to current regulatory standards, there is a limit to the population size able to be supported by available receiving waters. If wastewater treatment can remove 95% of the wastewater carbon content, and a river can support a flow of 2 g of biochemical oxygen demand (BOD) per m^3/s of flow, this corresponds to a population of approximately 400,000 people served by one sewage treatment works per m^3/s of flow. It is more difficult to make a similar calculation for wastewater nitrogen content, as this is strongly affected by local agricultural practices.

Taking England and Wales as a whole, population growth scenarios do not appear to be on a scale where the receiving capacity of the aquatic environment might be a limiting factor. However, some catchments are already potentially threatened at low flows: examples include the Tame at Minworth, northeast of Birmingham, the Soar at Leicester and the Avon at Coventry. Further, significant increase of population in a number of other cities would result in the need for wastewater effluent to be pumped to other catchments.

Environmental standards

Environmental standards have been one of the major drivers of recent investment and are prescribed by the European Union's Urban Wastewater Treatment Directive of 1991, which concerns discharges to the environment of municipal wastewater, the Water Framework Directive of 2000, which concerns water resources management, the Bathing Water Directive and the revised Bathing Water Directive of 2006. The Urban Wastewater Treatment Directive requires the collection and treatment of wastewater in all communities of greater than 2,000 inhabitants, and more advanced treatment for communities of greater than 10,000 inhabitants in designated sensitive areas. This is managed by requiring prior authorisation of all discharges of urban wastewater; by monitoring treatment plants and receiving waters; and by controlling the disposal and reuse of sewage sludge. Discharge consents are issued to the water companies for each treatment plant, specifying the levels of carbon, suspended solids, ammonia and phosphates permitted. Discharges of other substances are controlled, such as metals and pesticides. Most of the inland catchments in Britain are designated as sensitive to phosphates, which can cause eutrophication in freshwater bodies (Defra, 2012b).

Following the introduction of the above directives, substantial investment (Table 7.2) was required to ensure compliance, particularly at coastal locations, for example, Brighton and Stranraer (Defra, 2012b), where previously wastewater had been discharged to sea following only screening and primary settlement. While the most significant part of this expenditure has now been made, continuing investment will be required to replace ageing plants, to reflect population change (affects both the capacity and treatment level required to maintain environmental quality) and to comply with continuing tightening of environmental standards. This last aspect could be significant as new pollutant concerns are addressed.

Table 7.2 Recent capital expenditure (£M) on the U.K.'s wastewater industry (Source: Defra, 2012b)				
	England	Wales		Scotland
1990–2000	9,600	1,200	1996–2002	2,160
2000–2005	4,600	500	2002–2006	1,980
2005–2010	3,100	300	2006–2010	2,380
2010–2015	3,100	100	2010–2015	2,500
Total	20,400	2,100		9,020

Ageing infrastructure

Failure to replace ageing sewerage pipes can result in sewer collapses and groundwater contamination, whilst failure to reflect changes in volumetric demand can lead to sewer surcharging and flooding of foul water. This occurs because the sewer system conveys both domestic wastewater and surface runoff, and the volume to be accommodated can swell considerably during storms. Over 4,000 properties in England and Wales are at risk of foul water flooding from sewers (Mott MacDonald, 2011). Possible alternative responses to this capacity need are reduction in storm water transport, household demand reduction and decentralisation of treatment.

A programme in the US during the 1990s to reduce sewer flooding was achieved by storm water being separated from domestic wastewater in a number of cities, at a cost ranging between £5 million and £36 million per square kilometre (MWRA, 2011; Environmental Protection Agency, 1999; updated for inflation). Estimating the combined built up area in Britain to be 22,000 square kilometres (Ordnance Survey, 2014), and bearing in mind that 40% of the network comprises combined sewers (Defra, 2012a), this implies an expenditure of between £40 billion and £320 billion in storm water separation.

Thames tideway tunnel

London's sewer system was designed and built in the 1850s. Designed to accommodate both domestic sewage and surface runoff, the significant increase in London's population since that time means that there is no longer sufficient capacity in the system during periods of rainfall, and the mean annual discharge of untreated wastewater into the Thames is 20 million tonnes.

The solution is to build a 25 km, 7.2-m-diameter pipeline largely under the Thames, with a storage capacity of 1.6 million m^3, to transport the combined wastewater and surface runoff to Beckton sewage treatment works, and store excess in times of severe rainfall, with an estimated cost of £4.2 billion (2011 prices).

The tunnel, and a connecting tunnel in the Lee valley, is expected to be completed in 2023.

(Thames Tideway Tunnel Ltd, 2014)

The introduction of sustainable urban drainage systems (SuDS) in new developments is intended to reduce demand on the piped sewer network. These are a range of measures which permit the detention of runoff, and its absorption close to its source, ranging from reed beds, swales and infiltration ditches to permeable paving, and allowing for discharge directly to the ground, where possible.

An alternative strategy for reducing the demand for sewer capacity is the reduction in household water use, which implies less wastewater. Besides measures discussed in Chapter 6, domestic water use could be further reduced by the introduction of rainwater harvesting or grey water reuse, that is, by using rainwater runoff from roofs, or by reusing water from sinks, baths and showers for toilet flushing, estimated to make up 30% of water demand in Britain. However, responsibility for regular maintenance for such installations would have to be taken by householders. It is not clear whether this is the case in locations where these systems have been introduced, and if not, the system will not have the intended benefits (Moglia et al., 2013). Therefore, strategic choices about the future of wastewater systems involve trade-offs between the operational costs of wastewater treatment plants (which currently are lower, and more reliable, in large centralised plants) and the capital cost of construction and depreciation of sewer networks.

Cost and energy efficiency

The provision of wastewater treatment infrastructure is a problem of constrained optimisation: minimising the cost of service subject to achieving required public health and environmental standards. Within the context of existing wastewater treatment systems, a major factor in improving cost efficiency is reducing energy use. This is reinforced by Britain's carbon reduction commitment, which seeks a steady reduction of energy use by all large industrial users, with a target of reducing energy use to 20% of the 1990 level by 2050 (Climate Change Act, 2008). Financial incentives for the reduction of energy consumption are issued by the European Union and U.K. government in the form of tradable emissions allowances and Renewables Obligation Certificates to further reduce energy use and also to promote generation of renewable energy.

While introduction of discharge consents reflecting the seasonal impact on the environment and its use would reduce overall costs, options for reducing energy within the current system include reorganising plant processes to take advantage of economies of scale; introducing alternative, lower energy-cost treatment technologies; or increasing energy recovery from the wastewater itself at different stages in the processing. The recovery of resources, either in the form of energy or other nutrients (e.g. nitrogen N or phosphorus P) is also an option, which will be reflected in the prices that can be attained from those resources.

Given that the provision of wastewater treatment is constrained by demand, strategic options for wastewater treatment are driven by costs and efficiencies, and by environmental standards. Key costs for the wastewater industry are those of construction, both of plant and of sewerage, and of plant operation. Operating costs are strongly influenced by the cost of energy, and will increasingly be driven by the cost of carbon emission. Wastewater treatment is a net consumer of energy, primarily through the cost of treatment and to a lesser degree, pumping. It is, however, currently feasible to make the wastewater industry energy neutral or even a net exporter of energy by shifting, wherever possible, to technologies that generate energy during treatment, most obviously through methane recovery, but potentially also through microbial fuel cells. Recovery of resources in the wastewater besides energy and freshwater can also offset some of the operational cost, although the economics of doing so depend crucially on the value of those resources.

Barriers to innovation

Marlow et al. (2013) discussed the barriers to innovation and mechanisms for transition to a more sustainable wastewater infrastructure system, noting that while much is said about decentralisation of wastewater treatment, it has been adopted in few locations. Although their focus is on the value of treatment decentralisation, more general conclusions may be drawn from their work.

Barriers to new wastewater management solutions include both economic and management issues. Realities of a government-regulated private industry imply that service charges to the customer cannot be raised beyond what is publicly acceptable, thus limiting the rate of introduction and originality of new solutions that are costly. Proven technologies bring the benefits of reliability, operational efficiency and known cost within the constraints of that technology, while the introduction of new or unknown technologies may have unexpected performance and knock-on effects. Further, piecemeal replacement of assets tends to preserve the existing network structure. However, retaining existing infrastructure may prevent savings from being realised from potentially more innovative solutions.

Thus, since pre-existing systems have the benefit of design and operator experience, new solutions that can fit within existing systems have a better chance of working. Once a solution has been shown to be effective in one situation, it can be rapidly adopted elsewhere, using the benefit of experience gained in the initial installation. Thus, it is to be expected that innovative solutions will be introduced slowly in pilot systems for small communities giving experience at a low cost.

7.2.3 Technological options for wastewater infrastructure

Treatment technologies

In response to the above challenges there are a number of wastewater treatment technologies available. The aim of wastewater treatment is to remove unwanted substances from the wastewater, including biodegradable organic matter, contaminants such as metals and disease-causing organisms, so that the treated effluent can safely be returned to a river, lake or the sea. This does not necessarily involve sterilisation, but aims to reduce levels of contaminants and nutrients to levels where chemical, ecological and public health conditions in the receiving water are not significantly impaired.

Waste treatment is performed in a number of steps (Figure 7.4) including primary, secondary and tertiary treatment, which are progressively more costly, but also achieve progressively higher effluent quality. Primary treatment involves screening and settling processes, which remove inert solids and larger organic matter, typically amounting to about 30–40% of biological oxygen demand (BOD) in the original wastewater.

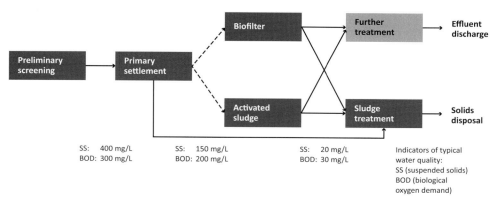

Figure 7.4 Schematic of current common practice in sewage treatment

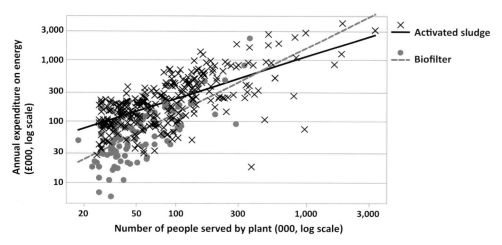

Figure 7.5 Expenditure on energy of wastewater treatment plants for sewage works serving more than 25,000 people, differentiated by secondary process type (Source: Ofwat, 2011)

Secondary treatment targets soluble biodegradable organic matter, promoting microbiological reactions in a manner identical to biodegradation processes in nature. The two most common secondary treatment processes used in Britain rely on bacterial populations to break down the organic matter in the waste. The first type is activated sludge, where the wastewater is actively aerated to increase rates of microbial degradation. Alternately, secondary treatment also can be performed using biofiltration processes such as trickling filters, where wastewater is passed slowly through porous media and bacteria attached to the media degrade the waste as it passes by.

Activated sludge tends to be expensive in energy because large amounts of air must be injected for active oxygenation (Figure 7.5). In contrast, aeration in biofilters is passive, which means less energy is needed, although carbon removal rates are sometimes lower and ground footprints are often larger than activated sludge systems. As a result of these traits, smaller volume treatment plants tend to use biofilter processes, whereas larger volume treatment plants (at least in the U.K.) use activated sludge processes, largely due economies of scale and superior effluent quality (Figure 7.3). In general terms, both activated sludge and biofilter systems produce high quality effluents if longer in-reactor contact times are used, although this also requires larger tanks or filters for the same population served.

If the dilution available in the receiving waters is small relative to the waste released (reflected in the discharge consent), tertiary waste treatments are sometimes needed. These most often further reduce the suspended solids remaining in the wastewater stream, or target nitrogen, phosphorus and other contaminants that are not readily removed by traditional secondary systems. As an example, removal of nitrogen compounds from wastes, most often ammonia, requires additional aeration to form nitrate, which can be subsequently biologically reduced under anoxic conditions to nitrogen gas. Further, phosphorus removal can also be performed from the wastes, either using chemical processes (e.g. addition of iron or alum to precipitate phosphates) or additional biological treatment.

Aerobic processes such as activated sludge or biofilters produce a large quantity of sludge, which is primarily composed of waste cells. This can be disposed to landfills (if local regulations allow) or can be further processed in a number of ways. Most treatment plants include AD, which not only reduces sludge volume, but also produces methane. This can be burned for energy on-site or fed into the power grid, depending upon gas quality. Waste sludge after AD is typically dewatered and recycled. If it is sterilised, it can be spread on land as a fertiliser or incinerated.

Economies of scale in wastewater treatment

It is observed throughout the world that there is an economy of scale in both capital and operational expenditure on wastewater treatment (Environmental Protection Agency, 1980; Friedler and Pisanty, 2006). Figure 7.6 shows the number of plants of different sizes currently in operation in Britain, and also the per capita waste processing cost of such plants. This economy of scale implies that based on treatment efficiency alone, it is always cheaper to process wastewater in large central plants, although this is tempered by the cost of pumping wastewater through long distances. However, small waste volumes currently processed in smaller plants (Figure 7.3) means that they have little effect on the overall treatment cost.

Therefore, as population and demand for waste treatment grow in the future, larger treatment plants are an option, which will in principle lead to increased process efficiency. The consequences for the existing portfolio of treatment plants of projected population growth during the first half of the twenty-first century is shown in Figure 7.7j. In principle, combining smaller wastewater treatment plants could lead to a further efficiency increase, and indeed water utilities have amalgamated several plants in recent decades, notwithstanding the additional capital cost of inter-connection. However, where effluent is transferred to a different catchment this approach will be limited by the capacity of local receiving waters for effluent disposal from very large plants.

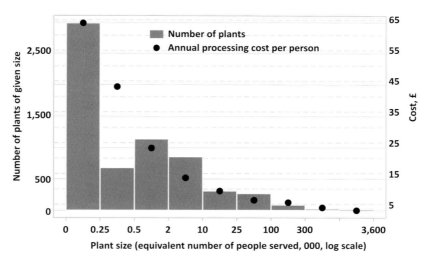

Figure 7.6 Current distribution of sewage treatment plant size, together with the cost of processing per person (Source: Ofwat, 2011)

Energy recovery

The primary route for recovery of the energy used in wastewater processing in Britain is through AD of concentrated sewage sludge. Anaerobic digestion (AD) serves the dual purpose of reducing sludge solid volume by approximately half, and producing biogas, which usually contains about 60% methane and 40% carbon dioxide, and can be burned to produce energy. As previously noted, approximately 75% of the sewage sludge produced in the U.K. is currently processed by AD (Defra, 2012b), recovering 140 MW of energy in England and Wales (Ofwat, 2011). This is equivalent to approximately 30% of the total energy cost of the wastewater service in England and Wales (Ofwat, 2011).

Thermal hydrolysis pre-treatment of sludge could significantly increase the biogas output of AD (Mills et al., 2014). In addition, increased application of AD across the water industry to ensure AD of sludge from all treatment plants serving at least 10,000 people would increase the proportion of energy costs recovered in the wastewater sector to 40%.

The recovered energy from AD of sewage sludge may be increased by the addition of household food waste. There are currently restrictions in the U.K. for co-digestion of sewage sludge and biodegradable solid waste because of incompatible legislation for the disposal of resulting digestates and the inevitable contamination of household food waste with debris such as plastic and glass. It is, however, practised in other countries, including Denmark and Germany (Iacodivou et al., 2012). Laboratory and pilot studies have shown that the addition of food waste to sewage sludge enhances methane production and combining the wastes can also improve process stability if the waste mix is maintained at stoichiometrically suitable ratios. Lebiocka and Piotrowicz (2012) found in a pilot plant that the optimum volumetric ratio is 75:25 sewage sludge to organic municipal solid waste, leading to an increase of 25%

in biogas production. The current production of food waste in Britain (12–20 Mt) is well in excess of the quantity (5 Mt) required to achieve this ratio (Defra, 2011). In addition to the above measures this would result in recovery of 50% of the energy used in the wastewater sector.

Nutrient recovery

The focus of wastewater treatment has shifted over the course of the twentieth century from the removal of substances detrimental to public health, to encompass the maintenance of the environmental health of waterways. The emphasis is beginning to broaden again, from the expenditure of energy for cleaning the wastewater to the recovery of valuable resources from the wastewater. Chief among these are the water itself, and energy within the wastewater, which has already been discussed. However, wastewater also contains nitrogen and phosphorus, both of which need to be controlled to prevent eutrophication of the receiving water bodies. However, nitrogen and especially phosphorus are globally diminishing resources for sustainable agriculture and a huge portion of our phosphorus is released in domestic wastes without consideration of recovery. This makes nitrogen and phosphorus in wastes of considerable present and future value. Further, the hydrocarbons butyric and propionic acid could be recovered from wastewater to provide even higher value waste-sourced products.

Around 80% of the sludge output from the wastewater processing in Britain is currently recycled to land (Defra, 2012b). This is a small proportion of the biomass currently spread to the land, so there is capacity for substantial increase in the event of population growth. However, the usable quantity will ultimately be constrained by variable phosphorus content and the possibility of heavy metals content, limited by the European Sewage Sludge Directive of 1986.

The chief emphasis of processing to remove (rather than recover) nitrogen content from wastewater is firstly in the reduction of ammonia content and secondly to convert as much as possible to gaseous nitrogen. Nitrogen is recovered from the air by the energy-expensive Haber–Bosch process, chiefly for fertiliser. It may be more economical to recover the nitrogen from wastewater effluent in a form more suitable for use as a fertiliser.

An example of the recovery of useful materials is struvite, a magnesium–ammonium–phosphate compound whose inadvertent precipitation in pipe networks is an expensive nuisance to be avoided. However, struvite can be precipitated and recovered in a controlled manner from anaerobic waste liquor streams to form a valuable slow-release fertiliser (Rahman et al., 2014). Since phosphorus would otherwise have to be removed from the wastewater stream by precipitation with chemical additives, the installation of a struvite plant, at a fractional additional capital cost, might save money in additive costs, quite apart from the sale value of the resultant fertiliser. One such plant, recently installed in Slough, is expected to recover its capital cost within ten years through reduced additive costs alone. However, although the phosphorus removal from anaerobic waste streams can be efficient, most of the phosphorus in wastewater still remains in sludge solids

(Jaffer et al., 2002). New technologies are under development to improve practical yields and also recover phosphorus earlier in the process train.

Freshwater reuse and recovery

There is an intrinsic linkage between treated wastewater released to the environment and freshwater supply. In fact, wastewater discharge consents are largely based on the recognition that treated wastewater ultimately becomes a water source. This is a situation so prevalent in Britain, which has many larger cities in close proximity to each other (e.g. the Midlands, or the Thames valley) where different cities use water in series as it passes down each valley.

Treated wastewater directly supplies 30% of Singapore's water needs (Public Utility Board, 2014), and is also used for agricultural irrigation in a number of parts of the world, in particular in the Middle East. However, use of treated wastewater for human consumption more usually involves groundwater recharge or discharge of treated wastewater into a reservoir or upstream of a river water intake. The pressure of population increase and climate change may make direct domestic reuse of treated wastewater a reality, particularly in the south and east of England. Achievement of an appropriate effluent quality requires tertiary treatment steps, including sterilisation using processes such as ozonation, chlorination, ultraviolet radiation and filtration.

Filtration is particularly useful, especially for groundwater-based reuse and recovery, because it can eliminate smaller particles that pass primary and secondary wastewater treatment processes. A filter size of less than 0.05 μm is needed to remove bacteria, while virus removal requires a filter finer than 0.0001 μm. A common filtration system for recycling drinking water is reverse osmosis (RO), which can remove bacteria, viruses and common chemical contaminants to below 0.0001 μm. Reverse osmosis is an expensive technology in terms of energy use, due to the pressure differential necessary to filter the water. However, if RO is preceded by coarse filtration, both energy use and RO membrane fouling can be reduced. Friedler and Pisanty (2006) reported an increase of 70% of operating costs for the addition of both nutrient removal and filtration over simple secondary treatment, and a 400% increase in costs if RO is also included.

A somewhat related treatment technology is membrane bioreactors (instead of secondary clarifiers), which are based on the same principles as biofilters and activated sludge. This technology is gaining rapid acceptance due to smaller cost and ground footprint, providing a more finely filtered effluent (<0.5 μm) than that arising from the combination of activated sludge and settling. While energy expenditure is higher, overall costs are now comparable to those of conventional activated sludge plants (Young et al., 2012), and earlier problems of membrane fouling are being overcome.

A number of comparisons have been made of the cost of reuse of treated wastewater and of desalination of brackish water. Desalination has until recently been more expensive than water reuse (Reddy and Ghaffour, 2007). However, the use of a common technology (i.e. RO) means that relative costs are converging, although the costs of both options are still higher than the current water supply in Britain. The use of desalination and direct reuse for water supply is discussed in Chapter 6.

7.3 Assessment of national wastewater systems

7.3.1 Modelling approach

The following section summarises the modelling approach used for strategy assessment. The national wastewater systems model is a simple annual tally of capacity, energy use and expenditure. The initial treatment capacity is that listed by EEA under the Urban Wastewater Treatment Directive, and involves locations and loads. The plants are matched to population in census districts, by proximity and balancing plant load with population served. Note that the balance cannot be perfect, as the load includes industrial waste, and not all population is served by wastewater treatment works. Initial treatment type is taken where possible from government sources (Ofwat, 2011), and otherwise drawn from binomial distributions to match proportions found in those government sources. Initial plant age is drawn from a uniform distribution ranging from 0 to the maximum age, taken here as fifty years (Friedler and Pisanty, 2006). From the point of view of plant age, plant with capacity over 30,000 person equivalents (PE) is split into five sub-plants of different ages, to represent past incremental development. Anaerobic digestion (AD) capacity is taken from Byrns et al. (2013), and split between the larger plant.

Using the population growth scenarios described in Chapter 3, treatment plant capacity is increased in line with population increase, using linear projections of current population growth rate in each district to the date of maximum plant age. In addition, it is assumed that each plant (or plant subsection) is to be replaced when it reaches the maximum age. Sewer length is increased with population growth, but capacity is not taken into account. AD plants are installed at all locations serving more than 10,000 people.

7.3.2 Strategy description

Strategies for wastewater treatment

Three strategies for wastewater treatment plants have been explored:

1. Business as usual: the existing plant are expanded in line with population in their catchment, as surplus capacity is exhausted.
2. Centralisation: when plant is due for expansion, it absorbs capacity from nearby smaller plant.
3. Hybrid processing: when plant is due for expansion, it is replaced in such a way as to convert as many streams as possible to series processing, consistent with the projected capacity requirement. Following the survey reported by Harrison et al. (1984), a unit of capacity of a single processing type is replaced with a quarter unit of biofilter in series with a quarter unit of activated sludge plant.

Table 7.3 Key data and sources used in strategy analysis. P is population served		
	Value used in calculations	Source
Pipeline and pumping station capital cost for amalgamating plant	£84, 820$P^{0.66}$	www.waterprojectsonline.co.uk
Sewer length (km)	0.00614P	Ofwat, 2011, table 17a
AD capital cost	£41P	www.waterprojectsonline.co.uk
Plant energy cost (activated sludge)	£61$P^{.737}$	Ofwat, 2011
Plant energy cost (biofilter)	£0.61$P^{1.08}$	Ofwat, 2011
AD net power generation	£1.28P	Ofwat, 2011
Sewer energy cost	£2, 383$P^{0.7}$	Ofwat, 2011
Plant operational cost	£52$P^{0.794}$	Ofwat, 2011
Other operational cost	£8P	Ofwat, 2011

It is assumed all new plant is designed for nitrification. This implies an increase in energy cost of all activated sludge plant by a factor of 1.5 compared with non-nitrifying plant, and an increase in volume (implying also energy cost increase) of all biofilter plant by a factor of 1.2 (representing an increase in both capital and operational expenditure). Where plants are in series, it is not expected that a biofilter is set up for nitrification. In addition, it is assumed that all new AD plants include thermal hydrolysis pre-treatment, resulting in increased energy production by a factor of 1.4. The following factors are not included in the calculations: (i) cost of final sludge disposal; (ii) cost of land; (iii) cost of pumping between amalgamated stations; and (iv) phosphorus recovery. Table 7.3 summarises the data and sources used in the strategy calculations.

In addition to the investment strategies discussed here, we also provide exploratory analysis on the potential for decentralisation as an alternative investment approach in Section 7.4.2 and lay out the basis for future strategies in Section 7.4.2 that align with the cross-sector analysis presented in Chapter 10.

Integrated strategies for wastewater treatment and sewer networks

We have combined the most appropriate treatment technologies with varying levels in the sewer network to develop integrated strategies for wastewater infrastructure systems.

Investment requirements in the different wastewater infrastructure strategies are summarised in Table 7.4. Investment requirements increase with population, and capital expenditure is expected to be primarily in the sewer system. Storm water separation in particular will be an expensive strategy.

Economies of scale in treatment will increase efficiency, strengthening the case for large plants in urban areas, and thus decentralisation will have a marginal effect on overall costs except in very small population centres. Furthermore, energy use is critical. There is potential for incremental improvements in energy efficiency, but, with technologies that are currently available, the wastewater sector should not be expected to become a significant net exporter of energy. In addition, stricter environmental standards could bring significant new

Table 7.4 Strategies for long-term investment in the wastewater sector, with costs to 2050 for medium population growth scenario

Strategy	Treatment plant	Sewers
Minimum intervention	Plant replacement and new capacity with current (or marginally improved) technology Total cost £40 billion	Low replacement rate of 0.3% per year Total cost £96 billion
Capacity expansion	Plant replacement and new capacity with current (or marginally improved) technology Total cost £40 billion	High replacement rate of 1% per year Total cost £146 billion
System efficiency	Operational and energy expenditure improvements due to technological change Total cost £36 billion	Moderate replacement rate of 0.5% per year Total cost £109 billion
System restructuring	Operational and energy expenditure improvements due to technological change, and increased efficiency of process due to storm water separation Total cost £36 billion	High replacement rate; Storm water separation Total cost £248 billion

energy costs. Finally, the future should see more efficient recovery of water and nutrients from our wastewater.

In summary, it is clear that significant changes in incurred costs cannot easily take place within the current wastewater infrastructure network due to long-term invested interests. However, there is significant potential for improvement in energy efficiency in wastewater processing, which should be acted upon at all opportunities. Regardless, given the sunk investment in existing infrastructure and the expected life of that infrastructure, changes in wastewater management will not likely be rapid in the near future.

7.4 Strategy analysis and discussion

7.4.1 Population and economic impacts on strategy performance

The economics of wastewater infrastructure in Britain have been modelled under the population scenarios shown in (Figure 7.7a), using current energy prices. Results for the three strategies are given in Figure 7.7. It can be seen that the capital expenditure on plant renewal (Figure 7.7b) and expansion is dwarfed by expenditure on sewer expansion (Figure 7.7c). This does not, however, include the cost of sewer renewal. With an estimated total replacement value in England and Wales of £180 billion and an

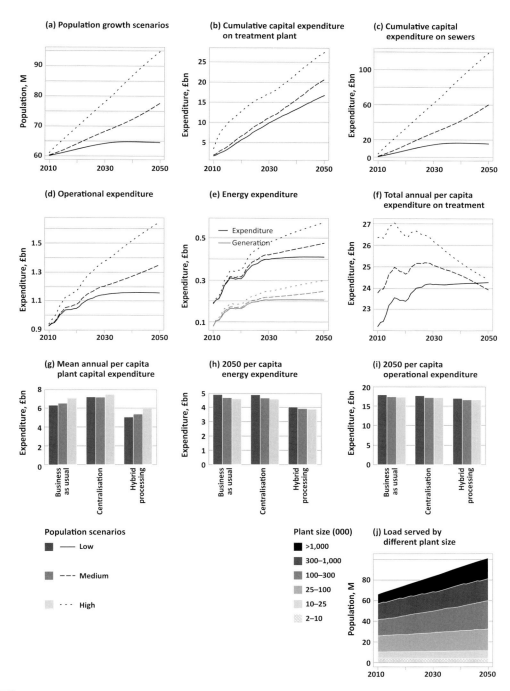

(a) Population growth scenarios

(b) Cumulative capital expenditure on treatment plant

(c) Cumulative capital expenditure on sewers

(d) Operational expenditure

(e) Energy expenditure

(f) Total annual per capita expenditure on treatment

(g) Mean annual per capita plant capital expenditure

(h) 2050 per capita energy expenditure

(i) 2050 per capita operational expenditure

Population scenarios

Low
Medium
High

Plant size (000)

>1,000
300–1,000
100–300
25–100
10–25
2–10

(j) Load served by different plant size

Figure 7.7 Forward projections of expenditure on wastewater infrastructure; (g), (h) and (i) show comparisons between different processing strategies: business as usual, centralisation and hybrid processing

assumed life expectancy of one hundred years, this would imply a necessary average additional expenditure on sewers of £1.8 billion per year, in addition to the estimated average annual expenditure on system expansion of £0.3 billion, 1.5 billion or £3 billion, depending on the population growth scenario. Storm water separation, with an estimated total additional cost of £180 billion, could cost more than the sewer expenditure, depending on the rate of implementation, although there are ancillary benefits such as reduced sewer-related flooding.

Combining the model estimates of plant capital expenditure and operational expenditure (Figure 7.7d), a slight decrease in annual per capita total expenditure is predicted. This is likely the effect of increasing plant size (Figure 7.7j), and the economy of scale. However, further centralisation has little effect on capital or operating expenditure, as it involves absorbing the smaller plants, which contribute little to the total operating cost (Figure 7.6). While moving to more efficient processing technologies can reduce the overall capital expenditure, the small reduction in energy expenditure is overshadowed by other operational costs.

7.4.2 Potential for decentralised wastewater systems

Apart from the three different processing strategies modelled above, we also explore the potential for decentralisation as an alternative strategy. Demand for sewer capacity can also be reduced by decentralisation of wastewater treatment. There has been considerable debate in the academic literature about the economics of centralisation and decentralisation of wastewater treatment. While there is substantial sunk cost in existing infrastructure, and concern about control of output quality for distributed systems, it is worthwhile to examine the circumstances under which decentralised treatment would be economically advantageous for new communities. Decentralised treatment not only has the potential benefit of reducing the requirement for a broader sewer infrastructure, but also increasing the flexibility of supply, making it much easier for provision to grow incrementally with population. More centralised service provision will result in installation of a greater unneeded surplus capacity.

As seen above, capital expenditure on wastewater treatment plant has economy of scale, of the order of a 0.7 power of population size. A comparison between the expenditure on sewerage and treatment plant hinges on the rate of growth of sewerage costs with population. It can be shown that for individual communities in the U.K., not only is it more economical to collect and centrally treat wastewater in towns of increasing size, but it is not economical to have more than one treatment works for a single community. This agrees with the results found by Maurer et al. (2010) for Swiss municipal sewer networks.

An analysis of groups of adjacent communities shows that there is a relationship between community size and the distance above which it ceases to be economical to treat wastewater centrally. Figure 7.8 shows the relationship between this distance and the community size, for both activated sludge and biofilter processing. It can be seen that the distance at which decentralisation becomes economical is smaller for cheaper processing; thus as per capita

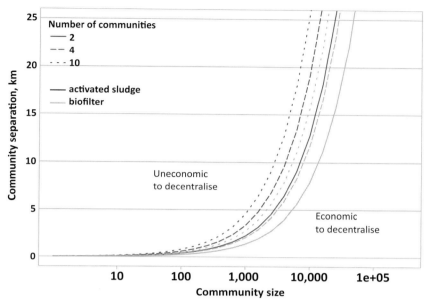

Separation distance between communities, above which decentralised wastewater treatment is not economic (Source: Ofwat, 2011)

processing costs are reduced, the value of treatment decentralisation also decreases. The distance is slightly greater if the communities are of unequal size. If more expensive centralised processing is compared with cheaper distributed processing, then the distance is somewhat smaller. In particular, it is economical to process the wastewater of a community of up to about 150 people at the scale of individual households only if the total cost of processing can be reduced by a factor of eight relative to a scaled-down centralised plant. Communities of this size represent only 10% of the U.K. population (ONS, 2012).

The economic benefit of decentralisation is in reduction of the sewerage network, so it would not be advantageous to pipe partially processed effluent to a central point. Thus, it is only reasonable to decentralise wastewater treatment if the effluent can be appropriately discharged to a watercourse without new sewer pipe provision. Regulations on the discharge of wastewater effluent through a soakaway to the environment depend on the capacity of the local ground to absorb the effluent (BSI, 2008), which is exceeded in densely populated areas.

For small communities, sludge would still have to be collected by road tanker as it is at present for septic tanks and other small installations. If other nutrients are recovered from the wastewater, these would also need to be collected by road. This has not been included in the calculation, and would reduce the distance at which centralisation becomes a more practical option.

The calculation has also not included technologies for resource recovery, which may only be viable at larger scale plant. However, it has been shown in small community experiments in a number of countries that the introduction of separated domestic wastewater, and in particular in urine separation at source, can lead the way to new wastewater treatment options and efficiencies not available with current single-source treatment (Larsen et al., 2013).

7.5 Conclusions

Implementation of new technologies in the water industry tends to be cautious due the high cost of plant construction, long design lives and risk aversion in the realm of public health. Further, teething problems with new technologies can result in lapses of treatment quality, for example, recent problems with screen rupture in integrated fixed-film activated sludge systems (Leaf et al., 2011), which can be exceedingly costly environmentally, financially and in public confidence. In addition, technologies which look promising in the lab or at pilot scale may not be suitable for operational use, if they are not resilient to vagaries of the wider environment, such as colder temperatures, variable feed organic loading rates or sudden volume changes due to large runoff events.

Despite this conservative approach, considerable gains are possible in energy efficiency and process effectiveness using current technologies, simply by optimising the efficiency of each stage of the process. In addition, combining current biofilter and activated sludge technologies in series (Harrison et al., 1984), or using hybrid or granule-based reactors (Germain et al., 2007; Ni, 2012) enables lower power requirements and increased capacity within the same plant. If treatment quality is to be maintained for series reactors, using activated sludge as the final processing stage makes most sense. First, it ensures equivalent effluent quality under a range of external conditions, but it can be costly in terms of aeration. Therefore, placing pre-treatment processes (e.g. a biofilter) before an activated sludge unit can reduce the organic loading on the aerobic unit, reducing the amount of aeration required to degrade the carbon. Obviously, capital investment is needed if pre-treatment is to be included, but operating costs will decline due to reduced aeration.

The use of sequenced reactors is increasingly common for wastewater processing in warmer climates (Chernicharo, 2006), where the pre-treatment reactor is frequently anaerobic. The anaerobic reactor has the potential for producing some methane and reduces organic loads on the aerobic system, reducing aeration requirements. However, anaerobic biological pre-treatment of raw wastewater tends not to be practical in cooler climates because slow growth rates of cold-adapted psychrophilic anaerobic bacteria make such processes more vulnerable to environmental shocks.

Another technology that may have promise for the distant future in wastewater treatment is microbial electrolysis cells, which can biologically treat the wastewater, providing hydrogen as an output (Heidrich et al., 2013). As with the methane from anaerobic processes, combustion could be used to offset the energy cost of treatment. However, this technology has yet to do more than recover energy required for the operation itself, and much more research and development is needed to demonstrate its true viability.

In the continuing drive to make environmental standards more stringent for wastewater treatment across Europe, there is considerable debate among governments and the water industry about the need for micropollutant treatment, such as endocrine disrupting compounds in human wastes. Reducing such compounds in treatment has potential merit, as they can impact aquatic life in receiving waters. However, existing treatment technologies were never designed to remove such compounds and process modifications for their removal will be exceedingly expensive. As an example, ozone treatment can effectively remove many such contaminants; however, it is energy intensive. Alternately, biological sequence reactors also have possible merit, although such systems do not readily treat wide ranges of micropollutants because each pollutant can require very specific treatment conditions; developing sequences that are universally effective is difficult. Therefore, it is wiser and more economical to reduce the use of such compounds in human activity rather than treat them in a dilute form in mixed wastes, which is a key policy decision that would make waste treatment sustainable into the future.

It is clear that the more stringent the imposed environmental standards, the greater are the capital and operational costs of the wastewater treatment, in particular their energy expenditure (e.g. Friedler and Pisanty, 2006). The joint necessities of satisfying environmental standards and reducing energy use in the face of population growth suggest the need for compromise, particularly in areas of high population density.

Acknowledgements

The authors are grateful to Sue Horsfall and Andrew Moore of the Northumbrian Water Group for providing technical guidance and helpful review of this chapter.

References

BSI (2008). Code of practice for the design and installation of drainage fields for use in wastewater treatment. BS 6297:2007+A1:2008.

Byrns, G., A. Wheatley and V. Smedley (2012). "Carbon dioxide releases from wastewater treatment: potential use in the UK." *Proceedings of the ICE-Engineering Sustainability* 166(3): 111–121.

Carbon Trust (2013). *Conversion factors–energy and carbon conversion guide*. London, UK, Carbon Trust.

Chernicharo, C. A. L. (2006). "Post-treatment options for the anaerobic treatment of domestic wastewater." *Reviews in Environmental Science and Bio/Technology* 5(1): 73–92.

CIWEM (2013). *A blueprint for carbon emissions reduction in the UK water industry*. London, UK, Chartered Institution of Water and Environmental Management.

Climate Change Act (2008). *Climate Change Act 2008*. London, UK, HM Government.

Defra (2011). *Anaerobic digestion strategy and action plan*. London, UK, Department for Environment, Food and Rural Affairs.

Defra (2012a). *National policy statement for waste water*. London, UK, TSO, Department for the Environment, Food and Rural Affairs.

Defra (2012b). *Waste water treatment in the United Kingdom – 2012*. London, UK, Department for Environment, Food and Rural Affairs.

Environmental Protection Agency (1980). *Construction costs for municipal wastewater conveyance systems: 1973–1978*. Washington, DC, USA, US Environmental Protection Agency.

Environmental Protection Agency (1999). *Combined sewer overflow management fact sheet–sewer separation*. Washington, DC, USA, US Environmental Protection Agency.

Friedler, E. and E. Pisanty (2006). "Effects of design flow and treatment level on construction and operation costs of municipal wastewater treatment plants and their implications on policy making." *Water Research* 40(20): 3751–3758.

Germain, E., L. Bancroft, A. Dawson, C. Hinrichs, L. Fricker and P. Pearce (2007). "Evaluation of hybrid processes for nitrification by comparing MBBR/AS and IFAS configurations." Water Science & Technology 55(8-9): 43–49.

Harrison, J. R., G. T. Daigger and J. W. Filbert (1984). "A survey of combined trickling filter and activated sludge processes." *Journal (Water Pollution Control Federation)* 56(10): 1073–1079.

Heidrich, E. S., J. Dolfing, K. Scott, S. R. Edwards, C. Jones and T. P. Curtis (2013). "Production of hydrogen from domestic wastewater in a pilot-scale microbial electrolysis cell." *Applied Microbiology and Biotechnology* 97(15): 6979–6989.

Iacovidou, E., D.-G. Ohandja and N. Voulvoulis (2012). "Food waste co-digestion with sewage sludge–realising its potential in the UK." *Journal of Environmental Management* 112: 267–274.

Jaffer, Y., T. A. Clark, P. Pearce and S. A. Parsons (2002). "Potential phosphorus recovery by struvite formation." *Water Research* 36(7): 1834–1842.

Larsen, T. A., K. M. Udert and J. Lienert (2013). *Source separation and decentralization for wastewater management*. London, UK, IWA Publishing.

Leaf, W. R., J. P. Boltz, J. P. McQuarrie, A. Menniti and G. T. Daigger (2011). "Overcoming hydraulic limitations of the integrated fixed-film activated sludge (IFAS) process." *Proceedings of the Water Environment Federation* 2011(11): 5236–5256.

Lebiocka, M. and A. Piotrowicz (2012). "Co-digestion of sewage sludge and organic fraction of municipal solid waste. A comparison between laboratory and technical scales." *Environment Protection Engineering* 38(4): 157–162.

Lofrano, G. and J. Brown (2010). "Wastewater management through the ages: A history of mankind." *Science of The Total Environment* 408(22): 5254–5264.

Marlow, D. R., M. Moglia, S. Cook and D. J. Beale (2013). "Towards sustainable urban water management: A critical reassessment." *Water Research* 47(20): 7150–7161.

Maurer, M., M. Wolfram and H. Anja (2010). "Factors affecting economies of scale in combined sewer systems." *Water Science and Technology* 62(1): 36–41.

Mills, N., P. Pearce, J. Farrow, R. B. Thorpe and N. F. Kirkby (2014). "Environmental and economic life cycle assessment of current and future sewage sludge to energy technologies." *Waste Management* 34(1): 185–195.

Moglia, M., G. Tjandraatmajda and A. K. Sharma (2013). "Exploring the need for rainwater tank maintenance: survey, review and simulations." *Water Science and Technology: Water Supply* 13(2): 191–201.

Mott MacDonald (2011). *Future impacts on sewer systems in England and Wales*. Report for Ofwat. Cambridge, UK.

MWRA (2011). MWRA completed CSO projects. Massachussetts Water Resources Authority.

Ni, B.-J. (2012). *Formation, characterization and mathematical modeling of the aerobic granular sludge: formation, characterization and mathematical modeling of the aerobic granular sludge*. Berlin, Germany, Springer-Verlag.

Ofwat (2011). June return. London, UK, Office of Water Services.

ONS (2012). National population projections, 2012-based. Office for National Statistics.

Ordnance Survey (2014). *Meridian 2 – user guide and technical specification*. Southampton, UK, Ordnance Survey.

Public Utility Board (2014). "NEWater." Retrieved from www.pub.gov.sg/water/newater/.

Rahman, M. M., M. A. M. Salleh, U. Rashid, A. Ahsan, M. M. Hossain and C. S. Ra (2014). "Production of slow release crystal fertilizer from wastewaters through struvite crystallization – a review." *Arabian Journal of Chemistry* 7(1): 139–155.

Reddy, K. V. and N. Ghaffour (2007). "Overview of the cost of desalinated water and costing methodologies." *Desalination* 205(1–3): 340–353.

Thames Tideway Tunnel Ltd (2014). "Thames Tideway Tunnel." Retrieved from www .thamestidewaytunnel.co.uk/.

World Health Organisation (2014). UN-water global analysis and assessment of sanitation and drinking-water. *GLAAS 2014 report*. Geneva, Switzerland, World Health Organisation.

Young, T., M. Muftugil, S. Smoot and J. Peeters (2012). "MBR vs. CAS: capital and operating cost evaluation." *Water Practice and Technology* 7(4).

Solid waste systems assessment

GEOFF V. R. WATSON, ANNE M. STRINGFELLOW, WILLIAM POWRIE,
DAVID A. TURNER, JON COELLO

8.1 Introduction

Wastes are defined in the Waste Framework Directive (European Parliament and Council of the European Union, 2008) as 'any substance or object which the holder discards or intends or is required to discard'. Over the last two to three decades, waste management in the industrialised world has gradually shifted from providing safe disposal of unwanted materials, often by entombing the waste in a sophisticated, engineered landfill, to recovering materials and value from that which is no longer needed through reuse, recycling, composting and energy recovery. In Britain, this shift has resulted in a 71% reduction in the amount of biodegradable municipal waste (BMW) going to landfill since 1995. Recycling and composting have increased from almost nothing in 1995 to nearly 45% of municipal waste treatment today and energy from wastes accounts for about a third of renewable energy generated (Defra, 2015a). This has required significant investment in infrastructure as well as sustained efforts to change the attitude of industry and consumers. Recent publications on resource security (Defra, 2012), resource efficiency (European Commission, 2011) and sustainable materials management (OECD, 2012) show that there is a move away from the linear view of resource management (extraction, manufacture, use, final disposal) towards a more circular view in which waste management becomes primarily a resource recovery operation and final disposal is necessary only for those materials from which further value can no longer be economically or technically extracted.

Material and value are recovered from wastes through recycling and composting (42%) and energy recovery (22%) with the remainder being landfilled (34%). (Figures are from Defra for 2012/2013 and are for local authority collected waste (LACW) in England.) Recycling accounts for most of this recovery and the rates for the most commonly collected materials (glass, steel, aluminium, dense plastics (e.g. plastic bottles) and paper, card and cardboard) are shown in Table 8.1. Garden wastes are recovered for composting and food or mixed food and garden wastes for in-vessel composting (IVC) or anaerobic digestion (AD). Other materials (e.g. tetrapaks and plastic film) are recovered more rarely. Energy is generally recovered by incineration from mixed wastes, from mixed wastes processed to produce fuels (solid recovered fuel (SRF) or refuse derived fuel (RDF)) for co-combustion or through the AD of biodegradable wastes (usually food waste or mixed food and green waste). Energy outputs are primarily electricity with heat, biogas (from anaerobic digestion

Table 8.1 Packaging recycling rates of the most commonly recycled materials (Source: Defra, 2015b for 2011)		
	Total packaging waste arising (tonnes)	Total recovered/ recycled (tonnes)	Recovery/recycling rate (%)
Paper	3,817,860	3,232,461	84.8
Glass	2,739,989	1,751,852	63.9
Aluminium	160,877	73,683	45.8
Steel	648,740	373,714	57.6
Plastic	2,515,809	609,910	24.2

or landfill) and syngas (from gasification or pyrolysis) becoming potentially more important in the future.

Waste infrastructure systems consist of transfer stations (for sorting, recovering and consolidating waste prior to onward carriage to processors or disposal), material recovery facilities (MRFs), recycling or other processing facilities (e.g. AD), landfill and incinerators and thermal treatment plant where waste is combusted usually to produce electricity. There are three main sub-systems: (i) collection; (ii) treatment and (iii) final disposal.

Throughout the developed world the delivery of waste infrastructure has been influenced by local and national governance and planning mechanisms. All European Union (EU) member states are required to prepare waste management plans in accordance with the Waste Framework Directive (European Parliament and Council of the European Union, 2008). As a result of implementation of the plans, the average amount of municipal solid waste (MSW) treated by recycling or composting in the EU28 has risen from 18% in 1995 to 42% in 2012 (Eurostat, 2014), with landfilling (34%) and incineration (24%) also being used. In some cases, countries have gone beyond the requirements of the Landfill Directive (European Parliament and Council of the European Union, 1999). For example, Germany (which, at 49%, has the highest recycling rate in Europe), banned sending non-pre-treated waste to landfill in 2005 and therefore has already stopped sending BMW to landfill. Sweden, Denmark and the Netherlands send only 1–2% of their waste to landfill, treating most waste by incineration (49–52% of waste) together with recycling and composting. Recent accession states, for example, Croatia, Slovakia, Bulgaria and Romania (and some older members, e.g. Poland and Greece) are still heavily reliant on landfilling (greater than 75%).

Countries outside the EU have also introduced waste management policies. The USA has no national waste management policy, but each state implements a plan under the Resource Conservation and Recovery Act (1976). This regulates hazardous waste management and provides guidelines for solid waste disposal, but does not set national recovery targets. Nonetheless, state targets have led to a reduction in waste generation and increased the average recycling rate to 35% in 2012, although landfill (54%) remains the major treatment option (Environmental Protection Agency, 2014). Canada and Australia have regional waste management policies aimed at achieving diversion from landfill and increasing recycling and recovery. South Australia, for example, has targets including a reduction of waste arising by 35% by 2020, and diversion of 70% of MSW, 75% of commercial and industrial waste and 90% of construction and demolition waste from landfill by 2015; the latter two diversion targets have been achieved, and MSW diversion is already at 58% (Rawtec Pty Ltd, 2014). China also has an ambitious target of 70% recycling of major waste types (metals, paper, plastics, glass, tyres, end of life vehicles (ELVs) and waste electrical and electronic equipment (WEEE)) by 2015 (Latcham, 2011). As a high percentage of waste in China is currently sent to landfill or waste to energy plants, a great deal of investment will be required to build the infrastructure needed to achieve this recycling target.

Cultural differences between nations can also affect the nature of such infrastructure. For example, Austria, Germany and the Netherlands, have invested heavily in recycling and composting, Denmark and Sweden have developed incineration and district heating and sanitary landfill remains common in Italy, Spain and the UK (CIWM, 2005). Nonetheless, in all advanced economies waste infrastructure systems share similar characteristics. This chapter explores possible future directions for solid waste infrastructure in Britain, making use of a new waste infrastructure systems assessment model.

8.2 Solid waste systems in Britain

8.2.1 System description

Over the last decade, there has been a shift away from the traditional view of wastes as materials which have to be disposed of, towards treating them as resources to be recovered. In Britain, despite this philosophical change, only about half of the annual 300 million tonnes of wastes have value recovered from them (mainly through material recovery, but also through energy recovery). Progress towards European targets on recycling and waste recovery has been driven primarily by government targets and taxation rather than an awareness of the value of recovered resources. It seems counter-intuitive that no wine and beer producers thought of reducing the weight of their glass bottles, even though it would clearly reduce their operating costs, prior to government intervention (e.g. WRAP 2008). More businesses are now becoming aware that they can reduce their costs by changing practices to minimise waste arisings and maximise reuse. This is likely to continue in the near future, alongside an increasing realisation that reducing waste leads to a reduction in both manufacturing costs and disposal costs.

Waste has traditionally been categorised by generating sector, for example, household (often used interchangeably with municipal solid waste (MSW)), commercial and industrial (C&I), construction and demolition (C&D), mining and quarrying and agricultural. Hazardous waste is categorised separately. For household waste, collection is from the kerbside or a bring site (e.g. bottle and textile banks; household waste recycling centre (HWRC)). Some C&I waste is collected along with green waste from parks and gardens or with household waste from the kerbside; this forms Local Authority Collected Municipal Waste (LACMW), formerly known as MSW. Licensed waste management companies collect the majority of the remaining commercial and industrial waste.

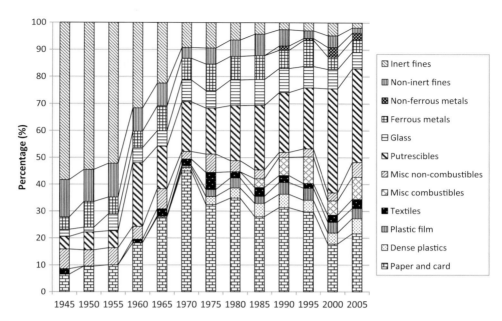

Figure 8.1 Variation in UK household waste composition 1945–2005 (modified from Brown et al., 1999)

Until 2010 MSW was defined as waste collected by local authorities (LACMW). This was changed (Defra, 2011b) following the modification of reporting waste arisings to the EU to ensure that the targets for the landfilling of biodegradable waste, as set out in the Landfill Directive (European Parliament and Council of the European Union, 1999), would be met. The arisings that are now reported to the EU consist of Local Authority Collected Waste (LACW) and the MSW-like component of C&I wastes which collectively are termed Municipal Waste (MW) (Defra, 2011b) and are roughly double MSW. The arisings modelled in this chapter are for MW.

The need to manage waste is primarily associated with population growth, settlement and concentration into cities, industrial development and high levels of production and consumption. Prior to the 1950s these were high levels of reuse, with small quantities of biodegradable material and a very large proportion of apparently inert materials such as ash and cinders in household waste (Figure 8.1). Thus, there was little perceived need to manage solid household wastes much beyond collection and sorting, with a small amount of largely inert material then being landfilled. The composition of waste began to change following the introduction of the 1956 Clean Air Act. Over the following decades the amount of ash being disposed of reduced considerably while plastics and other materials increased. Changing lifestyles and greater wealth meant that the amount of waste produced per household increased.

Figure 8.2 shows the British regional distribution of recycling, composting and energy recovery facilities and landfill. The first map shows the regional capacity of waste management facilities in Mt/year. The numbers show the annual permitted tonnage for non-hazardous landfill. *Recycling* is the capacity of MRFs and so underestimates recycling

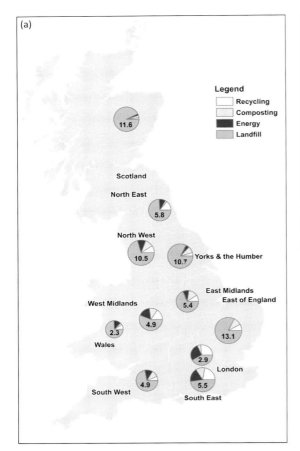
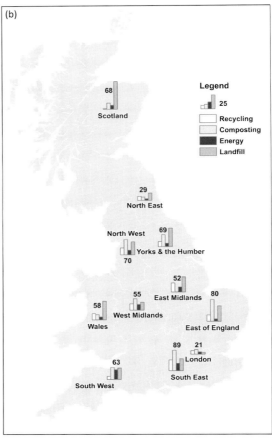

Figure 8.2 Maps of the English government office regions (GOR), Wales and Scotland. (A) The pie charts show the waste management capacity for each region in 2010, scaled so that the area represents relative total capacity in Mt/year. (B) shows the number of facilities as in (A), with the number showing the total number of facilities in each region. (Source: EA, SEPA and Natural Resources, Wales)

capacity as explained in Section 8.4.1. *Composting* is windrows and IVCs. *Energy* is EfW and anaerobic digestion. *Landfill capacity* is the annual tonnage allowed under the terms of their licence, not the amount permitted by the landfill directive, except for Wales where the figure used is for amount landfilled in 2010 in the absence of similar data. Other types of capacity are omitted for clarity but make up no more than 10% of the capacity. This mainly affects mechanical biological treatment (MBT) plant which typically will include recyclate recovery, biological processing and creation of refuse derived fuel (RDF). The second map shows the number of facilities in each region. It should be noted that this figure excludes MBT plant as these produce some or all of recyclates, compost and RDF and hence cross categories, but they account for less than 10% of the capacity. Figure 8.3 shows the flow of resources in Britain. The thick line at the base of the diagram is material which has been recycled or recovered.

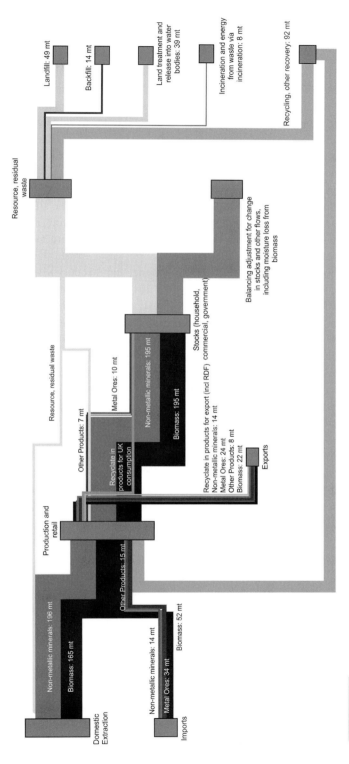

Figure 8.3 Sankey diagram of flow of resource in Britain, 2012, (excluding fossil fuels and energy carriers). The line at the bottom of the figure shows materials being recycled and reused. (Source: Defra, 2015a reused under the terms of the Open Government Licence)

Landfill: 49 mt

Backfill: 14 mt

Land treatment and release into water bodies: 39 mt

Incineration and energy from waste via incineration: 8 mt

Recycling, other recovery: 92 mt

Resource, residual waste

Balancing adjustment for change in stocks and other flows, including moisture loss from biomass

Resource, residual waste

Metal Ores: 10 mt

Non-metallic minerals: 195 mt

Biomass: 195 mt

Stocks (household, commercial, government)

Other Products: 7 mt

Recyclate in products for export (incl RDF)
Non-metallic minerals: 14 mt
Metal Ores: 24 mt
Other Products: 8 mt
Biomass: 22 mt

Recyclate in products for UK consumption

Exports

Production and retail

Other Products: 15 mt

Non-metallic minerals: 196 mt

Biomass: 165 mt

Non-metallic minerals: 14 mt

Metal Ores: 34 mt

Biomass: 52 mt

Domestic Extraction

Imports

8.2.2 Challenges and opportunities

EU Directives relating to waste, and British regulations transposing them, are the main drivers for reduction of waste generation and the increase in the recovery of resource from wastes. This is likely to continue for the foreseeable future; however, additional drivers such as low carbon and energy from renewables may gain importance. The possible banning of the disposal of all biodegradable municipal waste to landfill in the next decade may require new infrastructure. However, the rate at which MW is being sent to landfill has reduced faster than has been required either by the 1999 EU Landfill Directive or subsequent national policies (e.g. Waste Strategy for England (Defra, 2007)), although the rate of reduction has been slowing in recent years.

Britain's strategy for waste management follows the EU Waste Framework Directive waste hierarchy: prevent waste, prepare for reuse, recycle, recover energy from waste, and dispose to landfill as a last resort. The need to adhere to the hierarchy now applies to businesses as well as those who manage waste. Waste that is disposed of to landfills is currently (2014/2015) taxed at £80/tonne, and this cost will increase in line with the retail price index (RPI) thereafter (HMRC, 2014). The Landfill Tax has had the effect of making landfill the most expensive form of waste treatment in Britain, and hence making investment in other forms of treatment and disposal financially viable.

The 2007 Waste Strategy for England set targets for the recycling of household waste and reduction of waste going to landfill (Defra, 2007). Currently, local authorities are meeting targets for diversion of biodegradable waste from landfill, but it remains to be seen whether this will be achieved in the future. In terms of the recovery of resources from waste, Britain lags behind many other countries in the EU. This is in part because of the legacy of a large number of landfill sites and an historical reliance on this type of waste management, as well as a strong public antipathy towards any kind of waste combustion. There are very little data for C&I or C&D wastes, with the bulk of the data that are available coming from surveys of companies in a single region (e.g. northwest England) which is then extrapolated both spatially and temporally. Without reliable data, targets cannot be imposed or checked and this has been recognised in the 2011 National Infrastructure Plan (HM Treasury and Infrastructure UK, 2011) with The Responsibility Deal requiring sharing of C&I data by 2014. It is clear that the landfill tax has had and will continue to have an effect in reducing C&I disposal to landfill.

The British government has supported a number of initiatives to help increase recycling (e.g. the creation of WRAP – the waste and resources action programme), reduce the production of waste and promote the use of waste management treatment (e.g. through the Defra New Technologies Demonstrator Programme) such as mechanical biological treatment, mechanical heat treatment, anaerobic digestion and composting. Energy from waste plays a significant part in waste management and the production of RDF and SRF from waste will be increasingly important as the percentage of energy recovery from renewables increases in line with government policy. However, most waste disposal techniques still produce greenhouse gases (e.g. fossil carbon is burnt in incinerators and fossil fuels are used in the aerobic composting of wastes). An alternative approach may be to bury carbon rich wastes including both fossil and non-fossil carbon (e.g. plastics and paper) in carbon sinks. However, there are currently no policies that include this strategy, and the current European opposition to the use of disposing of waste to landfill makes this exceedingly unlikely in the near future.

Targets for materials recovery from MSW by recycling and composting plus overall recovery rates (including energy) were set in 2000, in the waste strategy for England and Wales (DETR, 2000) and updated in 2007 (Defra, 2007). Table 8.2 shows the targets for England, Scotland and Wales based on LACMW (formerly MSW) as well as the MW targets set out in 2011 legislation (House of Commons, 2011).

According to Environment Agency (EA) figures (Date, 2013, 2014a, 2014b; Holder, 2015), in 2012, 890,000 tonnes of U.K.'s waste (in the form of RDF) were shipped to continental Europe to be incinerated, rising to 1,590,000 tonnes in 2013. 2.37 million tonnes of RDF was exported in 2014 (Holder, 2015) and, based on exports from the first half of 2015, this is projected to rise to 2.73 million tonnes in 2015. The high levels of export are as a result of European incineration overcapacity and the consequent low/subsidised gate fees and transport costs.

A number of technological and management options exist to address the above challenges. For LACMW, collection is from the kerbside or HWRCs. Licensed waste management companies collect the majority of LAC&I waste. The main type of facility associated with waste collection is a transfer station or bulking facility. Kerbside collected recyclables are generally sent to MRFs for sorting prior to baling and sale. Black bag waste from kerbside collections is taken for further treatment or disposal. Final disposal, where needed, is almost exclusively to landfill. When waste is transported to large-scale waste treatment facilities for processing, the most common waste treatment facilities are:

- **Materials recycling facilities (MRFs):** of two types: dirty and clean. Dirty MRFs take black bag MSW whereas clean MRFs take dry, co-mingled recyclables. Outputs include recyclable fractions including paper, cardboard and metals. The non-recyclable residue may be sent for further processing, for example, IVC, used as refuse derived fuel, recovered to land (i.e. used to replace fertilisers for soil improvement), or sent to landfill.
- **Mechanical biological treatment facilities (MBT):** an extended MRF with an anaerobic or aerobic biological treatment stage to reduce the biodegradability of residual material. Outputs include recyclables, sometimes SRF or RDF, and treated residual waste which may go to landfill.
- **Composting facilities:** large-scale open windrows treating garden waste or IVC for food and green waste. Outputs include mature compost which may be used for soil improvement.
- **Anaerobic digestion plants (AD):** to treat food and green wastes. Outputs include digestate (which may be used for soil improvement), biogas (fuel) and non-fossil CO_2. The biogas is typically combusted on site to produce electricity and/or heat.
- **Energy from waste (EfW):** primarily incineration plants (AD is technically included in this broad terminology but often referred to separately as it is here). These may be combined with a MRF to recover recyclables prior to incineration. Outputs include electricity (and heat), recyclables, aggregate and ash as well as fossil and non-fossil CO_2 and nitrous oxides.

Table 8.2 Current waste treatment targets in Britain	
Location	Reduction targets and regulations
England	• Recycling & composting 40% by 2010; 45% by 2015; 50% by 2020 (Defra, 2007). • Recovery of value (above + energy recovery) 53% by 2010; 67% by 2015; 75% by 2020 (Defra, 2007). • Reduction of mass not reused, recycled or composted from 22 Mt in 2000 to 15.8 Mt in 2010 and 12.2 Mt in 2020 (Defra, 2007). • Reduction of BMW going to landfill (European Parliament and Council of the European Union, 1999). Where two figures are shown the former is for LACMW and the latter for MW (from House of Commons, 2011): • by 2010 no more than 11.2 million tonnes of BMW can go to landfill; • by 2013 no more than 7.46/14.51 million tonnes of BMW to landfill; • by 2020 no more than 5.22/10.16 million tonnes of BMW to landfill.
Scotland	*Landfill Directive (European Parliament and Council of the European Union, 1999) and the Waste Framework Directive* (European Parliament and Council of the European Union, 2008) *require that (note that, as above, where two figures are given, the former is for LACMW and the latter MW):* • by 2010 no more than 1.32 million tonnes of BMW can go to landfill; • by 2013 no more than 880,000/1,798,000 tonnes of BMW to landfill; • by 2020 no more than 620,000/1,258,000 tonnes of BMW to landfill; and • by 2020, 50% of household waste and similar to be reused or recycled. *Scotland's ambitious zero waste policy for MSW also requires that:* • MSW arisings do not increase after 2010; • EfW is limited to a maximum of 25%; • <5% waste to landfill by 2025; and • recycling & composting rates of at least 40% of MSW arisings by 2010, rising to 50% by 2013, 60% by 2020 and 70% by 2025.
Wales	*Comply with the Landfill Directive and landfill no more than 35% of its 1995 levels of BMW by 2020 (as above, where two figures are given, the former is for LACMW and the latter MW):* • by 2010 no more than 710,000 tonnes of BMW to landfill; • by 2013 no more than 470,000/919,000 tonnes of BMW to landfill; • by 2020 no more than 330,000/643,000 tonnes of BMW to landfill; and • by 2020, 50% of household waste and similar to be reused or recycled. *Zero waste targets are set out below:* • by 2025 all sectors should be recycling at least 70% of their waste; and • by 2050, the Welsh Assembly 'hope to have achieved zero waste' (Welsh Assembly Government, 2009).
EC	*Following a press release on 2nd July 2014, the following proposed targets for EU waste management have been announced (European Commission, 2014):* • by 2025 no recyclable waste (including plastics, paper, metals, glass and biodegradable waste) to be landfilled; • by 2030 recycling and preparing for re-use of municipal waste increased to 70%; and • increasing the amount of packaging waste recycled or prepared for re-use to 80% by 2030.

Any residual waste not recovered by treatment is diverted to landfill. It is notable that landfill operators currently make most of their profits from the sale of landfill gas (LFG) or the energy it produces. The amount of biodegradable wastes being landfilled is declining as a result of the EU Landfill Directive. It is very likely that future regulation will lead to decreased production of LFG and hence profits for operators, making landfill less attractive to investors. While this may not be problematic in the near future, there will be an on-going requirement for some landfill capacity to dispose of residual wastes unsuitable for other treatments and to deal with waste backlogs due to disasters (e.g. floods, terrorist attack) or treatment plant failure. There may be a need for a contingency of publicly funded and operated landfills to meet this need. Other potential options relatively new to Britain, but which may emerge in the future, are outlined below.

- **Mechanical heat treatment facilities (MHT):** an MRF where the mixed residual waste is heat treated to sanitise it. Outputs include recyclables, SRF and a residual waste fraction to landfill.
- **Gasification:** an advanced thermal treatment process. This may be combined with an MRF to recover recyclables or an MBT to recover recyclables and nutrients prior to gasification. Outputs include syngas (mainly CO and H_2) that is usually combusted on site to generate electricity (and heat), recyclables (if coupled with an MRF or MBT), soil improver (if combined with an MBT), slag and ash.
- **Pyrolysis:** an advanced thermal treatment. These may be combined with an MRF to recover recyclables or an MBT to recover recyclables and nutrients prior to treatment or use SRF as the feedstock. Waste is heated to high temperatures in the absence of oxygen. Outputs may include char (a carbon-rich solid fuel), liquid or gaseous fuels depending on processing temperature (Williams and Barton, 2011), recyclables (if coupled with a MRF or MBT), soil improver (if combined with an MBT) and ash as well as CO_2 and nitrous oxides.
- **Plasma arc gasification:** an advanced thermal treatment process. This may be combined with an MRF or MBT to recover recyclables and nutrients (in the case of MBT) prior to treatment or use SRF as a feedstock. Waste is heated in a low oxygen atmosphere using a plasma torch. Outputs include syngas which maybe combusted on site to produce electricity (and heat), recyclables (if coupled with an MRF or MBT), soil improver (if combined with an MBT) and vitrified slag.

8.3 Assessment of national solid waste systems

8.3.1 Modelling approach

Much of the modelling of waste infrastructure has examined the optimisation of treatment and transportation systems (e.g. Abou Najm & El-Fadel, 2004; Rathi, 2007; Anghinolfi et al., 2013). Although there has been some modelling to address future arisings or the requirement for new infrastructure, it is relatively uncommon. The U.K. government

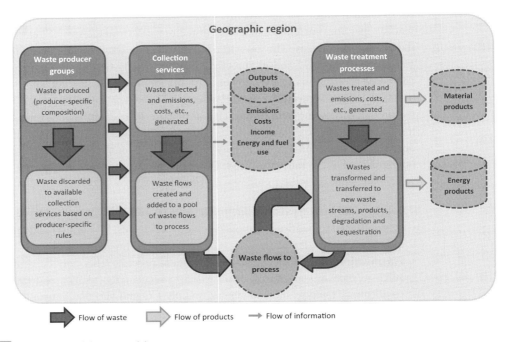

　　Schematic of solid waste model

(Defra, 2011c) examined four future scenarios for waste arisings/treatment to 2030. Fitzgerald (2010) modelled future waste arisings and treatments to determine the likelihood of reaching the biodegradable municipal waste (BMW) landfill diversion targets by 2020. She also investigated the requirement for new infrastructure under different treatment strategies and multiple waste arisings growth scenarios. Other national waste modelling has been carried out directly by government and various consultants, primarily to assess the likelihood of meeting waste management targets (e.g. Defra, 2013; Tolvik Consulting Ltd, 2011) or the probable future costs (e.g. SQWenergy, 2010). The EA developed a life cycle assessment tool (WRATE) to assist local authorities to adopt a more sustainable approach to the provision of waste management services.

In order to assess solid waste infrastructure at a national scale in Britain, we have developed a new solid waste infrastructure systems model, as depicted in Figure 8.4. The model is explained fully in Appendix D (available online at itrc.org.uk), and is briefly described here. Total annual waste arisings are calculated based on waste arisings per capita and three scenarios (high, medium and low) for population and economic growth discussed in Chapter 3. A waste arising decoupling factor is included to account for an expected future reduction in per capita waste arisings.

The total waste arisings are divided into waste types (e.g. cardboard, paper, WEEE and glass) based on pre-defined waste compositions for the different waste producer types modelled. The waste generated is allocated to waste collection processes, such as kerbside collections, HWRC and bring banks, using waste producer type-specific rule sets that vary depending on the collection processes available. Broadly, two types of waste producer are modelled: (i) householders and (ii) non-household MSW-like waste producers, with each further broken down into an average present day and optimal sorting variants. Householder waste producer discard rules have been developed based largely on Resource Futures' (2013) national MSW composition estimates for England, with variations in performance between collection setups and the optimal sorting producer type modelled based on a range of literature sources (e.g. WRAP, 2009a, 2009b, 2009c, 2011). Non-household MSW-like waste producer discard rules have been developed based on the work of Fitzgerald (2010) and Resource Futures (2013). This results in the production of realistic waste streams comprised of target materials and contaminants.

An exploration of the possible treatment and disposal paths available within the waste management network is carried out for each waste stream entering the system and aggregated financial and environmental impact performance metrics (e.g. £/t and CO_2e/t waste) are produced. These profiles form the basis of an optimisation process that seeks to allocate waste to the available waste treatment and disposal paths to meet a specified optimisation goal, such as minimising cost, minimising carbon footprint or maximising energy output or materials recovery.

8.3.2 Strategy description

Future strategies for solid waste infrastructure systems will combine, to differing extents, actions on the demand side (i.e. in waste arisings) ('D' options), more or less vigorous policy instruments ('S' options) and different combinations of technologies ('C' options) (see Table 8.3). We have combined these different aspects of solid waste infrastructure systems into four contrasting strategies, which are described in Table 8.3.

8.4 Strategy analysis and discussion

8.4.1 Impacts of alternative strategies on future waste arisings

The calculated arisings for MW in England are shown in Figure 8.5 for three population/ economy scenarios (low, medium and high growth) and four strategies. Figure 8.6 shows the arisings for the medium growth scenario only and it will be this scenario for which the remaining analyses will be shown. The strategies assessed are WE0 (Minimum Intervention), WE1 (New Capacity), WE2 (Closed Loop) and WE3 (Demand Reduction) with the waste arisings coupling factor taken to reduce annually at 2%, 1%, 3% and 4%, respectively. The arisings under the Closed Loop and Demand Reduction strategies decline with time because their annual reduction in waste arisings coupling factors compensate for the increase in waste due gross value added (GVA) growth.

Residual waste (i.e. the material not recycled or composted), will be treated using one of the residual waste collection/treatments previously discussed. The residual waste capacity must be such that nationally, the amount of residual waste landfilled complies with the

	Strategy name	Demand change	Policy instruments	Infrastructure technologies
WE0	Minimum intervention	SWD0 – partial decoupling	SWS0 – no change	SWC1 – thermal treatment increase SWC2 – bio treatment increase SWC3 – increase in recycling

Existing waste, reuse and recycling targets for household, C&I and C&D wastes are met by continuing the current trends and building new infrastructure, particularly EfW and MBT with AD plant as required. The landfill tax continues to improve performance of the waste sector and reduce landfilling.

	Strategy name	Demand change	Policy instruments	Infrastructure technologies
WE1	New capacity	SWD2 – increase in arisings or SWD0 – partial decoupling	SWS0 – no change or SWS1 – national planning	SWC1 – thermal treatment increase SWC2 – bio treatment increase SWC3 – increase in recycling

Developments in materials separation and recovery technologies mean that wastes require minimum source separation. Consumers disengage from concerns about waste & recycling but despite this, rates of recycling and composting/AD continue to rise as does the overall waste production. Materials left over from materials recovery are used for fuels in EfW plant. Landfilling is capped at 10% of arisings in 2025, reducing to 5% in 2030.

	Strategy name	Demand change	Policy instruments	Infrastructure technologies
WE2	Closed loop	SWD0 – partial decoupling or SWD1 – full decoupling	SWS1 – national planning and SWS2 – taxation and targets	SWC2 – bio treatment increase SWC4 – increase in reuse/recycling SWC6 – decrease in thermal treatment

A move to industrial symbiosis with the wastes from one process providing the raw materials for another. Waste is consciously eliminated from all stages by design and products are designed for reuse, refurbishment, repair and recycling (D4R). Landfill and incineration are largely phased out being retained primarily for disposal of hazardous wastes. Overall waste arisings drop. Landfilling is capped at 5% of arisings from 2030. The recycling & composting target increases to 70% in 2025 and 75% in 2030.

	Strategy name	Demand change	Policy instruments	Infrastructure technologies
WE3	Demand reduction	SWD1 – full decoupling	SWS1 – national planning and SWS2 – taxation and targets	SWC0 – no change SWC2 – bio treatment increase SWC4 – increase in reuse/recycling

Similar to the closed loop strategy but much less investment in infrastructure. A move from consumption to leasing with products designed for long life, easy repair and remanufacturing (D4R). Waste arisings reduced by increasing prices for waste disposal and increasing the involvement of the third sector in refurbishing of unwanted goods. Little investment in infrastructure and changes are driven by cultural and behavioural change. Landfill is capped at 10% of arisings in 2030. The recycling and composting target increases to 70% in 2030.

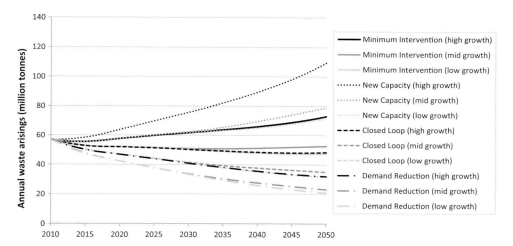

Figure 8.5   MW arisings for England (all growth scenarios)

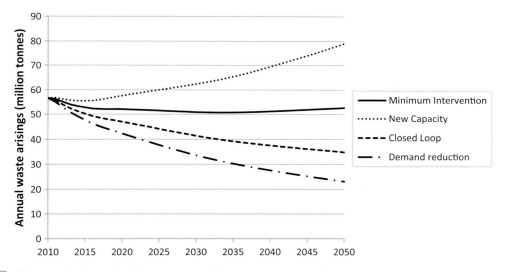

Figure 8.6   MW arisings for England (medium growth scenario)

Landfill Directive targets shown in Table 8.2. The residual MW arisings for England (medium growth scenario), have been calculated and are plotted in Figure 8.6 along with the current and consented (but not in planning stage) non-landfill, residual waste capacity (Eunomia, 2013) and the total residual capacity, including landfill. It should be noted that other reviews of future waste capacity (Defra, 2013; Centre for Environmental Policy, 2014) use a more complex approach than used here primarily because they were produced to determine the likelihood of England meeting the 2020 waste diversion targets (European Parliament and the Council of the European Union, 1999). The residual waste arisings are calculated on the basis that recycling and composting levels just meet the targets in Table 8.2

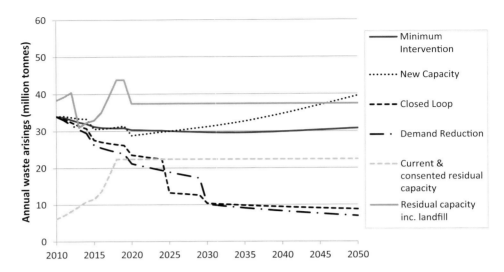

Figure 8.7 English MW residual arisings shown with the current capacity and that under construction and consented and the residual treatment capacity including the annual permitted landfill tonnage (assuming BMW is 68% of total)

(i.e. the worst case, compliant, situation) with the exception of the proposed EU targets, which are not included in the analyses reported here, given they have yet to be ratified and Defra's resistance to the 70% recycling target (Goulding, 2014). Strategy specific targets are also applied, as outlined in Table 8.3.

The capacities for recycling and composting are from the Environment Agency's 2010 Waste Infrastructure Report (EA, 2010), for MRFs and composting facilities. It should be noted that the capacities shown will not necessarily capture all recycling operations, for example, household recyclables that are collected through a kerbside sorted collection do not require an MRF. The same is true for household waste disposed of through HWRCs (which typically have a recycling rate of over 50% and account for about 10–20% of household waste) and through bring banks. The 2010 Waste Infrastructure Report stated that England has 10.8 Mt of permitted MRF capacity. However, according to Defra's 2009 C&I waste data, 22.9 Mt of C&I waste was recycled (of which about 12.9 Mt is likely to have been MSW-like C&I waste). In the same year, Defra also reported that 5.5 Mt of household waste was recycled. A proportion of both MSW and C&I waste is collected and dispatched directly to processors without passing through recycling facilities. Because the model uses total waste arisings and does not take into account the directly reprocessed recyclables, more recycling infrastructure will be built than is necessary.

Figure 8.7 shows the residual MW arisings (i.e. total arisings less the recycling and composting targets shown in Table 8.3) along with the current residual capacity including the annual landfill tonnage permitted by the Landfill Directive, and that reported by Eunomia (2013) as under construction or consented. The total residual capacity declines after 2020 because of the reduction in waste permitted to go to landfill under current waste management targets (Table 8.2).

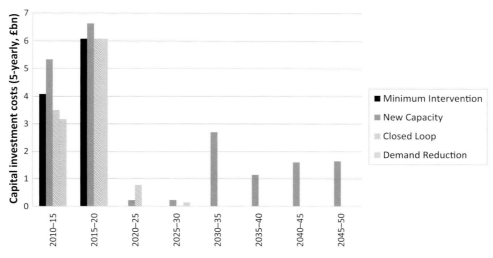

Figure 8.8 Five-yearly MW capital investment costs for England (medium growth scenario)

It is important to note that recycling and composting infrastructure will need to be built in order to reach the recycling and composting targets (see Figure 8.8 for details of capital expenditure). Where the waste arisings are greater than the residual capacity including landfill, new residual capacity will have to be built to ensure compliance with the Landfill Directive. In Figure 8.7, where the arisings line is above the residual capacity line, landfilling is required and building new infrastructure might reduce the treatment costs (due to the high level of the landfill tax). It can be seen that under current waste management targets:

- By 2025 there is sufficient residual capacity such that in the Demand Reduction and Closed Loop strategies there is no need to send any waste to landfill. It should be noted, however, that whilst there may be sufficient capacity for this, it may not be where it is needed.
- The New Capacity and Minimum Intervention strategies require significant amounts of landfill and after 2046, the New Capacity strategy will require the building of new residual waste infrastructure in order to comply with the landfill directive.

8.4.2 Investment costs for recycling, composting and additional capacity

Figure 8.8 shows the capital investment required to build sufficient recycling and composting and residual capacity (including landfill in the case of the Minimum Intervention strategy) for the current and strategy specific waste management targets as outlined in Table 8.3. In the years to 2018, investment is nearly identical, as the model assumes infrastructure will be built according to the Eunomia (2013) report, which describes capacity under construction and consented (this can also be seen in Figure 8.9). The capital expenditure is shown in the year that the plant comes on line, assuming new capacity is built if the previous year showed a shortfall, which implies accurate forecasting of demand.

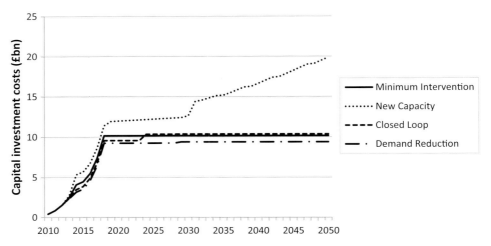

Figure 8.9 Cumulative investment required for MW treatment infrastructure for England

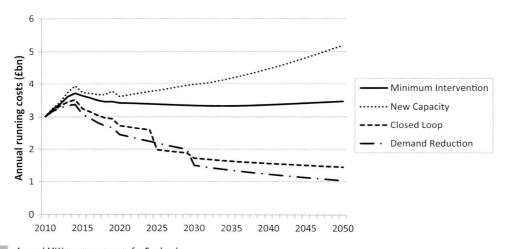

Figure 8.10 Annual MW treatment costs for England

As would be expected from the data shown in Figure 8.7. Figures 8.8 and 8.9 show that little investment is needed in the Demand Reduction and Closed Loop strategies. To some extent, this is due to the falling waste arisings resulting from the decoupling rate being greater than the GVA growth rate, though this does not include the capital cost of replacing life-expired plant.

Figure 8.10 shows that the annual treatment costs are initially similar but as the arisings diverge as a result of the different decoupling factors, the treatment costs also diverge until, by the end of 2050, the annual running costs for the solid waste sector are four times higher in the New Capacity than in the Demand Reduction strategy. It should be noted that despite the different treatment approaches embodied in each strategy, the per tonne treatment costs are not hugely different across strategies as can be seen in Figure 8.11. The reason for

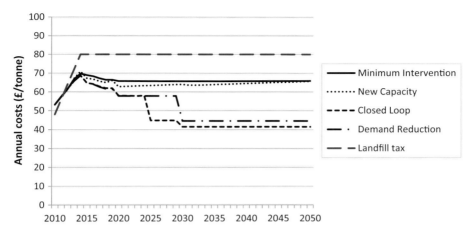

Figure 8.11 Per tonne costs of MW treatment for England for each strategy. Landfill tax per tonne is also shown

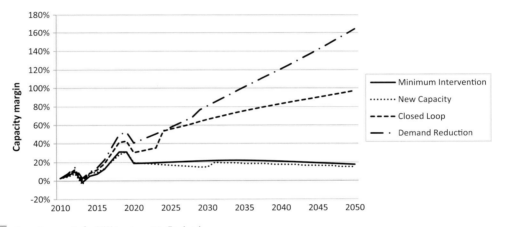

Figure 8.12 Capacity margin for MW treatment in England

this is, as Figure 8.7 suggests, that much of the waste infrastructure required for the next thirty-six years has been built or is in the process of being built at the moment, even though the population of England is projected to rise by almost a third by 2050.

Figure 8.12 shows the capacity margin for England for MW for each of the four strategies. Capacity is assumed to be the sum of residual treatment capacity, recycling and composting capacity and the mass of waste permitted to be landfilled. The capacity margin is taken as the capacity less the arisings divided by the arisings, that is, excess capacity per unit demand. Capacity margin gives some indication of the resilience of a system. It is clear from the graph that the strategies in which arisings reduce give the highest resilience.

The figures show that it is much more cost effective to adopt strategies that reduce waste than to continue with current (Minimum Intervention) or New Capacity strategies which add residual treatment capacity. In order to achieve these cost savings, it will be necessary to change behaviour. Over the last fifteen years the increases in recycling and composting

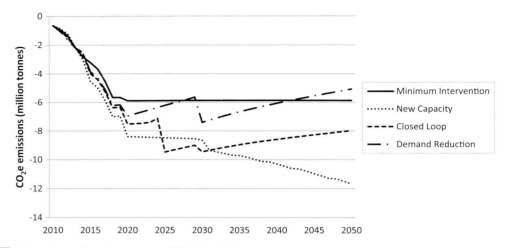

Figure 8.13 CO_2e emissions from the treatment of MW in England

and the concomitant decrease in landfilling has been driven primarily by: (i) The European Landfill and Waste Framework directives; (ii) The landfill tax and landfill tax escalator; and (iii) The 2000 and 2007 Waste Strategies.

Research has showed that a move to strategic regional planning authorities coupled with integration of planning across waste types and compensation for communities hosting waste facilities (and other strategic infrastructure) could significantly improve planning, remove the potential biases of some local authorities and ensure efficiencies of scale are accessed (CIWM, 2005). An ad-hoc version of this has already happened in, for example, the Southeast 7 – a group of five County Councils and two unitary authorities working in partnership (Defra, 2011a).

8.4.3 Energy and carbon performance of alternative strategies

Figure 8.13 shows the emissions due to the treatment of the wastes in each of the four strategies. The assumptions made, and the calculation method for emissions, are shown in Appendix D (available online at itrc.org.uk). It should be noted that negative emissions represent emissions 'savings' incurred as a result of the recovery of energy and secondary materials through waste management activities. The method is based on lifecycle assessment (LCA), and according to Ekvall et al. (2007), 'LCA models that calculate the environmental burdens per kg or tonne of waste generated allow for environmental comparisons of different options for dealing with this waste, but not for analysis of changes in the quantities of waste generated. They are inadequate for the identification and assessment of waste prevention strategies'. The consequences of this are that the strategies and scenarios which produce most waste also produce the most CO_2e 'savings'. Figure 8.14 shows the same data presented as CO_2e emissions per tonne of waste treated, showing that the Demand Reduction and Closed Loop strategies produce significantly larger per tonne emissions savings. As stated above, LCA cannot calculate the effects of avoided arisings, however an estimate of these

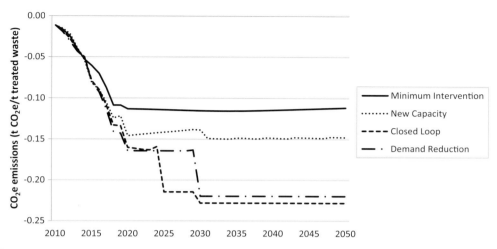

Figure 8.14 CO_2e emissions per tonne of MW treated

can be made using existing data (van Santen, 2010). From this it can be estimated that the CO_2e savings associated with avoided waste is about 1.5 tCO_2e/t waste (see Appendix D – available online at itrc.org.uk) which is substantially greater than the savings due to the waste processing techniques as shown in Figure 8.14. To put this into context, on the basis of the assumptions made as shown in Appendix D, the 56 million tonnes difference in waste arisings between the New Capacity and Demand Reduction strategies in 2050, results in a saving of almost 80 Mt of CO_2e, compared to the 12 Mt due to the recycling and waste used as fuel in the New Capacity strategy.

8.5 Conclusions

Although the recent trends of a reduction in per capita waste arisings are continued with all strategies, these reductions are swamped by increases in population and GVA in the New Capacity and Minimum Intervention strategies leading to significant rises in waste arisings and hence the infrastructure needed to manage it (in the case of the New Capacity strategy, this will roughly double the cost of capital investment to £20 billion by 2050). In the Demand Reduction and Closed Loop strategies, the rate of reduction is greatest and is sufficient to counteract some or all of the growth due to population increase and economic growth and so actual waste arisings fall. The Minimum Intervention strategy does not demand significant capital expenditure over that of the Demand Reduction and Closed Loop strategies, primarily due to the significant infrastructure build in the early years common to all strategies but also due to a significantly greater reliance on landfill. Redesigning packaging (e.g. reducing the mass of steel in cans or glass in bottles) may go some way to reducing waste but there will also need to be a radical reassessment of attitudes to food and food waste as well as consumption more generally. It is clear from the analysis presented here that reducing waste will be far more cost-effective and more environmentally beneficial than trying to build capacity to process the potentially large volumes of waste arisings likely to occur with future socio-economic change. Investing money in education and re-engineering and design for resource efficiency will be far more effective and achievable than spending money on building waste treatment capacity.

References

Abou Najm, M. and M. El-Fadel (2004). "Computer-based interface for an integrated solid waste management optimization model." *Environmental Modelling & Software* 19(12): 1151–1164.

Anghinolfi, D., M. Paolucci, M. Robba and A. C. Taramasso (2013). "A dynamic optimization model for solid waste recycling." *Waste Management* 33(2): 287–296.

Brown, K. A., A. Smith, S. J. Burnley, D. J. V. Campbell, K. King and M. J. T. Milton (1999). Methane emissions from UK landfills. *Report prepared by AEA Technology for the Department of the Environment, Transport and the Regions*. Abingdon, UK.

Centre for Environmental Policy (2014). *Waste infrastructure requirements for England*. London, UK, Imperial College.

CIWM (2005). Delivering key waste management infrastructure: lessons learned from Europe. Report by SLR Consulting for the Chartered Institute of Waste Management.

Date, W. (2013). "Sharp rise in RDF exports during 2012." Retrieved from www.letsrecycle.com/news/latest-news/energy/sharp-rise-in-rdf-exports-during-2012.

Date, W. (2014a). "RDF export market looks set to grow in 2014." Retrieved from www.letsrecycle.com/news/latest-news/energy/rdf-export-market-looks-set-to-grow-in-2014.

Date, W. (2014b). "RDF exports top 1.5 million tonnes in 2013." Retrieved from www.letsrecycle.com/news/latest-news/energy/rdf-exports-top-1.5m-tonnes-in-2013.

Defra (2007). *Waste strategy for England 2007*. London, UK, Department for Environment, Food and Rural Affairs.

Defra (2011a). *Government review of waste policy in England 2011*. London, UK, Department for Environment, Food and Rural Affairs.

Defra (2011b). "Local authority collected waste – definition of terms." Retrieved from www.gov.uk/local-authority-collected-waste-definition-of-terms.

Defra (2011c). Scenario-building for future waste policy. *Final report Research Project WR1508*. London, UK, Department for Environment, Food and Rural Affairs.

Defra (2012). *Resource security action plan: making the most of valuable materials*. London, UK, Department for Environment, Food and Rural Affairs.

Defra (2013). *Forecasting 2020 waste arisings and treatment capacity*. London, UK, Department for Environment, Food and Rural Affairs.

Defra (2015a). *Digest of waste and resources statistics – 2015 edition*. London, UK, Department for Environment, Food and Rural Affairs.

Defra (2015b). "Packaging waste statistics." Retrieved from www.gov.uk/government/policies/reducing-and-managing-waste/supporting-pages/packaging-waste-producer-responsibility-regimes.

DETR (2000). *Waste strategy 2000. Department for Environment, Transport and the Regions*. London, UK, HMSO.

EA (2010). *Waste infrastructure report 2010*. Bristol, UK, Environment Agency.

Ekvall, T., G. Assefa, A. Bjorklund, O. Eriksson and G. Finnveden (2007). "What life-cycle assessment does and does not do in assessments of waste management." *Waste Management* 27(8): 989–996.

Environmental Protection Agency (2014). *Municipal solid waste generation, recycling, and disposal in the United States: facts and figures for 2012*. Washington, DC, USA, US Environmental Protection Agency.

Eunomia (2013). *Residual waste infrastructure review: high-level analysis – issue 4*. Bristol, UK.

European Commission (2011). *Roadmap to a resource efficient Europe*. Brussels, Belgium.

European Commission (2014). "Higher recycling and packaging targets for 2030." Retrieved from http://ec.europa.eu/unitedkingdom/press/frontpage/2014/14_67_en.htm.

European Parliament and Council of the European Union (1999). Council Directive 1999/31/EC on the landfill of waste. Official Journal of the European Communities. Brussels.

European Parliament and Council of the European Union (2008). Council Directive 2008/98/EC on waste. Official Journal of the European Communities. Brussels, Belgium.

Eurostat (2014). News release 48/2014 – "In 2012, 42% of treated municipal waste was recycled or composted", Eurostat Press Office.

Fitzgerald, J. (2010). Waste treatment infrastructure requirements for municipal solid waste and commercial/industrial waste in England 2010 – 2020. London, UK, Imperial College. MSc.

Goulding, T. (2014). "Defra denies 'U-turn' on waste activities." Retrieved from www.letsrecycle.com/news/latest-news/waste-management/defra-denies-2018u-turn 2019-on-waste-activities.

HM Treasury and Infrastructure UK (2011). National Infrastructure Plan 2011. HM Treasury. London, UK.

HMRC (2014). *Landfill tax briefing – March 2014*. HM Revenue & Customs. London, UK.

Holder, M. (2015). "RDF export begins levelling off." Retrieved from http://www.letsrecycle.com/news/latest-news/rdf-export-begins-levelling-off/.

House of Commons (2011). *The Landfill (Maximum Landfill Amount) Regulations 2011*. UK Statutory Instruments. London, UK.

Latcham, R. (2011). "China sets out recycling goal of 70%." Retrieved from www.mrw.co .uk/8622174.article.

OECD (2012). Sustainable materials management – a synthesis. Working Party on Resource Productivity and Waste, Organisation for Economic Co-operation and Development.

Rathi, S. (2007). "Optimization model for integrated municipal solid waste management in Mumbai, India." *Environment and Development Economics* 12(1): 105–121.

Rawtec Pty Ltd (2014). South Australia's recycling activity survey 2012–13 financial year report. Report prepared for Zero Waste SA, Government of South Australia.

Resource Conservation and Recovery Act (1976). (PL 94–580, Oct 21, 1976) United States Statues at Large, 90, pp. 2795–2841.

Resource Futures (2013). National compositional estimates for local authority collected waste and recycling in England, 2010/11. Report prepared for Defra, EV0801. Bristol, UK.

SQWenergy (2010). Meeting Scotland's zero waste targets: assessing the costs associated with new waste management infrastructure. Report for the Scottish Government.

Tolvik Consulting Ltd (2011). The future of landfill? 2011 Briefing Report.

van Santen, A. (2010). Waste prevention actions for priority wastes – economic assessment through marginal abatement cost curves. Report prepared by ERM for Defra. Oxford, UK.

Welsh Assembly Government (2009). Towards zero waste. *A consultation on a new Waste Strategy for Wales*. Cardiff, UK.

Williams, P. T. and J. Barton (2011). "Demonstration scale flash pyrolysis of municipal solid waste." *Proceedings of the ICE – Waste and Resource Management* 164(3): 205–210.

WRAP (2008). *Lightweight wine bottles*. Waste and Resources Action Programme. Banbury, UK.

WRAP (2009a). *Choosing the right recycling collection system*. Waste and Resources Action Programme. Banbury, UK.

WRAP (2009b). *The composition of municipal solid waste in Wales*. Waste and Resources Action Programme. Cardiff, UK.

WRAP (2009c). *Kerbside recycling in Wales: environmental costs*. Waste and Resources Action Programme. Cardiff, UK.

WRAP (2009d). *MRF quality assessment study*. Waste and Resources Action Programme. Banbury, UK.

WRAP (2011). *Kerbside collection options: Wales*. Report prepared for WRAP Cymru, by Eunomia Research, Resource futures and HCW Consultants. Cardiff, UK.

Digital communications and information systems

EDWARD J. OUGHTON, MARTINO TRAN, CLIFF B. JONES, RAZGAR EBRAHIMY

9.1 Introduction

Digital communications and information systems infrastructure is comprised of a variety of communication and computation systems which provide the transmission, processing and storage of digital information. This includes: (i) legacy and fibre-optic cable networks; (ii) mobile, satellite and wireless networks; and (iii) data storage and processing centres. The acronym ICT is often used to describe these systems, though in other contexts ICT can refer to a much broader class of information and communications technologies. Here we use the shorthand ICT to refer to digital communications and information infrastructure systems. In comparison to other physical infrastructure sectors, ICT is relatively new and very rapidly evolving, so is less completely understood from a system-of-systems perspective. ICT infrastructure has become increasingly embedded in nearly all economic and social activities, including for operation of other infrastructure networks.

Because of the complexity and rapid innovation in ICT systems, their future is highly uncertain and analysing the systems' prospective direction is challenging. Yet, issues surrounding digital connectivity are high on the global policy agenda. In many advanced economies in Europe, North America and beyond, significant increases in ICT capacity have been delivered by inducing competition within industry combined with a light-touch, market-led regulatory approach. Innovation has provided new technologies and has helped serve growing consumer demand, which is itself largely driven by innovations in consumer technologies and business practices.

Globally, the telecommunications sector has increased dramatically over the past decade to serve almost three billion Internet users, amounting to over 40% of the world's population (ITU, 2014a). Investment in competitive markets by network operators is underpinned by analysis comparing prospective revenues with investment costs (Tselekounis and Varoutas, 2013) so is determined by population density, topology and the capital cost of new technologies (Gotz, 2013). Population density is an important factor in serving local telecommunication markets because, like other infrastructure systems, large fixed-capital investments need to be spread over many consumers with the purchasing power to acquire service subscriptions. However, if left to its own devices the market would likely be considered to be socially unjust and not provide all the potential welfare benefits, as the majority of investment would likely flow into pockets of dense, wealthy and well-educated urban areas. This would particularly leave remote rural or other uneconomically viable locations

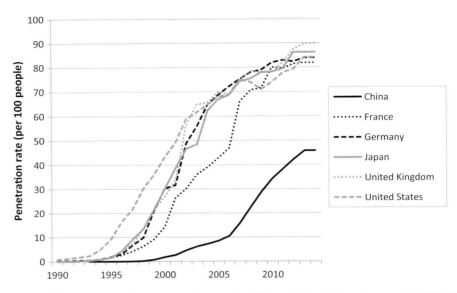

Figure 9.1 Global demand for internet services – penetration rate (per 100 people) for the top six economies 1990–2014 (Source: World Bank, 2013; ITU, 2014a)

at a severe disadvantage. Indeed, the economics and regulation of ICT infrastructure have a profound impact on the supply and capacity of the ICT system.

Total revenue in the telecommunications sector alone in 2012 was estimated at being 2.7% of world GDP ($1.88 trillion), with the largest increases being seen in developing nations (ITU, 2014b). Investment flows in the whole of the ICT infrastructure system are driven by four key aspects which affect global supply and demand dynamics. These are: (i) markets and competition; (ii) policy, regulation and government; (iii) availability of content, applications and services; and (iv) education and skills. The first two areas have the largest impact on supply, while the latter two areas have the largest impact on demand. In particular, population density and the ability for consumers to pay for, access and utilise the digital connectivity services enabled by ICT infrastructure have the largest influence. The demand for Internet services is illustrated in Figure 9.1 and shows that the US maintained the largest Internet penetration rate throughout the 1990s. However, other advanced nations began to catch up in the first decade of the twenty-first century as the U.K., Germany and Japan performed strongest out of the OECD countries. Although in 2014 China had a moderate penetration rate, due to its vast population it had over 632 million Internet users over 80% of whom were using mobile devices (CNNIC, 2014).

Around the world there has been a shift in the focus of policy from basic access to *access quality*. For example, the European Commission's Digital Agenda for Europe has the aspiration of providing Internet access speeds to all EU citizens of 30 megabits per second (Mbit/s), with over 50% of citizens subscribing to a connection over 100 Mbit/s, by 2020. In Europe, the EU Structural and Rural Development Funds have been utilised to roll out fibre to poorly connected places to help achieve this target. Equally, in the US access quality issues were addressed in the American Recovery and Reinvestment Act 2009 which required the Federal Communications Commission to start drafting a National Broadband Plan, which ultimately aimed to deliver 100 million American households with 100 Mbit/s access by 2020. Similar stimulus plans can be seen around the world with notable cases including Australia, Finland and Japan (Zhen-Wei Qiang, 2010).

The composite ICT Development Index (ITU, 2014b) is comprised of metrics for ICT readiness (reflecting networked infrastructure and access), intensity (reflecting use of ICTs) and impact (reflecting the result/outcome of efficient and effective ICT use). Denmark leads the table in 2014 taking the top spot from the Republic of Korea which ranked first in 2013. Many of the Nordic countries rank highly (Sweden, Iceland, Finland and Norway) due to their long tradition of comprehensive state aid broadband policies, supplemented by strong involvement by municipalities and energy companies (Briglauer and Gugler, 2013). This contrasts strongly with the U.K.'s more laissez-faire regulatory approach. Britain has moved into fifth position, moving up two places since 2013. All top ten countries are from Europe and Asia and the Pacific (the aforementioned countries plus The Netherlands, Luxembourg and Hong Kong) and almost two-thirds of the top thirty are European.

9.2 Digital communications and information systems in Britain

9.2.1 System description

ICT infrastructure is comprised of communication systems which include fixed and mobile telephony, broadband, television and navigation systems, and computation including data and processing hubs. However, the ICT sector can be viewed as a system-of-systems because of the interdependencies which exist between four key entities: (i) networked element providers (including connecting devices such as tablets and smartphones, to telecoms switches and transmission systems); (ii) network operators (covering telecoms, cable TV and satellite networks); (iii) content, application and service providers; and (iv) final consumers (Fransman, 2010). The ICT sector differs significantly from other infrastructure sectors because rapid growth and frequent change due to new technologies make it harder to analyse and forecast over the long term. Additionally, the sector is mostly commercially driven with private providers responding rapidly to changes in technology and consumer demand. The sector also has a strong international dimension marked by dependence on global markets. Compared to other sectors, artefacts of the ICT system are generally smaller, less expensive and have shorter lifetimes, thus infrastructure expansion can mostly be made in a rapid fashion without constructing large, physical objects. However, some artefacts such as information technology (IT) systems can be considered large, expensive and have long development times. Furthermore, ICT systems are subject to 'generation' upgrades, whereby development and deployment is significant and costly.

Figure 9.2 illustrates the inter-connection of some of the key components in the ICT infrastructure system. A variety of ways are used by businesses and consumers to access digital communications including legacy copper-based methods, Fibre to the Cabinet (FTTC),

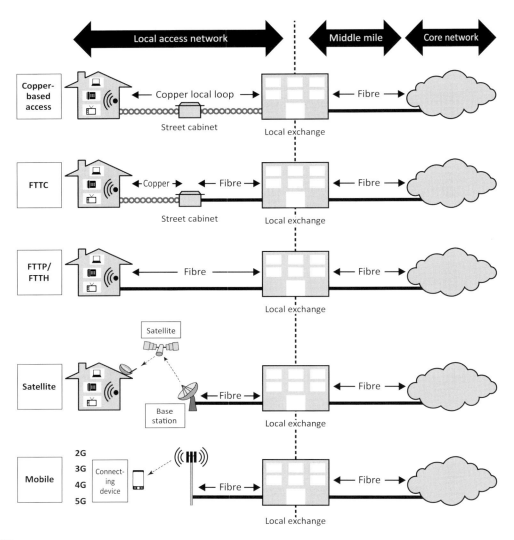

Figure 9.2   Simplified schematic of the ICT infrastructure system

Fibre to the Premises (FTTP), Fibre to the Home (FTTH), commercial leased lines or satellite and mobile technologies. As illustrated in Figure 9.2, FTTC involves laying fibre between the telephone exchange and the street cabinet, and using copper or coaxial cable for the final connection. FTTP and FTTH involve laying fibre all the way to the business or consumer's premises, providing the highest speeds but at the greatest expense. The core communications network is used to carry data around Britain and beyond, and the middle mile (referred to as the backhaul) connects local areas into this core network. The local access network (also known as the 'final mile') usually links premises to a street cabinet and then the local telecoms exchange. Much of the debate around bandwidth issues have traditionally been focused on fixed broadband infrastructure in the final mile, as copper-based solutions struggled to cope with the demands of new bandwidth-heavy content, applications and services. Fibre has traditionally been considered as necessary for a modern, high-capacity ICT system and is being rolled out towards the premises via FTTC but newer technologies are starting to enter the market which have the potential to deliver faster and faster speeds over traditional copper infrastructure.

The use of mobile for voice communications still plays a critical role with near ubiquitous saturation in advanced and developing nations. When a call is made a microwave transmission is sent from the user's device to a local base station. It is then sent via fixed connectivity to the main network and another base station where it can finally be channelled again by microwave to the receiver. Equally, wireless technologies are also becoming increasingly critical for data transfer. Particularly where the economic costs of delivering fibre are unviable (such as remote rural areas), both satellite and 4G mobile broadband have a potential role to play in improving data transfer rates in comparison to traditional copper-based access. In contrast, wireless access (specifically 4G) can provide a complementary way of accessing data in areas with good quality fixed connectivity (in dense urban areas). Transmission is understood in a similar way to mobile telephony whereby data transmission is sent to local base station wirelessly and sent by fixed connectivity from there. Each of the evolving components will now be examined in turn.

Computational hardware – modern computers are built from integrated circuits that create the equivalent of transistors on tiny areas of (mainly) silicon. The industry trends (e.g. hardware speed) are well described by Moore's law whereby the number of transistors that can be placed inexpensively on an integrated circuit doubles approximately every two years. Currently available computation resources are considerably more advanced than the necessary requirements needed to operate the infrastructure sector.

Bandwidth – the network bandwidth capacity is the measurement of bit-rate transmission in a digital network. The bandwidth available influences the speed of transmission and therefore the tasks users are able to undertake. The available bandwidth to end-users is one of the critical underlying constraints in the ICT sector and receives considerable policy attention because of worries over a 'digital divide' between those who have sufficient access to bandwidth, and those who do not. As with silicon chips, the growth in available bandwidth has been exponential, to the extent that annual global IP traffic will pass one zettabyte (equivalent to 250 billion DVDs) by the end of 2016 (Cisco, 2014). In many countries we currently see moves to roll fibre out towards end-users via FTTC, FTTP and FTTH, although some have questioned the multi-billion dollar subsidies provided by some governments to expedite this process (Kenny and Kenny, 2011). The issue is whether there should be a greater focus on improving connectivity at either the top end or the bottom end of the market, as the economic benefits of increased connectivity might be overstated.

Radio spectrum – wireless communications and broadcasting infrastructure depend on available radio frequency spectrum to transfer data. The regulation and use of radio spectrum can have significant implications for the functionality of wireless communications and broadcasting infrastructure. Demand for radio communications has grown from the variety of applications which have emerged across mobile communications, sound and television broadcasting, aviation, railway and maritime transport, defence, medical electronics, emergency services, remote control and monitoring, radio astronomy and space research.

Radio spectrum is a finite resource regulated by government and therefore it is crucial that frequency usage is maximised for both consumers and operators in the allocation of licences (Haykin et al., 2009). Consequently spectrum management reform is a topical issue to prevent applications using their allocation inefficiently and to ensure new technologies such as mobile-broadband can access a greater range of available radio spectrum (Medeisis and Holland, 2014; Minervini, 2014). By some accounts less than 14% of radio spectrum is truly busy at any given time (Rubenstein, 2007). Due to the finiteness of radio spectrum for use in wireless communications a key challenge in the coming decade is to balance private sector and government spectrum usage (Marcus, 2014). Various options to address this are available, for example, by increasing the efficiency of spectrum via improved technology, spectrum reuse or shared spectrum access (Weiss and Jondral, 2004; Blackman et al., 2013; Madden et al., 2014).

Satellite communications and navigation – data can be transferred between a dish antenna at a user's premises, a ground station and a satellite usually in geostationary orbit to provide connectivity. Additionally, Global Navigation Satellite Systems (GNSS) provide position, navigation and timing services via satellites for local electronic receivers. The main advantage of satellite services are that they require little terrestrial infrastructure, thus services can be available in remote areas with poor or no communications infrastructure (Maral and Bousquet, 2011). The most common consumer services include satellite television, radio, broadband, telephony and positioning (Richharia, 2001). The satellite industry is a subset of both telecommunications and space industries, with very high involvement of non-commercial organisations. For example, of the 986 satellites in orbit only 37% are commercial communications satellites representing 4% of telecommunication revenues, and 61% of space industry revenues (SIA and Futron, 2011). But growth is expected to continue. Euroconsult (2011) estimates 1,145 satellites to be built and launched for the period 2011–2020, an increase of about 51% compared to the previous decade (2001–2010). This is fuelled by expansion of satellite services, by both government and commercial organisations. Globally, the demand for navigation services is growing as a consequence of new applications used in cellular and data networks, emergency services and road, air, maritime and rail transport.

Information technology and data processing – dependence on IT has accelerated during the last two decades. In combination with the rise of advanced digital communication networks, information processing systems and services are taking more important roles. IT projects (especially large-scale ones) are emerging as significant parts of various infrastructure sectors and government. National governments apply IT in a variety of ways including online public services (tax collection or benefits payments) and business intelligence systems to process information and assist with decisions, and day-to-day operations (e.g. finance, human resources, other systems) (NAO, 2011). Information technology also plays an important role for physical infrastructure services, such as congestion charging, or the next generation energy network or smart grid (Liebenau et al., 2009). The IT landscape is changing for other organisations and SMEs as well. The recent emergence of cloud computing is providing greater business efficiency and lower upfront start-up costs, as well as providing altogether new economic opportunities. By using the cloud, organisations can minimise local ICT infrastructure and can scale their systems on demand to provide increased operational flexibility.

9.2.2 Challenges and opportunities

The ICT infrastructure system is especially critical in advanced service-led economies like Britain where it serves over 22.6 million fixed residential and SME broadband subscribers (>80% of households) (Ofcom, 2014a) and more than 83.2 million active mobile connections (Ofcom, 2014b). In 2014, 38 million adults (76%) in Britain accessed the Internet every day on average, with 68% using a mobile phone, portable computer and/or handheld device (ONS, 2014b). Firms are categorised as part of the digital economy if they utilise a set of inputs (in production and distribution) and outputs (via products and services), which are underpinned by ICT. While many advanced economies also have exceeded an 80% Internet penetration rate in 2014, Britain has one of the fastest growing digital economies with around 270,000 active companies (14.4% of all companies) (Nathan et al., 2013). Conservative estimates show that Britain leads the G20 nations with its digital economy accounting for 8.3% of total GDP in 2010, increasing to 12.4% by 2016 (BCG, 2012).

Britain is one of the top economies in Europe for the proportion of sales arising from e-commerce. The adoption of ICT in businesses has led to 20% of total turnover in 2013 being attributable to e-commerce, amounting to an estimated £557 billion (ONS, 2014a). This is due to strong growth in computer to computer e-commerce between businesses using standardised data formats known as Electronic Data Interchange (EDI), which makes up 65% of total e-commerce sales, with website sales making up the remainder (35%). This translates to an increase in demand for communications services. Traffic statistics over the London Internet Exchange's (LINX) network routers, which interconnect Britain's Internet Service Providers (ISPs), show that data traffic has increased by more than six times since 2008 (London Internet Exchange, 2015). Conservative traffic estimates show that in 2014 almost 1.5 Terabytes of traffic was exchanged per second between U.K. ISPs. Given the economic and societal implications of ICT systems, it is important to understand the system's changing demand profile, and what the current coverage and future capacity requirements of the system are.

To explore these implications we use Britain as an example of an industrialised nation pursuing rapid ICT investment focusing on issues of capacity and demand. In this section, many of the available statistics are for the U.K., which we use as a proxy for Britain.

9.3 Assessment of national digital communications and information systems

9.3.1 Future demand for digital communication and information services

In Figure 9.3, the six key indicators that Dauvin and Grzybowski (2014) identified as determining digital demand have been supplemented by additional indicators relating to, among others, the penetration of devices and services among the population and businesses. In essence, consumers and businesses are readily able to take advantage of new ICT because

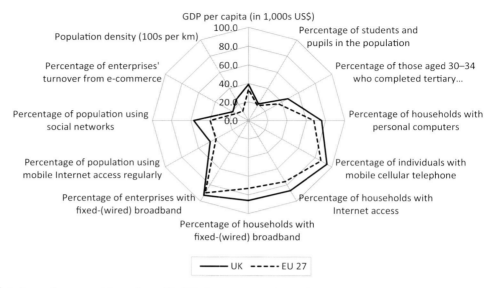

Figure 9.3   Indictors of present and future demand for ICT infrastructure – UK versus the EU27 (Source: Eurostat, 2014; ITU, 2014a; OECD, 2014; ONS, 2014a; ONS, 2014b). Note: number of households removed for scaling purposes

of high purchasing power and the ability to spend money on new devices and associated content, applications and services.

The device statistics demonstrate that ownership of personal computers (76% population) and mobile phones (94.2% population) are among the highest in Europe, leading to the fast diffusion of new applications and services. Education levels are high in Britain as a large percentage of individuals have completed tertiary university-level education which naturally helps businesses realise the real economic advantages of new technologies (47.1% vs. 35.8% for the EU 27). This is related to and compounded by that fact that Britain has one of the largest Knowledge Intensive Business Services (KIBS) sectors in Europe (Corrocher and Cusmano, 2014), bolstering demand for digital connectivity.

In terms of the adoption of digital connectivity services, Britain outperforms the EU average with a very high overall percentage of Internet users (87%). The large majority of users have access to fixed broadband connections at home (86%) as does practically every business (93%). But this dynamic is changing as more users move to mobile access, with nearly half the British population already accessing the Internet via their mobile regularly. Figure 9.3 shows that 57% of the population uses social networking which is the highest in Europe opening up new opportunities for marketing, Big Data and new associated economic activities. Population density is also an important factor because it takes into consideration a nation's overall agglomeration factor, which in the case of telecommunications is important because it makes new technologies more economically viable to deliver to market due to economies of scale. The combination of these demand determinants translates to considerable growth in total data traffic in the long term. For example, future projections show aggregate business and consumer traffic will be three times larger in 2018 for both fixed and mobile connectivity compared to 2013 (Figure 9.4).

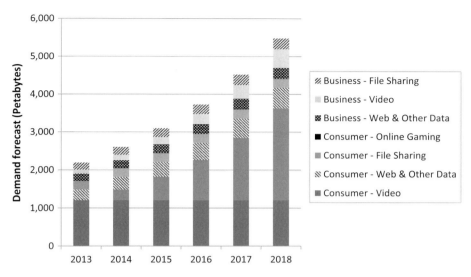

Figure 9.4   Aggregate demand forecast for fixed and mobile business, and consumer traffic in U.K., 2013–2018 (Source: Cisco, 2015)

Consumer video will be the primary driver of increasing traffic volume (Cisco, 2015) especially as we transition from high-definition television to 4K television.

The ongoing increase in networked devices will affect both traffic volume and the amount of bandwidth demanded as these devices increase from a total of 279 million in 2012 to over 460 million by 2017 (6.8 networked devices per capita) (Cisco, 2015). However, as well as increases in traffic and devices, we will also see changes in how users choose to connect. For example, IP traffic from PCs will have decreased by 25% by 2017, while the proportion of IP traffic from mobile portable devices such as tablets and smartphones will have increased by 19% (comprising 21% of aggregate UK IP traffic) (Cisco, 2015). Among the 83.2 million active mobile connections in Britain across all mobile network operators in 2014 (approximately 1.3 mobile connections per capita), the total uploaded and downloaded mobile data traffic was approximately 44,300,000 GB (equivalent to 9.4 million DVDs) (Ofcom, 2014b). Per active SIM, total data demand increased by 53% in 2014, which is similar to the increase seen in 2013. This has been driven by continued growth in the smartphone and tablet PC markets, and the subsequent consumption of video which accounts for 39% of mobile traffic. The verdict is still out as to whether Mobile Network Operators (MNOs) should see wireless networks as complementary, disruptive, substitutionary or competitive technologies (Hüsig, 2013). We could see a displacement effect away from mobile data services if we continue to see rapid increases in the number of WiFi hot spots in Britain (Ofcom, 2014b). Certainly for mobile operators there are clear advantages of this shift in demand away from cellular 3G or 4G LTE to Wi-Fi, as it reduces the capacity constraints placed on their networks (Telecoms.com Intelligence, 2014).

Current demand for Very High Bandwidth Connectivity (VHBC) services is driven by the substantial volumes of data recorded from large numbers of customer transactions, financial trading and intensive use of imaging and video data (CSMG, 2013). This is in

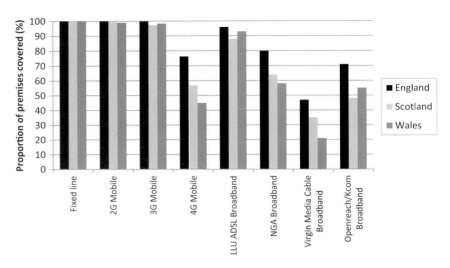

Percentage of the population covered by communications infrastructure by region (Source: Ofcom, 2014a)

addition to the increasing use of remote, off-site data centre services and cloud computing. Certain economic sectors are driving this increase, including growing trading volumes in financial services, the desire for richer content and HD video in media, the aggregation of large numbers of CCTV channels to prevent fraud and theft, and more data intensive research programmes in higher education and industrial R&D.

Looking beyond 2017, It has been projected that demand for mobile broadband may grow between 23 and 297 times over the period 2012–2030, with an eightyfold increase the mid-case scenario (Real Wireless, 2012). This analysis was based on current trends and an analysis of devices and user mobility. This increase in demand can be attributable on the one hand to unlimited or increased data usage allowances with mobile phone contracts, and on the other the rollout of faster 4G networks enabling the streaming of more data intensive applications and services.

9.3.2 Investment in new capacity for digital communication systems

Britain occupies a top ten spot in most global ICT-related rankings, having high population coverage and capacity by fixed and mobile ICT infrastructure (Figure 9.5). Recent estimates show that basic broadband of 2 Mbit/s has been reached at 97% of premises. Moreover, 85% of premises are able to achieve a broadband speed of 10 Mbit/s. Speeds greater than 30 Mbit/s are now available to 75% of premises with 21% having taken-up next generation access (NGA) services exceeding 24 Mbit/s. However, around 3% of premises fall below the British government's current minimum target download speed of 2 Mbit/s (Ofcom, 2014b).

Overall, 78% of premises have access to NGA and this is split between two main network operators – BT and Virgin Media. As a proportion of those premises with NGA access, a total of 34% is served by BT, 9% is served by Virgin Media and the final share of 35% of premises have access to NGA services from both operators (Ofcom, 2014b). In addition to delivering 100% coverage of basic broadband, the Department for Culture, Media and Sport (DCMS) which has overview of Broadband Delivery UK (BDUK) is aiming to achieve 95% superfast broadband coverage by 2017, which is defined as providing download speeds in excess of 24 Mbit/s.

Equally, mobile communications networks have achieved 2G (voice and low data rate) coverage for 99% of premises by at least one mobile network operator, and 97% by all three providers. For 3G (voice and slower data services) coverage, 84% of premises are served by all four network mobile operators, although this falls to 75% in Scotland and 65% in Wales. With the rollout of 4G (voice and fast data services), 35% of premises now have coverage from all three of the main licence-holding mobile network operators – Vodafone, EE and O2. In terms of coverage from at least one provider, 72% of premises can gain 4G access.

The existing coverage and capacity issues exist in the least profitable markets. This includes those premises in low-density areas where fixed infrastructure upgrades might be unviable because network operators cannot achieve the economies of scale necessary to justify large capital investments. Some have claimed that the statistics purveyed on network coverage to date are misleading (BBC, 2014). For example, even if FTTC is enabled some premises located far away from the telephone exchange still might not achieve the 2 Mbit/s target. Utilising national data the extent of fixed broadband and mobile communications network coverage and capacity is examined, which illustrates variations in broadband infrastructure. Using postcode level data comprised of sync speed measurements for fixed broadband connections from 2014, the percentage of postcodes with residential premises not receiving basic broadband of 2 Mbit/s has been calculated (Figure 9.6a).

Figure 9.6a illustrates that many of the areas with the highest percentage of premises without basic broadband (> 10%) are in Scotland and Wales. These areas are lagging behind their contemporaries particularly in southern England for access quality. For example, the approximate area from Nottingham to Southampton, and from London to Bristol all have more than 90% of premises with access to basic broadband. Digital connectivity north or west of this area is generally poorer.

Figure 9.6b illustrates the number of postcodes with NGA where fibre has been deployed to the cabinet. This deployment is hence highly correlated with speed improvements. However, disparities between achievable speeds and NGA can arise due to the geography of the settlement pattern and the use of copper in the final mile. For example, northern Scotland and Wales have poor NGA deployment in many areas where it can be well below 50% coverage. While this has a strong consequential impact on speed in Wales, Scotland still achieved reasonably acceptable fixed broadband speeds when compared with other areas. Population density and the spatial positioning of premises around the exchange have a significant impact on these results as it affects the length of the copper loop used in the final mile. For example, Scotland may have low density settlements clustered around telephone exchanges, allowing acceptable fixed broadband speeds over copper. Figure 9.6c illustrates the temporal change in connection speed seen across different platform technologies between 2010 and 2014. The average increase in connection speed to 18 Mbit/s in 2014 has been driven by dramatic increases in speeds via cable and NGA.

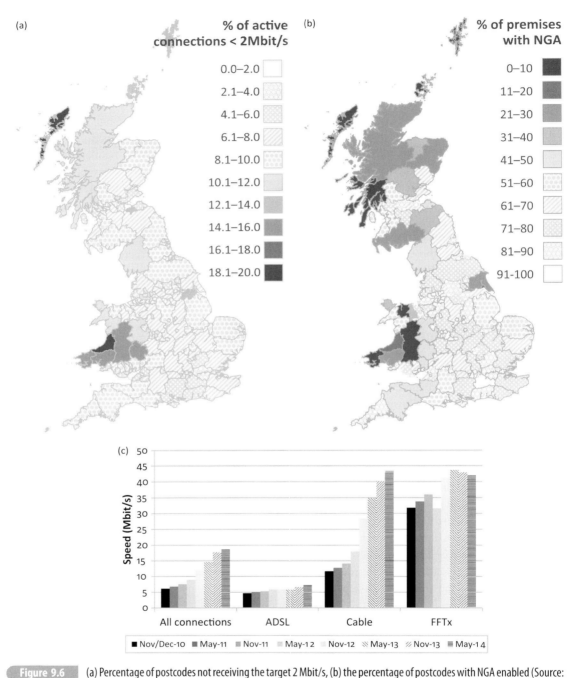

Figure 9.6 (a) Percentage of postcodes not receiving the target 2 Mbit/s, (b) the percentage of postcodes with NGA enabled (Source: Ofcom, 2013a) and (c) the average speed by connection type (Source: Ofcom, 2014a). Note: Next generation access (NGA) services have the potential to deliver download speeds of 30 Mbit/s or more

In terms of overall core network capacity, modern technologies such as wavelength-division multiplexing (WDM) have enabled ISPs and Very High Bandwidth Connectivity (VHBC) users to multiply by several times the bandwidth transmissible in communications networks (e.g. over fibre), without necessarily expanding the physical network infrastructure. But for most residential and business premises using fixed broadband, it is the final copper part of the connection between the consumer and the local street cabinet and exchange (the final mile), that remains the bottleneck in the system. The barrier to improved delivery lies in the economics of NGA access because of the large costs associated with digging and laying fibre in the final mile. Hence, wireless and satellite technologies are starting to play a bigger role in delivering digital connectivity to remote users. In the future, we may turn to the new fixed technologies being tested which can deliver fibre-like speeds over existing copper infrastructure. BT – the incumbent operator with the largest market share – is planning to upgrade its network via the deployment of G.fast technology which is claimed to enable 'ultra-fast' speeds up to 500 Mbit/s within the next decade. The main cable operator Virgin Media, recently announced a £3 billion investment plan to upgrade and extend their network to pass another four million businesses and households, offering speeds up to 152 Mbit/s.

Meanwhile in mobile communications networks, many coverage and capacity issues still exist at the bottom end of the market. These range from some areas having no voice or data signal at all (labelled by Ofcom as 'complete not-spots'), to some areas which have coverage but lack competition as they are served by only one or two operators. Current spectrum policy aims to improve competition and coverage by (a) ensuring extensive coverage obligations for mobile licences and (b) making sure more spectrum is available in the sub-1 GHz range due to its good propagation characteristics (Ofcom, 2014b). For example, obligations attached to some licenses required the network operator to roll out indoor coverage to 98% of the population, with at least 95% being covered in each individual U.K. nation by 2017. Britain wants to be one of the European front-runners for the supply and capacity of 4G mobile communications as they are expected to underpin future economic and societal change. Indeed, there is already significant interest in attempting to move beyond 4G to 4G+ to improve mobile broadband speeds. It is hoped that 5G will begin to be deployed by 2020 onwards to enhance mobile deployment (Ofcom, 2013b). However, to ensure Britain remains competitive in this area, there needs to be a continued and strong commitment by industry and government to a digital communications strategy which bolsters the delivery of new capacity and coverage across both the fixed and mobile domains. The following section explores the broader implications of large-scale uptake of ICT on both infrastructure performance and the changing nature of service demand.

9.3.3 Future connectivity and the digital divide

With regards to investment in this sector there are two opposed schools of thought developing in the Britain. Interventionists believe that Britain needs a substantial one-off public investment programme across a range of fixed and mobile broadband technologies to deliver the critical infrastructure necessary for the continued growth and development of Britain's digital economy. This desired multi-billion pound investment stimulus would include the delivery of gigabit fixed broadband connectivity to both businesses (FTTP) and consumers (FTTH). Those in favour of the interventionist approach to 'future-proof' the supply of digital connectivity particularly include Small and Medium Sized Enterprises (SMEs) in the high-tech sector that see potential for a substantial change in this infrastructure on their day-to-day business operations (Tech London Advocates, 2015). Many consumers, particularly in remote rural areas which are unviable for private sector investment, are also in favour of support for improving fixed and mobile digital connectivity.

On the other hand, the opposing viewpoint questions the evidence around the actual economic and social impacts of more bandwidth. For example, one study analysed 156 household demand profiles and estimated the median household will only require bandwidth of 19 Mbit/s by 2023, whilst the top 1% of high usage households will demand 35–39 Mbit/s (Kenny and Broughton, 2013). Hence, some argue that there is no guarantee that delivering gigabit speeds will translate to larger economic and social outcomes as the current supply practically meets our demand needs already. Also, where average monthly data downloaded is analysed in relation to average download speed, data usage increases significantly between 0 Mbit/s and 10 Mbit/s due to bandwidth being a constraining factor (Ofcom, 2014b). But past this point the correlation between downloaded data and average speed is less pronounced in comparison. Hence, there is greater importance in ensuring that there is universal access of 10 Mbit/s to all, as opposed to delivering speeds which well exceed this point. Moreover, if the average user does not need speeds in excess of, for example, 30 Mbit/s (which will be almost ubiquitously delivered in the next few years with the committed roll-out of FTTC), then there is no justification for costly public sector intervention to supply the types of speeds delivered via FTTP/FTTH. The applications and online services we use simply do not require the bandwidth delivered by these costly technologies. In any case with G.fast and other new technologies having the potential to deliver fibre-like speeds over copper we might simply be able to exceed our bandwidth requirements at a fraction of the cost without having to put fibre in the local access network.

Additionally, too often in this debate the efficiency benefits of upgrading connectivity to businesses are applied to household consumers, when in reality there is patchy evidence to suggest that added bandwidth leads to more economic growth (Kenny and Kenny, 2011). More evidence-based analysis needs to be undertaken to (i) understand the amount of band-width required for satisfactory usage of current and future online applications and services, (ii) understand the applications and online services being used in countries with widespread FTTP/FTTH to examine the return on investment and (iii) provide more examples of where high bandwidth connections actually provide business and social opportunities which could not be achieved with the bandwidths we already obtain today (Kenny, 2015).

9.4 Analysis and discussion

9.4.1 Interdependence with infrastructure, energy and transport services

Since the 1990s, ICT has been integrated with practically all infrastructure activities, changing the way in which assets are operated, infrastructure services are delivered and how

Table 9.1 Global projections for future power consumption in digital communication systems (Source: Pickavet et al., 2008; IEA, 2009)

	Power consumption in 2008 (GW)	Growth rate (% p.a.)	Power consumption in 2020 (GW)
Data centres	29	12	113
Network equipment	25	12	97
PCs	30	7.5	71
TVs	44	5	79
Total	168		433
Worldwide electricity	2,350	2.0	2,970
ICT fraction	7.15%		14.57%

infrastructure services are demanded. For decades, the largest degree of interdependency in the national infrastructure system resulted from the demand for energy from the transport, waste, water and communications sectors. But as ICT is used in more activities it equally has risen to underpin the functionality of other infrastructure sectors. As physical infrastructures become more sophisticated and loaded to capacity they become more interdependent and rely on ICT. ICT often has a supporting role for physical infrastructures, so it is important to identify possible capacity limitations and vulnerabilities within ICT that could affect the future development of infrastructures. Additionally, ICT will become increasingly diffuse across all infrastructure sectors influencing supply-side efficiency and demand-side management (DSM) with unknown implications for long-term systems performance. There is a particular need to understand how ubiquitous low-cost ICT might change the long-term physical performance of infrastructure systems and the nature of future service delivery. In the following we explore how increasingly diffuse ICT will impact upon energy and transport services.

The digital communications sector is growing rapidly and having an increasingly large impact on energy consumption. Table 9.1 provides a simple projection of future power consumption from ICT based on five categories of equipment: (i) data centre equipment comprises computing, storage and network equipment in data centres and additional supporting equipment including Heating Ventilation and Air Conditioning (HVAC) and Uninterruptable Power Supply (UPS); (ii) network equipment includes switches and routers, modems and home gateways. Network interface cards are not considered here since they are accounted for in either data centres or PCs; (iii) PCs comprise laptops and desktop computers; (iv) televisions include additional equipment like DVD players; and (v) other devices include telephones, mobile phones, printers, fax machines, etc. Worldwide electricity consumption for these five categories are based on methods and data in Pickavet et al. (2008) and used to predict a business as usual projection for 2020. This has been compared with global energy electricity consumption levels (IEA, 2009). Table 9.1 shows that ICT consumes about 7% of global electricity consumption and is set to double to 14% by 2020. Additionally, power consumption is fairly equally distributed across categories and therefore all need to be tackled in future energy reduction strategies (Vereecken et al., 2010).

The predicted future growth in the number of connected devices and bandwidth demanded will drive this energy demand upwards. However, simple strategies can be used to reduce ICT related energy consumption. Research on energy-aware backbone networks has demonstrated that by turning off spare devices whose capacity is not required to transport off-peak traffic, it is possible to easily achieve more than a 23% energy saving per year (Chiaraviglio et al., 2009). Moreover, around 8–22% energy demand could be reduced from cellular network infrastructure if network operators switched off redundant base stations during periods of low traffic (e.g. night time) (Oh et al., 2011). Yet to date, energy costs have played a limited role in the planning, management and regulation of communications infrastructure, even as we move towards the delivery of Next Generation Networks (NGNs) (Coomonte et al., 2013).

Additionally, ICT has the potential to reduce energy use in the transport sector. The availability and increased penetration of superfast broadband infrastructure by 2024 has been estimated to save 2.3 billion kilometres in annual commuting, and 5.3 billion kilometres in annual business travel (SQW, 2013). Governments around the world are now taking action to reduce the negative impacts of transport on climate change, energy and environment. A central part of these strategies are the mass deployment of ICT enabled smart transport systems. The idea of intelligent or 'smart' mobility is where travellers are able to plan and execute their journeys seamlessly and optimise the full range of services across all modes and infrastructures. This is enabled by intelligent transportation systems (ITS) that provide a set of strategies for advancing transportation safety, mobility and environmental sustainability. For example, large-scale ICT deployment can increase traffic speeds during peak times by $\sim$12%, mitigate total passenger vehicle GHG emissions by $\sim$16% and reduce local air pollution (NOx and PM2.5) by $\sim$2–12% by 2050 (Tran, 2013). This is made possible by the rapid advancement and integration of ICT into the management and operation of the transportation system across all modes and infrastructures.

In the past twenty years, the highest growth has occurred in the area of intelligent vehicles (42%), followed by transit (37%), management and operations (29%), freight (23%) and roadway operations (21%) (USDoT, 2013). It is anticipated that ITS technologies have the potential to revolutionise surface transportation by connecting vehicles, infrastructure and passengers via wireless devices and other real-time information dissemination applications. This will allow drivers, operators and commuters to send and receive real-time information about transport options, potential hazards, road conditions and all other means of information to optimise mobility services. Changes in mobility will also create the potential for reducing peak traffic demand during rush hour periods, which could reduce the negative environmental impacts associated with capacity-stricken transportation systems (White et al., 2010). However, the expected demand for infrastructure services can be changed considerably by these shifting work and lifestyle patterns. Consequently, this requires greater consideration for how ICT systems should be adapted to support shifting economic and social activities (Helbing, 2013).

9.4.2 Smart cities and lifestyle change

There is now global anticipation for so called 'smart cities' underpinned by investment and mass roll-out of smart grid networks and related technologies, which have the potential to connect urban energy and transport systems, and to some extent influence lifestyle change. Governments around the world are implementing policies to encourage investment into smart grid architecture. Smart grid networks potentially have an enabling role for low-carbon energy technologies, including electric vehicles, variable renewable energy sources and managing demand response. It is believed that there is potential to increase overall system efficiency by better matching energy demand and supply through improved data monitoring and information feedback. For example, network operators will get more detailed information about supply and demand improving management of the system such as shifting demand to off-peak times (Cheng et al., 2014). Chapter 11 analyses the economic and energy impacts from large-scale investment into offshore renewables to meet rising electricity demand enabled by mass deployment of smart grid technologies in Britain.

New research is considering how the smart grid concept can be extended to broader energy and societal contexts termed the 'smart community' or 'smart city'. This involves integrating several energy supply and use systems within a region to optimise operations, while maximising the integration of renewable energy resources (from large-scale wind to micro-scale PV) and coupled with residential energy management systems. This would also include existing infrastructure systems, that is, electricity, water, transportation, gas, waste and heat and future systems, that is, hydrogen and electric vehicle charging (IEA, 2014). Not only will this transform existing physical systems, but consumer engagement will also change, as many households could become small-scale electricity producers through decentralised on-site energy conversion and storage technologies (PV, wind, electric batteries, etc.). The term 'prosumers' has recently emerged to describe customers who consume and produce electricity and may provide energy services. Demand-side management and demand response will also increase the role of the consumer to not only use electricity, but also support system operation. However, stakeholders also need to take seriously the increased potential of cyber attacks, and the implications for future governance and regulation of information and privacy rights.

Indeed, the social impact of technologies on everyday lifestyle is becoming more and more evident such that everyone feels the need to be connected to the Internet. The greater flexibility over working and living patterns, provided by ICT connectivity, is changing how we routinely move around our environment (Sayah, 2013). These new patterns appear to have many advantages, although the evidence is not always clear cut (Wilks and Billsberry, 2007). The potential benefits of these changes range from new and more efficient forms of business organisation, through to improved work-life balance for workers.

9.4.3 Risks of ICT infrastructure failure

As ICT systems have become smaller and less expensive, they have become more pervasive. This has led to an increasing reliance on the interaction between people and the design of dependable ICT systems. It is essential to note that as 'dependability' has increased, so has dependence. ICT is now pervasive in systems that are financially, militarily or safety critical. Any complex system such as ICT can fail, but today failure can arise from myriad factors related to hardware, software or malicious attack.

As the role of ICT has been increasing within national infrastructure and general usage, it has become a more important target for various threats, such as hackers, viruses, identity thieves, etc. The general threat of hacking and cyber espionage (e.g. banking fraud and identity theft) has been a persistent concern for the general public and organisations. Cyber-crime has been estimated to cost as much as $1 trillion per year globally (Cabinet Office, 2010). An important issue with tackling cyber-security threats is that the response is usually fractured and uncertain. There is need for development of standards of best practice, continuity planning and risk management. While it is recognised that unexpected threats cannot be eliminated completely, response mechanisms within organisations or government must contain high levels of communication, capability and agility to successfully address these risks (Cornish et al., 2011). Chapter 14 focuses on governance issues arising from increasing infrastructure interdependencies due to pervasive uptake of ICT.

The performance of ICT systems under stress whether from component failure or malicious attack can be unpredictable and compromise services for multiple users. Risks of failure also increase where ICT systems are working close to capacity (i.e. above 40%) (CST, 2009). The resilience of ICT networks is improved by their nature and availability: multiple networks and/or ICT services (e.g. wired or wireless) are available to switch across multiple links simultaneously and re-route dynamically in real time. The issues of communications infrastructure resilience are being recognised in academia, government and industry. Involved parties are acting to maintain and improve resilience, as well as react quickly to failures and reassess ICT infrastructure provision (Horrocks et al., 2010). Legislation to improve security and resilience in the communications sector is now being developed around the world. Chapter 12 focuses on risk and cascading failures in interdependent infrastructure networks, which in some instances will be amplified or mitigated by ICT.

9.5 Conclusions

Superficially, ICT infrastructure appears to be categorically different to the other infrastructure considered in this volume. The long-term investment planning challenges that are faced by other sectors seem to be less relevant for digital communications, where provision of new infrastructure is highly adaptive and driven by the market and by technological innovation, in particular in consumer technologies. However, we have demonstrated that there are long-term strategic planning issues, in particular relating to coverage in areas where providers may be reluctant to incur the costs of adequate service provision. Moreover, whilst demand for digital communication services has grown enormously, and will continue to do so in coming years, it is not guaranteed to continue growing at the same rate. For example, this demand growth has been driven by consumer video and there could be a point at which this tapers off, as consumers are constrained in the amount of time they can spend watching video and because the benefits of higher video resolution begin to diminish. Furthermore, whilst capacity constraints are not an immediately significant planning issue for fibre in the core network, there are capacity issues in the access network, particularly for mobile connectivity. Moreover, there needs to be considerations about the limited allocation of spectrum for wireless usage. With regard to fibre, there are significant cost implications for commercial and residential delivery in the 'final mile'. While there is an argument for delivering fibre based around 'future proofing', there does not appear to be substantial evidence to suggest that there should be public sector intervention to achieve this. Besides, new technological innovations have proved they can achieve ultra-fast broadband speeds over traditional copper cable which will more than satisfy medium-term bandwidth demand. In light of this, there should be more focus on ensuring that everyone can achieve 10 Mbit/s (which Britain is on its way to achieving as the committed investment in FTTC is rolled out over coming years), and a greater role for evidence-based public sector intervention.

Whilst ICT capacity will not likely limit the growth of physical infrastructures, this chapter indicates there are important outstanding issues concerning risk and resilience in the system. Moreover, the human dimensions of dependence on ICT along with the long-term performance of networked systems require greater understanding. The largest impacts of ICT on society are still to come, but much of the above discussion indicates increasing interdependency between the built environment and human activity enabled by ICT. In the future, our physical infrastructure systems may shift from historically capital intensive physical assets prone to lock-in, towards more modular and flexible systems adaptive and responsive to human needs. This may prompt a step change in future demand and service delivery business models, which are further explored in Chapter 10 where we develop strategies to assess long-term transformational change of the current system.

References

BBC (2014). "Rural broadband: how the money will be spent." Retrieved from www.bbc.co.uk/news/technology-26338920.

BCG (2012). *The internet economy in the G20 – the $4.2 trillion growth opportunity.* Boston Consulting Group.

Blackman, C., S. Forge and R. Horvitz (2013). "Liberating Europe's radio spectrum through shared access." *info* 15(2): 91–101.

Briglauer, W. and K. Gugler (2013). "The deployment and penetration of high-speed fiber networks and services: Why are EU member states lagging behind?" *Telecommunications Policy* 37(10): 819–835.

Cabinet Office (2010). *A strong Britain in an age of uncertainty: the national security strategy.* London, UK.

Cheng, X., X. Hu, L. Yang, I. Husain, K. Inoue, P. Krein, R. Lefevre, Y. Li, H. Nishi and J. G. Taiber (2014). "Electrified vehicles and the smart grid: the ITS perspective." *IEEE Transactions on Intelligent Transportation Systems* 4(15): 1388–1404.

Chiaraviglio, L., M. Mellia and F. Neri (2009). Reducing power consumption in backbone networks. IEEE International Conference on Communications, 2009. ICC '09.

Cisco (2014). "VNI forecast highlights." Retrieved from www.cisco.com/web/solutions/sp/vni/vni_forecast_highlights/index.html.

Cisco (2015). The zettabyte era: trends and analysis.

CNNIC (2014). *Statistical Report on Internet Development in China*. Beijing, China, China Internet Network Information Center.

Coomonte, R., C. Feijóo, S. Ramos and J.-L. Gómez-Barroso (2013). "How much energy will your NGN consume? A model for energy consumption in next generation access networks: the case of Spain." *Telecommunications Policy* 37(10): 981–1003.

Cornish, P., D. Livingstone, D. Clemente and C. Yorke (2011). *Cyber security and the UK's critical national infrastructure. A Chatham House Report*. London, UK, The Royal Institute of International Affairs.

Corrocher, N. and L. Cusmano (2014). "The 'KIBS engine' of regional innovation systems: empirical evidence from European regions." *Regional Studies: The Journal of the Regional Studies Association* 48(7): 1212–1226.

CSMG (2013). *Research on very high bandwidth connectivity*. CSMG. London, UK.

CST (2009). *A national infrastructure for the 21st century*. Council for Science and Technology. London, UK.

Dauvin, M. and L. Grzybowski (2014). "Estimating broadband diffusion in the EU using NUTS 1 regional data." *Telecommunications Policy* 38(1): 96–104.

Euroconsult (2011). Satellites to be built and launched by 2020, World Market Survey. A Euroconsult Research Report.

Eurostat (2014). "Glossary: e-commerce." Retrieved from http://ec.europa.eu/eurostat/statistics-explained/index.php/Glossary:E-commerce.

Fransman, M. (2010). *The new ICT ecosystem: implications for policy and regulation*. Cambridge, UK, Cambridge University Press.

Götz, G. (2013). "Competition, regulation, and broadband access to the internet." *Telecommunications Policy* 37(11): 1095–1109.

Haykin, S., D. J. Thomson and J. H. Reed (2009). "Spectrum sensing for cognitive radio." *Proceedings of the IEEE* 97(5): 849–877.

Helbing, D. (2013). "From technology-driven society to socially oriented technology – the future of information society – alternatives to surveillance." Retrieved from http://futurict.blogspot.co.uk/2013/07/from-technology-driven-society-to.html.

Horrocks, L., J. Beckford, N. Hodgson, C. Downing, R. Davey and A. O'Sullivan (2010). Adapting the ICT sector to the impacts of climate change – final report. *Defra contract number RMP5604*. AEA group.

Hüsig, S. (2013). "The disruptive potential of PWLAN at the country-level: the cases of Germany, the UK, and the USA." *Telecommunications Policy* 37(11): 1060–1070.

IEA (2009). *World energy outlook 2009*. Paris, France, International Energy Agency.

IEA (2014). *Energy technology perspectives 2014: harnessing electricity's potential*. Paris, France, International Energy Agency.

ITU (2014a). "ITU statistics." Retrieved from www.itu.int/en/ITU-D/Statistics/Pages/stat/default.aspx.

ITU (2014b). *Measuring the information society 2014*. Geneva, Switzerland, International Telecommunication Union.

Kenny, R. (2015). Exploring the costs and benefits of FTTH in the UK. A report for NESTA. London, UK.

Kenny, R. and T. Broughton (2013). Domestic demand for bandwidth – an approach to forecasting requirements for the period 2013–2023. A report for the Broadband Stakeholder Group.

Kenny, R. and C. Kenny (2011). "Superfast broadband: is it really worth a subsidy?" *info* 13(4): 3–29.

Liebenau, J., R. Atkinson, P. Kärrberg, D. Castro and S. Ezell (2009). *The UK's digital road to recovery*. LSE Enterprise & The Information Technology and Innovation Fund.

London Internet Exchange (2015). "LINX traffic stats." Retrieved from www.linx.net/pubtools/trafficstats.html.

Madden, G., E. Bohlin, T. Tran and A. Morey (2014). "Spectrum licensing, policy instruments and market entry." *Review of Industrial Organization* 44(3): 277–298.

Maral, G. and M. Bousquet (2011). *Satellite communications systems: systems, techniques and technology*. Chichester, UK, John Wiley & Sons.

Marcus, M. J. (2014). "The challenge of balancing private sector and government spectrum use." *Wireless Communications, IEEE* 21(3): 8–9.

Medeisis, A. and O. Holland (2014). *Cognitive radio policy and regulation: techno-economic studies to facilitate dynamic spectrum access*, Cham, Switzerland, Springer International Publishing.

Minervini, L. F. (2014). "Spectrum management reform: rethinking practices." *Telecommunications Policy* 38(2): 136–146.

NAO (2011). *Information and communications technology in government – landscape review*. London, UK, National Audit Office.

Nathan, M., A. Rosso, T. Gatten, P. Majmudar and A. Mitchell (2013). Measuring the UK's digital economy with big data. Growth Intelligence and The National Institute of Economic and Social Research.

OECD (2014). "OECD Broadband portal." Retrieved from www.oecd.org/sti/broadband/oecdbroadbandportal.htm.

Ofcom (2013a). "Facts & figures." Retrieved from http://media.ofcom.org.uk/facts/.

Ofcom (2013b). *Infrastructure report 2013*. London, UK, Ofcom.

Ofcom (2014a). *Communications market report: UK*. London, UK, Ofcom.

Ofcom (2014b). *Infrastructure report 2014*. London, UK, Ofcom.

Oh, E., B. Krishnamachari, X. Liu and Z. Niu (2011). "Toward dynamic energy-efficient operation of cellular network infrastructure." *Communications Magazine, IEEE* 49(6): 56–61.

ONS (2014a). "E-commerce and ICT activity." Retrieved from www.ons.gov.uk/ons/guide-method/method-quality/specific/business-and-energy/e-commerce-and-ict-activity/index.html.

ONS (2014b). *Internet access – households and individuals, 2014: statistical bulletin*. Cardiff, UK, Office for National Statistics.

Pickavet, M., W. Vereecken, S. Demeyer, P. Audenaert, B. Vermeulen, C. Develder, D. Colle, B. Dhoedt and P. Demeester (2008). Worldwide energy needs for ICT: the rise of power-aware networking. 2nd International Symposium on Advanced Networks and Telecommunication Systems, Mumbai, India.

Real Wireless (2012). Capacity techniques for wireless broadband. *Real Wireless*. Pulborough, UK.

Richharia, M. (2001). *Mobile satellite communications: principles and trends*. Boston, MA, USA, Addison-Wesley Longman Publishing Co., Inc.

Rubenstein, R. (2007). "Radios get smart." Retrieved from http://spectrum.ieee.org/consumer-electronics/standards/radios-get-smart.

Sayah, S. (2013). "Managing work–life boundaries with information and communication technologies: the case of independent contractors." *New Technology, Work and Employment* 28(3): 179–196.

SIA and Futron (2011). State of the satellite industry report.

SQW (2013). *UK broadband impact study*. Cambridge, UK, SQW.

Tech London Advocates (2015). *Joining the dots: building the infrastructure for London tech*. London, UK, Tech London Advocates.

Telecoms.com Intelligence (2014). Industry Survey 2014. Telecoms.com.

Tran, M. (2013). *Impact Study on Intelligent Mobility*. London, UK, InnovITS.

Tselekounis, M. and D. Varoutas (2013). "Investments in next generation access infrastructures under regulatory uncertainty." *Telecommunications Policy* 37(10): 879–892.

USDoT (2015). "Intelligent transport systems joint program office benefits database." Retrieved from www.itsbenefits.its.dot.gov/.

Vereecken, W., W. V. Heddeghem, D. Colle, M. Pickavet and P. Demeester (2010). Overall ICT footprint and green communication technologies. ISCCSP 2010, Limassol, Cyprus.

Weiss, T. A. and F. K. Jondral (2004). "Spectrum pooling: an innovative strategy for the enhancement of spectrum efficiency." *Communications Magazine, IEEE* 42(3): S8–14.

White, P., G. Christodoulou, R. Mackett, H. Titheridge, R. Thoreau and J. Polak (2010). *Impacts of teleworking on sustainability and travel*. Social Sustainability in Urban Areas: Communities, Connectivity and the Urban Fabric. A. Manzi, K. Lucas, T. Lloyd-Jones and J. Allen (eds.) London, UK, Earthscan.

Wilks, L. and J. Billsberry (2007). "Should we do away with teleworking? An examination of whether teleworking can be defined in the new world of work." *New Technology, Work and Employment* 22(2): 168–177.

World Bank (2013). World Bank open data.

Zhen-Wei Qiang, C. (2010). "Broadband infrastructure investment in stimulus packages: relevance for developing countries." *Info* 12(2): 41–56.

PART III

INTEGRATIVE PERSPECTIVES FOR THE FUTURE

Assessing the performance of national infrastructure strategies

MARTINO TRAN, JIM W. HALL, ROBERT J. NICHOLLS, ADRIAN J. HICKFORD

10.1 Introduction

In Part II of this book (Chapters 3–9) we have developed a case study of systems analysis and assessment for Britain. In Chapter 3, we developed scenarios of future population and economy, which could be used in a consistent way across sectors. In Chapters 4–9, we explored strategies for provision of infrastructure services in energy, transport, water, wastewater, solid waste and digital communications sectors, respectively. Here we bring these results together in a system-of-systems perspective and a national infrastructure assessment.

We achieve consistency across sectors in this assessment by using the same scenarios from Chapter 3 and by analysing consistent cross-sectoral strategies for infrastructure provision. Our approach for developing cross-sectoral strategies was presented in Chapter 2. In that chapter, we explained how our strategy generation process combined a top–down perspective, beginning with three high-level policy dimensions (investment level, commitment to demand management and commitment to environmental objectives) alongside a bottom–up perspective that assembled strategies for each infrastructure sector. We have now presented (in Chapters 4–9) the range of possible strategies in each infrastructure sector and evaluated their performance with our system models. Let us begin by restating the four cross-sectoral strategies that were presented in Chapter 2:

Minimum intervention (MI): takes a general approach of minimal intervention, reflecting historical levels of investment, continued maintenance and incremental change in the performance of the current system. There is no long-term vision to reduce future demand or implement more stringent commitments to environmental policies; instead, the focus is on short-term incremental improvements at a sector level, and thus fails to account for sectoral interdependencies.

Capacity expansion (CE): focuses on planning for the long term by increasing investment in infrastructure capacity. Priority is given to the expansion of physical capacity to alleviate pinch-points and bottlenecks. There is no long-term vision to reduce future demand. Provision of new capacity will provide opportunities for introduction of some more environmentally benign technologies.

System efficiency (SE): focuses on deploying the full range of technological and policy interventions to optimise the performance and efficiency of the current system targeting both supply and demand. This implies targeted investments to increase capacity at severe bottlenecks in the shorter term, but the medium to long term vision is to invest heavily in maximising the throughput of the current system, without high investments in CE. Improvements in the efficiency of plant and vehicles enable significant increases in the through-put of the system with relatively modest capacity investments. There may be a limit to the extent to which these efficiency improvements can address expected demand pressures from population and the economy.

System restructuring (SR): focuses on fundamentally restructuring and redesigning the current mode of infrastructure service provision, deploying a combination of targeted centralisation and decentralisation approaches. This will require long-term investments aiming to utilise a wide range of technological innovation, incorporating policy incentives, changes in demand, and integrated planning and design.

These strategies will be used for our cross-sectoral evaluation in this chapter. In our analysis of these strategies we bring together quantitative assessments of the energy, transport, water supply, wastewater and solid waste infrastructure systems. We exclude digital communications from this analysis because, as we saw in Chapter 9, quantified assessment of that system on extended timescales is less meaningful, though Chapter 9 did set out some future trends that we envisage materialising in each of the cross-sectoral strategies. Furthermore, we anticipate different levels of uptake of information and communications technology (ICT), with it being stronger in the SE and SR strategies.

10.2 Cross-sectoral performance metrics

Our aim in this chapter is to develop an evaluation of infrastructure systems that applies across sectors and can be used to provide an overview of future system performance, in the context of different exogenous scenarios and national infrastructure strategies. Infrastructure performance is clearly a multi-attribute construct, since we are interested in many aspects of infrastructure performance (Francis and Bekera, 2014).

The primary consideration in evaluating infrastructure system performance relates to the capacity of the system to provide users (customers) with the services that they need. This directly leads to an interest in the capacity of the system to deliver infrastructure services (e.g. energy and water), as compared to the demand for those services. However, this framing in terms of capacity and demand immediately raises questions from an economic perspective. Such a perspective would argue that neither the demand nor the supply are, in general, a fixed quantity. They are functions, whose slope is determined by elasticities and from which the total amount of service delivered can be computed.

It is indeed important to remember that infrastructure capacity and demand are not fixed quantities. Increasing demand for rail travel will encourage train operating companies to run more trains (increase capacity) and/or increase prices (to suppress demand). These dynamic relationships between supply, demand and price are enacted in our models of the energy and transport sectors, where they are most relevant. In the transport sector, demand for transport is deterred by the effects of congestion increasing journey times (a cost) and suppressing demand.

Nonetheless, in energy, transport and all of the other sectors, it is reasonable to think in terms of a maximum quantity of infrastructure service that could be delivered for a given network configuration (i.e. without further investments). There is a maximum amount of electricity that a given set of power plants and grid configuration can generate and deliver. There is a maximum yield of water that a water supply system can deliver to customers. Effectively, beyond this quantity, the supply becomes price inelastic. Even in transport, there are a maximum number of vehicles, trains, aircraft or ships that a given stretch of road or rail, airport or port can cope with. On a stretch of road, increasing demand beyond that maximum number of vehicles/hour will reduce speeds and lead to congestion. We therefore argue that it is meaningful to think of the maximum capacity of an infrastructure system to deliver infrastructure services, and this is an important performance metric for an infrastructure system. Whilst for most of our infrastructure sectors (energy, water, wastewater and solid waste) it is reasonable to aggregate (sum) the service delivery over the whole network, for transport there needs to be some spatial averaging to account for the fact that some links (road, rail) or nodes (ports, airports) are very heavily utilised whilst others, typically in remote areas, may be very sparsely utilised. We use a passenger-weighted average across all links/nodes in the network to generate our aggregate capacity metric.

Demand for, and use of, infrastructure services is clearly influenced by a range of different factors. Unlike capacity, it is not reasonable to think in terms of people's or businesses' maximum capacity to consume infrastructure services. Instead we use empirical data on demand, in the context of each of our infrastructure strategies. Each of those strategies contains assumptions about demand, both as a factor of population and as a consequence of the demand management actions that have been adopted as part of the strategies. The SE and SR strategies contain a more ambitious package of measures on the demand side than the MI strategy, which, in turn, contains more demand side measures than the CE strategy that is dominated by supply-side investments.

As a consequence of these different strategies for demand management (including pricing and regulatory measures), and in the context of the socio-economic scenarios, we generate projections of infrastructure service delivery, that is, the amount that the infrastructure system is projected to be used. Services delivered are calculated by sector specific models and are therefore measured by sector level metrics as follows: energy (MWh/yr), water supply and wastewater (ML/yr), transport (VKm/yr) and solid waste (Mt/yr).

Whilst Service Delivery (SD) is an interesting metric in its own right, it is even more informative to decision-makers when combined with the metric of Capacity. By comparing these two quantities, that is, the amount of infrastructure service that a system can deliver (Capacity, C) with the amount that it is projected to actually deliver (Service Delivery) we generate a non-dimensional metric of Capacity Margin. Capacity Margin (CM) is defined as the ratio between total available capacity and service delivery

$$\text{CM} = \frac{C - \text{SD}}{\text{SD}} \times 100\%. \tag{10.1}$$

CM assesses the difference between available capacity and demand at a given time for a given sector, representing security of supply and system redundancy. Capacity Margins are a well-developed approach used for power system planning to determine the level of over-capacity in the electricity network to meet peak demand (RAE, 2013). In water resources planning, there is also a long tradition of engineering hydrology in simplifying the water supply problem to compare a single value of supply, or yield (Law, 1955), with annual demand. The concept is less familiar in the context of transport, wastewater and solid waste, but here we generalise the CM approach to these other national infrastructure sectors.

Because of the need for spatial averaging in the transport sector, we adapt the metric to represent Capacity Utilisation (CU), which measures use of each link/node in the transport network. The CU of each link/node is the predicted throughput (in terms of vehicles per hour, trains per hour, airport passengers per year) as a percentage of the maximum capacity of the link/node for the given infrastructure configuration. The aggregate figure is the passenger-weighted combination over all links/nodes. The aggregate figure for the whole transport system is the passenger-weighted combination over the three networks considered, accounting for the current mode share (road $\sim$90%, rail $\sim$9% and air $\sim$1%). Here we assume that this split does not change dramatically in the future, although this could be changed with development of additional strategies for future analyses. Because of spatial heterogeneity across the network, it is not realistic to suppose that the network will ever be operating at full capacity in all places. For comparison with the Capacity Margin for other sectors, we take $CM_{transport} = 60\% - CU$.

Cumulative Investment (CI) quantifies the cost of providing the physical infrastructure in a given strategy. This metric captures the cumulative system costs for service provision, including capital investments and operational costs. For this analysis, the investment in 2010 is determined for each sector and referenced against national statistics (HM Treasury, 2013). This provides a base year from which per annum investment can then be modelled in subsequent years based on the strategy assumptions discussed above. Starting in 2010, CI can then be summed across sectors giving an indicator of the total economic performance of the entire national infrastructure system over a selected time period, which we set here at 2050. The CI costs discussed in the following section includes capital and operational costs for each sector with the exception of transport due to data limitations. However, operational costs are small in proportion to capital expenditure for the transport sector in comparison to operational costs in other sectors, for example, energy.

Carbon emissions (CO_2) is an indicator of the environmental performance of the system, and the externalities of service provision. Although there are multiple environmental impact indicators specific to each sector, carbon emissions are common across sectors, and are also closely linked to other environmental pollutants (e.g. NO_x, SO_2), especially for the energy and transport sectors. Carbon emissions can also be summed across sectors, providing an indicator of total system performance and can be contrasted and compared to the economic performance of the national infrastructure system. Carbon emissions are also a key indicator to assess national infrastructure performance against national climate policy targets.

Each sector also has Sector Metrics (SM), which are specific to each sector and quantify different aspects of service quality, capacity and performance (see Chapters 4–9 for sectoral assessments, and online Appendices for technical details). These compliment the cross-sectoral analysis, allowing a deeper analysis of each strategy at different spatial

Table 10.1 National infrastructure system attributes and cross-sectoral performance metrics

	National infrastructure system attributes	Cross-sectoral performance metric(s)	Issues captured
1.	Security of supply	Capacity margin/utilisation (CM/CU)	System redundancy, frequency and extent of supply shortages
2.	Provision of infrastructure services	Services delivered (SD)	Scale of provision of infrastructure services to society and the economy
3.	System Costs	Cumulative investment (CI)	Capital investments, operational costs
4.	Externalities	Carbon emissions (CO_2)	Environmental impact; contribution to achieving carbon targets
5.	Sector-specific Service Indicators	Sector-specific metrics (SM) – see Chapters 4–9	Service quality, e.g. transport (congestion), water (environmental quality), energy (peak demand), etc.

and temporal resolutions. The combination of the above cross-sectoral metrics and sector-specific metrics capture the full range of national infrastructure attributes (Table 10.1) providing a quantitative basis for comparing alternative strategies.

10.3 Assessment of cross-sectoral strategies

For each of the strategies (MI, CE, SE and SR), we describe their future performance in qualitative terms and report on the quantitative metrics that were introduced in Section 10.2. The assessment is primarily done in the context of the central scenario for population growth, economic growth, energy prices and climate change. Sensitivity analysis with respect to other scenarios has been described in the sector chapters (Part II).

10.3.1 Minimum Intervention (MI) strategy

The MI strategy is based on a minimum plausible level of intervention, through capacity investment or demand-side measures. The focus is upon on-going maintenance and operation of currently available infrastructure. Investment follows current trends with no major future investment to expand or modify the existing system. There is no long-term vision to reduce future demand or commitment to environmental policy. This strategy focuses on short-term incremental improvements at the sector level, and does not account for increasing sectoral interdependencies. Advanced technologies such as ICT, new policies (incentives and penalties) and integrated planning and design are not adopted to alter conventional capacity provision, or influence end-use demand.

Figure 10.1 shows the performance of the MI strategy, based on CI, per annum carbon emissions (CO_2), capacity margin (CM) and total services delivered (SD). From 2010 to 2050, CI reaches nearly £670 billion, which is dominated by increasing energy supply capacity to meet growing demand. Water supply and wastewater is 31% of total investment by 2050, with minor additional investment into the transport network and solid waste infrastructure. The main consequence of this low investment strategy is poor systems security of supply, shown by long-term declines in capacity margins across all sectors, and lower quality of services shown by a near flattening of service provision out to 2050. This is also coupled with poor environmental performance with a 36% increase in carbon emissions reaching 387 Mt/yr in 2050 from 2010 levels. Without substantial additional investments across all sectors, the MI strategy will have negative implications for meeting Britain's infrastructure security of supply, service quality and energy and climate policy targets.

At the sector level, for energy there is no significant strengthening of climate policies resulting in a fossil fuel dominated mix with carbon emissions rising from 190 Mt in 2010 to nearly 265 Mt in 2050. Investment primarily targets energy security. However, without incentives for increased efficiency and demand reduction, capacity margins (based on gross capacity margins assuming full import) decrease from 53% in 2010 to 26% in 2050. Consequently, existing long-term increasing trends in demand continue with population and economic growth (based on the central growth scenario defined in Chapter 3). This is shown by a steady increase in electricity service provision reaching 505 TWh/yr in 2050 from 369 TWh/yr in 2010. The energy supply sector changes slowly, with continued dominance of large-scale investments by large companies with no significant investment in nuclear or carbon capture and storage (CCS). Investments in renewables continue as costs fall, but capacity increases slowly. Power sector investment continues to rely largely on combined cycle gas turbines (CCGTs) with gas supplies from imported, but diverse, sources. Heat remains largely dependent on gas, although with continued efficiency improvements. Transport fuel supply remains largely oil dependent with slow penetration of biofuels and electricity (Baruah et al., 2014; Eyre and Baruah, 2015).

The transport sector continues on its current trend heavily dependent on fossil fuels. There is an increase in services delivered from 500 billion VKm in 2010 to 700 billion VKm in 2050 largely due to population growth. However, infrastructure security of supply is poor shown by a decline in network capacity to 42% in 2050 from 47% in 2010. Networks become more congested with no incentives for mode shifting away from private passenger vehicles. With minor future investment, road networks continue to deteriorate as carbon emissions increase 34% reaching 124 Mt in 2050 due to continued reliance on the existing fossil fuel based vehicle stock.

For the water supply sector there is minimal intervention at a national scale with no new inter-company transfers, limited leakage reduction and modest demand reduction. Investment options target modest reservoir enhancement and groundwater exploitation, limited capacity of desalination plants and no water re-use. To meet increasing demand from population growth, starting in 2010, CI reaches £57 billion in 2050. Per capita demand follows historical trends shown by the increase in services delivered from 4,240 GL/yr in 2010 to 5,370 GL/yr in 2050 with no demand reduction. With minor infrastructure investments the provision of water supply remains largely unchanged from the

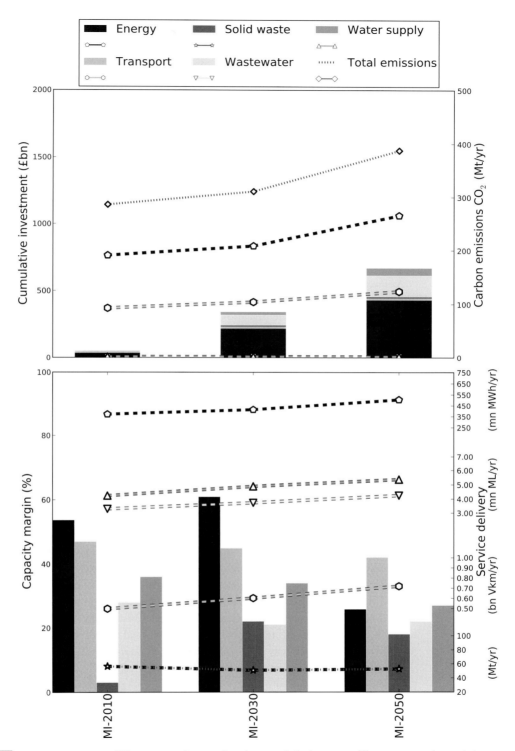

Figure 10.1 Minimum intervention (MI) strategy performance based on cumulative investment (£), per annum carbon emission (CO_2), capacity margins (%) and service delivery: Energy (MWh/yr), water supply and wastewater (ML/yr), transport (VKm/yr), solid waste (Mt/yr) for 2010–2050. Note: bars relate to the left hand, and dashed lines relate to the right hand y-axis, respectively

existing system configuration with carbon emissions remaining stable at ~2.3 Mt/yr out to 2050.

In the wastewater sector there is also little structural change with current management strategies persisting out to 2050. The per capita volumetric demand for wastewater services, the biological oxygen demand of sewage and the chemical oxygen demand of sewage remain constant, corresponding to no change in the consumptive behaviour of consumers. To meet population growth, CI reaches £156 billion in 2050 primarily for maintaining the sewer system. But without per capita demand reduction and minor efficiency gains, capacity margins decrease from 28% in 2010 to 22% by 2050. Sewerage service providers maintain the existing sewer network, extending and enhancing where necessary to meet the growth in demand in line with current consumer behaviour and marginal efficiency gains.

For the solid waste sector, existing waste, reuse and recycling targets for household, commercial and industrial (C&I) and construction and demolition (C&D) wastes are met by continuing the current trends and building new infrastructure, particularly energy from waste (EfW) and anaerobic digestion (AD) plants as required. From 2010, CI reaches £13 billion in 2050. There is a steady improvement in the performance of the waste sector and the amount of waste being landfilled continues to fall due in part to the continuing increases in landfill tax. As a result, the capacity margin increases from 3% in 2010 to nearly 18% by 2050, while maintaining the current level of services of around 53 Mt/yr in 2050.

10.3.2 Capacity Expansion (CE) strategy

This strategy focuses on provision of new infrastructure to meet increasing demand and to avoid capacity constraints in the future. Priority is given to the expansion of physical capacity to alleviate pinch-points and bottlenecks soon after they are identified. As expected, this strategy involves high capital costs. It can be less robust to future uncertainties unless optionality can be built-in. This strategy is effective at meeting demand in the short to medium term, but performs poorly over the long term due to physical limitations in capacity expansion, lock-in to current technology and design and lack of long-term vision to reduce or redistribute demand. There is also marginal commitment to environmental policy causing trade-offs between increasing capacity but poor environmental performance over the long term. The overall structure and function of the infrastructure systems does not change radically, with continued investment into conventional technology and design, and little forward planning to address increasing sector level interdependencies.

Figure 10.2 shows the cross-sectoral performance of the CE strategy. From 2010, total CI reaches £1.8 trillion with transport overtaking energy supply by 2030 and dominating total investment reaching £900 billion in 2050. There are also significant CI over the same period for solid waste (£25 billion), water supply (£93 billion), wastewater (£208 billion) and energy (£527 billion). As a result, over the medium term, systems security of supply is maintained across sectors with capacity margins stabilising despite increasing demand from population and economic growth. However, without demand reduction measures, energy and water security begins to decline by 2050. Services continue to rise over the long term especially for transport with a near doubling of vehicle kilometres (VKms) by 2050 from

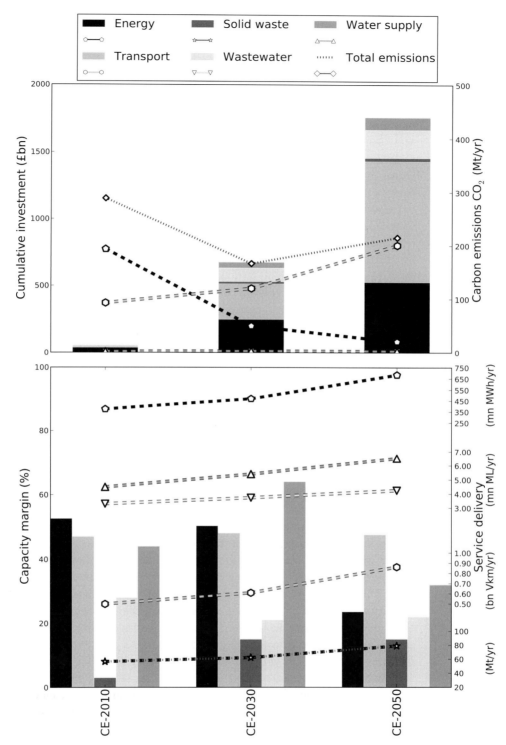

Figure 10.2 Capacity expansion (CE) strategy performance based on cumulative investment (£), per annum carbon emission (CO_2), capacity margins (%) and service delivery: Energy (MWh/yr), water supply and wastewater (ML/yr), transport (VKm/yr), solid waste (Mt/yr) for 2010–2050. Note: bars relate to the left hand, and dashed lines relate to the right hand y-axis, respectively

current levels. But this has negative implications for environmental performance. Over the medium term this strategy reduces carbon emissions from major investments into nuclear power generation, but over the long term emissions increase from the massive expansion of transport infrastructure and rising demand. Consequently, carbon emissions reach 215 Mt in 2050, similar to 2010 levels despite high investment into nuclear power generation. Consequently, emission reductions made in the energy sector are lost due to increasing transport CE. This demonstrates the importance of deploying a harmonised investment strategy that accounts for cross-sectoral performance and interdependency.

Large-scale CE in the energy sector is driven by security of supply concerns rather than meeting climate policy targets. As a result, there are large investments into indigenous resources and technology, first shale gas and then nuclear power reaching £530 billion by 2050. This significantly reduces carbon emissions from 193 Mt in 2010 to 21 Mt in 2050. However, existing long-term trends in demand continue with upward pressures from population and economic growth offset by some energy efficiency gains, but only limited improvements in regulatory standards, and tax incentives. This is shown by a near doubling of electricity services delivered combined with a decline in capacity margins to 23% in 2050 from a current 52% in 2010. Therefore, while CE is effective in the medium term it does not achieve long-term energy security or environmental benefits.

The transport sector attempts to build its way out of congestion, with large-scale road building and widening programmes, airport and seaport expansion and construction of additional railway lines. As a result, CI in transport reaches £900 billion and services nearly double reaching 862 billion VKm by 2050. Large-scale construction is determined by conventional cost-benefit analysis with externalities given low priority. Consequently, despite major carbon reductions in the energy sector, there is an overall increase in carbon emissions due to transport CE reaching 200 Mt in 2050. There are no incentives for demand reduction, or mode shifting as a result the expansion of transport networks continues to release latent demand, which is offset to some extent by peak travel. This is shown by the capacity margin remaining relatively stable around 47% over the long term, but as network congestion increases the quality of transport services declines.

The water supply sector also focuses on building large-scale, long-term investments in physical CE to meet increasing demand with CI from 2010, reaching £94 billion by 2050. This strategy involves no demand reduction and medium leakage reduction with an emphasis on reservoir enhancement, new groundwater and no regulated limit on the number of desalination plants. Water re-use is not explored as an option but inter-company transfers are allowed between all nearby water companies. As a result, water services delivered increases from 4,490 GL/yr in 2010 to 6,520 GL/yr in 2050. But with no demand reduction and minor investment into network efficiency, water security declines over the long term with the capacity margin dropping to 32% in 2050 from 44% in 2010. Carbon emissions also increase by nearly 40% reaching 3 Mt in 2050, but is relatively minor compared to the transport and energy sectors.

The wastewater sector invests heavily to meet increasing population growth with capital expenditure primarily for the sewer system with CI reaching £208 billion by 2050. Storm water separation, in particular is an expensive strategy. For treatment plants, replacement and new capacity with current (or marginally improved) technology is £40 billion; and for sewers there is a high replacement rate of 1% per year costing £146 billion. Services delivered and capacity margins are stable over the long term in line with population growth, and carbon emissions while doubling to 2 Mt in 2050 are relatively minor compared to other sectors.

For the solid waste sector, developments in materials separation and recovery technologies mean that wastes require minimum source separation. At the household level, this means two bins – food/green wastes and everything else. Consumers disengage from concerns about waste and recycling but despite this, rates of recycling and composting continue to rise as does overall waste production. Due to an over-reliance on technology and capacity expansion, there is no incentive for demand reduction with total services reaching 79 Mt/yr in 2050 from current levels of around 57 Mt/yr in 2010. To meet increasing population growth, CI from 2010 reaches nearly £25bn in 2050. A potential positive benefit from the increase in generated waste is that the materials left over from materials recovery are used for fuels in energy from waste (EfW) plants. As a result, this strategy has potential negative emissions of 12 Mt in 2050.

10.3.3 System Efficiency (SE) strategy

The SE strategy focuses on optimising the performance of the current system. This relies on deploying the full range of currently available technological innovations (ICT), and policies (incentives/penalties) to increase supply-side operational efficiency, and influence end-use demand. There is targeted investment to increase capacity at severe bottlenecks in the short term, but the medium- to long-term vision is to invest heavily to maximise throughput of the current system, without massive investments into CE. There is an important strategic shift in reframing the provision of infrastructure services by identifying and prioritising economic trade-offs and synergies between supply and demand. This reframing is strongly influenced by environmental policy and industry innovation to reduce carbon emissions along the entire supply chain, and increasing forward planning to capitalise on sectoral interdependencies. There are important trade-offs between this strategy and CE, which will likely perform better in the short term in alleviating bottlenecks. SE will be more competitive over the medium term since it can meet aggregate demand at less cost than physical expansion through steady operational efficiency improvements, and demand reduction. However, without a fundamental change in system design, performance of this strategy over the long term is less robust if improvements reach the limits of efficiency that can be extracted from the current system configuration, or is ultimately outpaced by long-term demand growth. In general, this approach focuses on deploying the full range of technological and policy interventions to increase efficiency of the current system targeting both supply and demand.

Figure 10.3 shows the performance of the SE strategy. From 2010 to 2050, total CI reaches £756 billion with energy supply comprising the largest proportion (55%). However, CI into water supply and wastewater combined is also significant, with far less invested into transport compared to CE. This strategy focuses more on demand reduction over the long term especially in energy, water and solid waste. As result systems security of supply remains relatively stable over the long term despite increasing population growth. For the

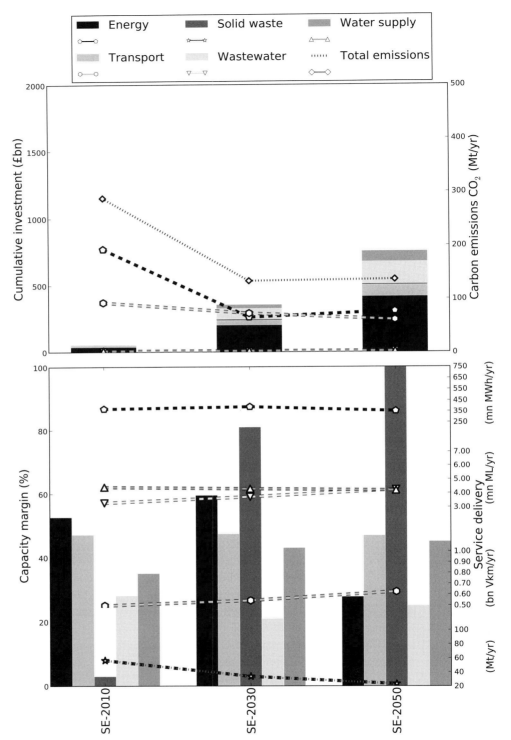

Figure 10.3 System efficiency (SE) strategy performance based on cumulative investment (£), per annum carbon emission (CO_2), capacity margins (%) and service delivery: Energy (MWh/yr), water supply and wastewater (ML/yr), transport (VKm/yr), solid waste (Mt/yr) for 2010–2050 Note: bars relate to the left hand, and dashed lines relate to the right hand y-axis, respectively

solid waste sector, with rapid per capita demand reduction, there is significant additional capacity by 2050. Despite decreasing electricity demand in the energy sector, capacity margins decline over the long term due to increasing reliance on intermittent renewables and phasing out of fossil fuel generation. Over the medium term this strategy performs well. By 2030, annual carbon emissions decline by 46% reaching 134 Mt through efficient transport networks, high investments into renewable power generation and levelling demand trends in the water supply sector. However, over the longer term, without a major restructuring of infrastructure service provision, efficiency gains could be lost from population growth shown by an upward trend in carbon emissions in the energy sector reaching 77 Mt/yr in 2050 from 65 Mt/yr in 2030.

For the energy sector, concerns over energy security are reflected in greater emphasis on individual and small-scale action, with more reliance on local resources, in particular solar energy and locally produced biofuels. Existing long-term trends in demand are reduced as upward pressures from population and economic growth are offset by higher efficiency heating systems, for example, heat pumps and combined heat and power (CHP), moderate improvements in building fabric energy efficiency and mass roll out of smart meters. This is stimulated by a combination of government policies and rising awareness of energy security driving local action. As a result, total electricity services stabilise over the long term at around 353 TWh/yr in 2050 and capacity margins reach 27%, which is an improvement over CE with no demand reduction and a capacity margin of 23%. Along with deployment of distributed generation via solar and biofuels there is greater investment into wind power. This results in CI of £417 billion and a 60% reduction in electricity-related carbon emissions reaching 77 Mt/yr in 2050 compared to current levels.

To increase network efficiency the transport sector relies heavily on advanced information and communication technologies (ICT) to enhance transport operations with a high level of embedded technology. Intelligent mobility measures include efficient road vehicle routing, based on real time traffic information enhanced by vehicle positioning systems; automated 'platoons' of vehicles on trunk roads to increase capacity utilisation and potentially increase maximum permitted speeds and use of hard shoulder running; real-time road pricing based on enhanced traffic information; cooperative traffic management systems; flexible pathing and moving block signalling on the railways; and smart logistics to optimise freight movements by all modes. Large-scale deployment of intelligent transport systems (ITS) requires significant CI reaching £83 billion by 2050, but is ten times less than a CE approach. Increasing the throughput of existing transport networks also reduces carbon emissions by 35% and with marginal increases in services delivered a current capacity margin of 47% is maintained over the long term.

For the water supply sector this strategy emphasises increased efficiency with measures to reduce the need for additional infrastructure. There is a strong focus on demand reduction and a high level of leakage reduction. Limiting infrastructure expansion also requires placing a limit on the capacity of desalination plants allowed and restricts new water re-use facilities or inter-company transfers. Targeting demand reduction and increasing the efficiency of the existing water delivery system results in CI of £80 billion with total water services delivered decreasing slightly to 4,200 GL/yr by 2050. In contrast to CE, water security in this strategy increases with a capacity margin of 45% in 2050 from 35% in 2010 despite continued population growth over the long term. Environmental performance also improves with carbon emissions of 2 Mt/yr in 2030 and levelling off out to 2050.

For the wastewater sector, economies of scale in treatment have increased efficiency, and thus decentralisation is not a viable approach except in very small population centres. Energy use becomes important with potential for incremental improvements in energy efficiency. However, the wastewater sector does not become a significant net exporter of energy. Cumulative investment reaches £165 billion in 2050. The bulk of investment is for sewers where there is a moderate replacement rate of 0.5% per year costing £109 billion; and investment for treatment plants, operational and energy expenditure improvements due to technological change cost £40 billion. Wastewater services delivered increase with population growth reaching 4,252 GL/yr and a capacity margin of 25% by 2050.

In the solid waste sector there is a move from consumption to leasing with products designed for long life, easy repair and remanufacturing (D4R). Waste arisings are reduced by increasing prices for waste disposal and greater involvement of the third sector in refurbishing of unwanted goods. There is little investment in infrastructure with increasing systems efficiency through demand reduction driven by cultural and behavioural change. There is a high emphasis on enhanced consumer education and deployment of information programmes. As a result, waste arisings decrease by nearly 60% in 2050 from current levels. All new infrastructure build is completed by 2030 costing nearly £20 billion, and by 2050 capacity margins exceed 100% due to major reductions in per capita waste arisings. There are also negative carbon emissions for this strategy of 5 Mt/yr, which is around half of the waste recovery for energy from CE.

10.3.4 System Restructuring (SR) strategy

This strategy focuses on fundamentally redesigning the current infrastructure system to improve total system performance. There is an important strategic shift in reframing the provision of infrastructure services from one of physical CE to the uninterrupted flow of goods and services. This results in identifying trade-offs and synergies along the entire service delivery chain and capitalising on sectoral interdependencies. This strategy leverages the full range of technological innovation, policy incentives and integrated planning and design through maximum use of ICT for operational planning and demand-side management (DSM). There is a strong commitment to environmental policy coupled with long-term investment to incentivise new service delivery models. The general approach is to deploy a balance of centralised and decentralised delivery mechanisms depending on regional and temporal trade-offs, such as determining where and when benefits from economies of scale can be achieved versus reducing long-term demand, or increasing capacity in the short term versus improving environmental performance over the long term.

There are important trade-offs between this strategy and CE and SE. CE will be competitive in the short term and SE in the medium term, with some exceptions in transport where CE can have long lead times compared to efficiency gains through targeted consumer behaviour. SR could deliver regionally dispersed benefits in the short to medium term through decentralisation. However, large-scale benefits may not be realised until the long term when economies of scale can be achieved. But this will require sustained investment to restructure the current delivery system, large-scale diffusion of advanced supply-side technologies, and major shifts in demand away from current consumption patterns. However, this strategy is the most robust against long-term trends in demand growth, where the other strategies fall short.

Figure 10.4 shows strategy performance for SR. From 2010 to 2050, total CI reaches £968 billion driven by high investments into renewables and offshore energy supply infrastructure along with smart grid architecture, district heating and domestic demand management. There are also significant investments in the wastewater sector due to replacing existing sewer systems. In contrast, there are marginal investments into transport capacity expansion due to major changes in travel demand patterns over the long term stimulated by incentives to mode shift and reduce vehicle travel. Demand trends in the water sector also levels off requiring lower capacity investments in the long term. As a result, this strategy is the most resilient with stabilised or increasing capacity margins across all sectors by 2050. In addition, due to major changes in end-use demand and a focus on new service delivery models, this strategy is the only one to achieve long-term reductions in carbon emissions to 59 Mt by 2050.

Restructuring the energy sector requires a strong emphasis on meeting climate policies while balancing energy security concerns (Skea et al., 2010). A combination of centralised national incentive programmes and decentralised delivery mechanisms are deployed. This requires significant CI of £540 billion by 2050. Social acceptance of the need to decarbonise the energy system leads to strengthened national government policy, and greater decentralised action. Concerns about energy security continue and are addressed by large investments in energy efficiency, with a mix of low carbon energy sources, including microgeneration driven by significant carbon prices. Low carbon electricity generation and biomass technologies are widely adopted reducing grid carbon emissions to 10 Mt by 2050, a 95% reduction from current levels. Renewable technologies capture a high market share in the electricity supply mix, with gas-fired generation used at low load factors to provide continued flexibility in the face of high levels of intermittency. There is greater reliance on electrification of transport and domestic building services shifting away from natural gas. However, smart meters are rolled out and used effectively for both demand response and reduction. District heating plays a significant role, with local authority led initiatives in cities, largely supplying urban centres. Most schemes are based on large commercial and/or public sector heat loads, but also supply neighbouring residential areas, where the heat density is sufficient to justify this (Kennedy et al., 2014). Heat sources come from waste to energy schemes and waste heat from power generation, with some residual gas-firing (Eyre and Baruah, 2015). As a result, this strategy reduces total electricity demand by nearly 30% and has the highest system security of supply with a capacity margin of 70% by 2050.

Restructuring the transport sector requires major industry and policy incentives for mode shifting, vehicle kilometre reductions and diffusion of alternative fuelled vehicles (Anable et al., 2012; Tran et al., 2014). Without major road network capacity expansion and related infrastructure works, this strategy requires CI of £40 billion, half the amount for CE in 2050.

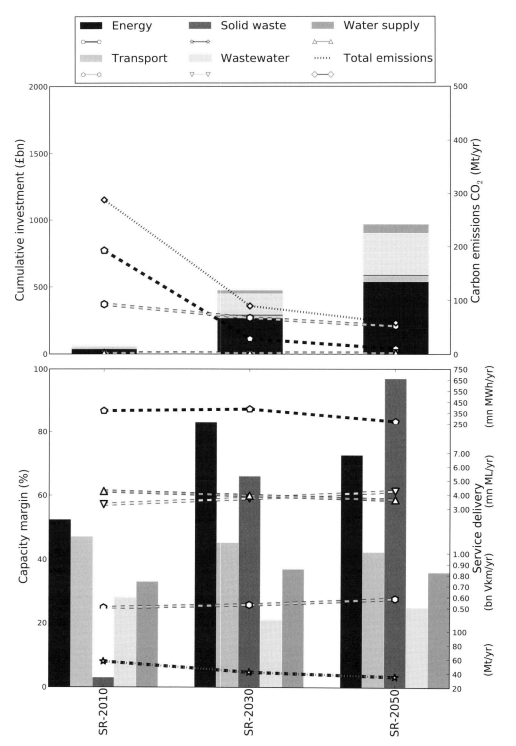

Figure 10.4 System restructuring (SR) strategy performance based on cumulative investment (£), per annum carbon emission (CO_2), capacity margins (%) and service delivery: energy (MWh/yr), water supply and wastewater (ML/yr), transport (VKm/yr), solid waste (Mt/yr) for 2010–2050. Note: bars relate to the left hand, and dashed lines relate to the right hand y-axis, respectively

A national program of measures to influence and alter travel behaviour and freight logistics using a variety of 'smarter choice' interventions to promote more sustainable travel are deployed (Cairns et al., 2004). Decentralised business models supported by large-scale ICT deployment and 'the internet of things' promotes travel substitution, influences workplace travel plans and deploys targeted discounts and promotional material to promote an increase in cycling, walking and public transport. Consequently, intra-zonal road congestion is reduced despite long-term population growth. Overall, a larger portfolio of mobility options supported by increasing urban densification and new business models that promote mobility services rather than vehicle ownership reduces vehicle travel. As a result, total vehicle kilometres increases marginally over the long term by 20% while maintaining a capacity margin of 42% in 2050. Reductions in fossil fuel based vehicle use results in a 43% reduction of carbon emission reaching 52 Mt in 2050 relative to current levels.

The water supply sector deploys a combination of national policies to promote consumer awareness to limit final demand, and stimulates innovative new delivery options. There is also significant investment to increase operational efficiency targeting network leakage reduction and high allowances for inter-company transfers and water re-use with a limit on the capacity of desalination plants. This requires nearly £67 billion in CI by 2050. However, the strong policy focus on demand reduction reduces final water consumption 15% reaching 3,640 GL/yr in 2050 relative to 2010, and maintains water security despite long-term population growth. This is shown by a capacity margin of 36% by 2050 similar to current levels. Over the same period, carbon emissions increase slightly to 2.2 Mt but are relatively minor compared to other sectors.

Restructuring the wastewater sector focuses on efficient recovery of water and nutrients from wastewater. There are also significant improvements in energy efficiency in wastewater processing. This is particularly important since stricter environmental standards increase energy costs. However, given the sunken investment in existing infrastructure and the expected life of that infrastructure, changes in wastewater management do not change rapidly in the near future. The combination of maintaining the existing infrastructure while increasing operational efficiency results in CI of £309 billion by 2050. Most of this investment is for the high replacement rate and storm water separation for sewers, along with increasing operational and energy expenditures for treatment plants. Process efficiency gains are made however due to storm water separation. Service delivery increases in line with population growth and the current capacity margin of around 25% is maintained over the long term.

Major changes in the solid waste sector are underpinned by a move towards industrial symbiosis where the wastes from one process provide the raw materials for another. This is combined with dematerialisation where raw material use is reduced while maintaining similar product functionality. Waste is also eliminated from all stages by design, and products are designed for reuse, refurbishment, repair and recycling (D4R). Extended producer responsibility (EPR) is also widely mandated (Lindhqvist, 2000; OECD, 2012). Landfill and incineration are largely phased out being retained primarily for disposal of hazardous wastes. These initiatives result in CI of £12bn and generate negative carbon emissions of nearly 10 Mt by 2050. There is also a 40% reduction in waste arisings resulting in major additional capacity of nearly 100% over the long term. These structural changes are supported by major behavioural shifts away from consumerism towards leasing and product service systems (Mont, 2002; Baines et al., 2007).

10.4 Total system performance

Figure 10.5 compares the comparative performance between the four cross-sectoral strategies from 2010 to 2050. There are important differences in CI between strategies with a range of £669 billion to £1.8 trillion by 2050 for MI and CE, respectively. Over the same time, CI for SE is £756 billion and SR is the second highest reaching £968 billion. Investment is dominated by energy supply across all strategies, except for CE where massive investments are made in transport infrastructure, resulting in steadily increasing road usage and carbon emissions. Conversely, SE has the potential to incur only half the investment cost as CE while emitting 36% less carbon emissions. Not surprisingly, SR incurs substantial investment costs over the long term, second only to CE, but provides the basis for a major configuration of infrastructure during the twenty-first century.

There are also important differences in future environmental performance between strategies. In 2050, the best performing strategy is SR with carbon emissions of 59 Mt/yr compared to nearly 388 Mt/yr for MI. In the medium term, CI in SE and MI are similar around £350 billion by 2030. However, with relatively smaller increases in investment compared to other strategies, SE achieves greater carbon emissions reduction by 2050. Importantly, all strategies improve environmental performance by 2030 except for MI. While environmental performance gains are made over the medium term, by 2050 only SE and SR appear to be robust against long-term population and economic growth trends, with only SR showing a continued decline in emissions.

In terms of system security of supply, MI is the least secure with declining capacity margins across all sectors except for solid waste, where there has been historical decoupling of per capita waste arisings. In the medium term, CE has the largest positive security effects in the water supply sector, but due to limited demand reduction has marginal security effects for all other sectors. CE also has a large impact on releasing latent demand in the transport sector with a near doubling of services measured by vehicle kilometres, and is also a major source of carbon emissions. Consequently, our results suggest that the conventional approach of physical capacity expansion to meet rising demand and maintain system security of supply is not a robust long-term strategy. In contrast, SE actively targets demand side measures combined with retrofitting existing infrastructure networks to increase throughput of the system. As a result, capacity margins improve and services are generally maintained to 2030 with aggregate demand falling in energy and solid waste by 2050. However, from 2030 onwards there is a slow increase in total carbon emissions due to long-term population growth exceeding efficiency gains. Therefore, SR appears to be the most robust strategy against long-term population and economic growth. It is able to maintain current capacity margins and service delivery levels across most sectors with aggregate demand reductions in energy, water supply and solid waste by 2050.

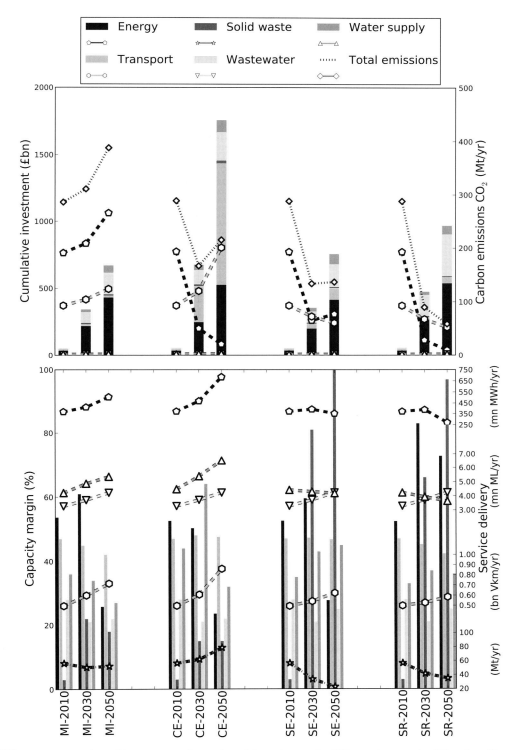

Figure 10.5 Comparative analysis of the cross-sectoral strategies: minimum intervention (MI), capacity expansion (CE), system efficiency (SE), system restructuring (SR). Multi-attribute performance based on cumulative investment (£), per annum carbon emission (CO_2), capacity margins (%) and service delivery: Energy (MWh/yr), water supply and wastewater (ML/yr), transport (VKm/yr), solid waste (Mt/yr) for 2010–2050. Note: bars relate to the left hand, and dashed lines relate to the right hand y-axis, respectively

Importantly, it is the only strategy that sees an increase in energy security over the long term due to major demand reductions and a balanced investment portfolio of new low carbon technologies and energy efficiency.

In the short to medium term, the key message is that all sectors need to invest heavily to improve system security of supply, continue service provision and reduce carbon emissions to meet economic and environmental policy goals. However, over the long term, each strategy performs quite differently allowing a distinction of the policy choices that are available.

10.5 Conclusions

This chapter has examined cross-sectoral performance of four distinct and contrasting strategies for infrastructure in Britain, integrating energy, transport, water supply, wastewater and solid waste sectors. Major innovations in ICT infrastructure are also implicit in these strategies and examined in Chapter 9. Here we demonstrate how national-scale cross-sectoral strategies for infrastructure systems can be constructed, analysed and appraised in terms of a range of performance metrics. The results demonstrate how there are genuine choices in infrastructure provision for the twenty-first century. Importantly, we show how the policy choices we make now can influence the future performance of the system.

The MI strategy delivers incremental change to the overall system, and marginal impacts on long-term performance. Capacity provision increases incrementally with minor prioritisation of regional demand trends to meet short-term demand growth. Demand continues to rise with increasing capacity constraints across all regions in the medium to long term. Investments increase following historical trends, and carbon emissions steadily rise due to insufficient investment resulting in poor quality services and continued rise in demand.

The CE strategy delivers large-scale enhancement of the current system, with improved performance over the short to medium term. However, it is less robust over the long term due to physical capacity limitations, increasing demand and lock-in to conventional technology and design. Capacity rapidly increases over the short to medium term across most regions, but is followed by demand increases in the most densely populated areas over the long term. Demand is met in the short to medium term, but over the long term there are increasing bottlenecks due to physical limits in capacity expansion. Investment increases dramatically over the entire period. Carbon emissions reduction may be achieved in the medium term from improved infrastructure performance, but rise over the long term due to increasing build and population growth. Sectoral interaction occurs but without a coordinated investment approach that targets cross-sectoral interdependencies, performance gains in one sector (energy) are lost in other sectors (transport).

The SE strategy delivers moderate change to the current system depending on where and when efficiency gains can be achieved at least cost. The strategy performs well over the medium term, but may be less robust over the long term if continued demand growth outpaces efficiency improvements. Capacity provision increases over the short to medium term but far less than for CE over the long term. Demand is met in the short to medium term, but over the long term there could be bottlenecks in the highest growth regions due to long-term demand trends surpassing efficiency improvements. Investments also increase substantially over the short to medium term, but decline over the long term. Carbon emissions decrease in the medium term due to efficient transport networks and high investments in renewable power generation, but over the long term begin to rise.

The SR strategy delivers a fundamentally restructured national infrastructure system over the long term. It could perform unevenly in the short to medium term depending on specific regional and local circumstances, but it is likely to be the most robust strategy across all regions over the long term due to major reductions in demand, but also faces a high degree of investment uncertainty. Capacity provision increases over the short to medium term but is distributed unevenly across regions depending on where economies of scale can be achieved. Demand steadily increases in the short to medium term resulting in bottlenecks in high growth regions, but reduces significantly over the long term across most regions. Investments increase substantially over the entire period without pay back until the medium to long term. Carbon emissions could decouple from economic growth over the long term due to the major structural redesign and sustained reductions in per capita demand.

These four national infrastructure strategies capture many of the major choices that society faces, but they are by no means exhaustive, and many other variants could be constructed and analysed using these methods. The four strategies presented here could be seen as an initial scoping, which might guide future strategy selection and analysis. Nonetheless, the results demonstrate the contrasting possible outcomes and emphasise that over a forty-year timescale, the decisions we make and the strategies that we employ today will have an important role in determining the multiple attributes of our future infrastructure systems. Chapter 11 continues this analysis and looks at examples of possible interactions between infrastructure sectors.

References

Anable, J., C. Brand, M. Tran and N. Eyre (2012). "Modelling transport energy demand: A socio-technical approach." *Energy Policy* 41(Feb): 125–138.

Baines, T. S., H. W. Lightfoot, S. Evans, A. Neely, R. Greenough, J. Peppard, R. Roy, E. Shehab, A. Braganza, A. Tiwari, J. R. Alcock, J. P. Angus, M. Bastl, A. Cousens, P. Irving, M. Johnson, J. Kingston, H. Lockett, V. Martinez, P. Michele, D. Tranfield, I. M. Walton and H. Wilson (2007). "State-of-the-art in product-service systems." *Proceedings of the Institution of Mechanical Engineers, Part B: Journal of Engineering Manufacture* 221(10): 1543–1552.

Baruah, P., N. Eyre, M. Qadrdan, M. Chaudry, S. Blainey, J. W. Hall, N. Jenkins and M. Tran (2014). "Energy system impacts from heat and transport electrification." *Proceedings of Institution of Civil Engineers – Energy* 167(3): 139–151.

Cairns, S., L. Sloman, C. Newson, J. Anable, A. Kirkbride and P. Goodwin (2004). *Smarter choices: changing the way we travel*. London, UK, Department for Transport.

Eyre, N. and P. Baruah (2015). "Uncertainties in future energy demand in UK residential heating." *Energy Policy* 87: 641–653.

Francis, R. and B. Bekera (2014). "A metric and frameworks for resilience analysis of engineered and infrastructure systems." *Reliability Engineering & System Safety* 121: 90–103.

HM Treasury (2013). *National Infrastructure Plan 2013*. London, UK, HM Treasury.

Kennedy, C. A., N. Ibrahim and D. Hoornweg (2014). "Low-carbon infrastructure strategies for cities." *Nature Climate Change* 4(5): 343–346.

Law, F. (1955). "Estimation of the yield of reservoired catchments." *Journal of the Institution of Water Engineers* 9: 467–493.

Lindqhvist, T. (2000). Extended producer responsibility in cleaner production: policy principle to promote environmental improvements of product systems, Doctoral Dissertation, University of Lund.

Mont, O. K. (2002). "Clarifying the concept of product–service system." *Journal of Cleaner Production* 10(3): 237–245.

OECD (2001). *Extended producer responsibility: a guidance manual for governments*. Paris, France, Organisation for Economic Co-operation and Development.

RAE (2013). GB electricity capacity margin. A report by the Royal Academy of Engineering for the Council for Science and Technology. London, UK.

Skea, J., P. Ekins and M. Winskel, Eds. (2010). *Making the transition to a secure low-carbon energy system*. Earthscan.

Tran, M., C. Brand and D. Banister (2014). "Modelling diffusion feedbacks between technology performance, cost and consumer behaviour for future energy-transport systems." *Journal of Power Sources* 251(Apr): 130–136.

Quantifying interdependencies: the energy–transport and water–energy nexus

MARTINO TRAN, EDWARD A. BYERS, SIMON P. BLAINEY, PRANAB BARUAH,
MODASSAR CHAUDRY, MEYSAM QADRDAN, NICK EYRE, NICK JENKINS

11.1 Introduction

Planning infrastructure investment has historically been done in each infrastructure sector in isolation. However, infrastructure sectors are becoming increasingly interdependent due to technological advancements, climate policy and increasing cross-sector demands. Infrastructure interdependencies introduce layers of complexity, uncertainty and risk to systems planning and design. Infrastructure is rapidly shifting from unconnected structures to interconnected networks. This shift has important implications for the resilience and sustainability of the system. Even small, temporary failures can have significant effects on economic productivity. In the long term, these risks intensify as systems become larger and increasingly interdependent. The combined effect of ageing infrastructure, growing demand (nearing capacity limits) from social and economic pressures, interconnectivity and complexity can weaken infrastructure resilience and sustainability (Francis and Bekera, 2014).

Importantly, changing patterns of demand can influence different infrastructure sectors in similar ways, providing a further source of interdependence in the long term. For example, reducing household water use can alleviate pressures on fresh water supply while decreasing the energy used for domestic water heating. Moreover, one infrastructure sector can be a major component of demand for another sector, where the transport sector, for example, represents 34% of energy demand in the Britain, whilst electricity generation is responsible for 32% of all non-tidal water abstractions (Defra, 2009). Future demand for infrastructure services will also impact upon the environment creating interdependency between sectors. Investment in water treatment may lead to quality improvements in rivers and coastal areas, but could then be offset by thermal cooling water discharges from future electricity production. While large electric vehicle fleets have the potential to mitigate environmental and carbon emissions from transport, this is dependent upon a commitment to investment in low carbon power generation. National infrastructure consumes over half the energy supply in most advanced economies and is a major source of carbon emissions (IPCC, 2014). Therefore, carbon emissions reduction in one sector may be offset by increases in another sector. Meeting increasing environmental standards and broader sustainability objectives therefore requires investment decisions that account for sector interdependencies.

Extensive research has been done on the failure of interdependent critical infrastructure to deliver essential goods and services over the immediate to short term with important consequences for security and resilience (Ouyang, 2014). This issue is addressed in Chapter 12 of this book. However, less is known about how strategic decision-making can influence sectoral interdependencies and the resulting long-term consequences. Here, we show how climate mitigation policy in one sector can be dependent upon investment decisions taken in other sectors. Specifically, we assess interdependencies that emerge from pursuing electricity as an alternative energy vector (Williams et al., 2012) and the cascading demand effects on the transport, energy and water sectors. Large-scale electrification has become an important climate mitigation pathway but will increase interdependency between sectors (energy, transport, buildings and water sector) for which the long-term sustainability impacts are not well understood. To explore these implications we provide two case-studies focusing on (i) energy–transport and (ii) water–energy interactions in Britain. In doing so, we provide fully integrated modelling results on energy, transport and water sector demand interdependencies and assess the implications for climate mitigation, economy and environment. As with Chapter 10, the assessment is primarily done in the context of the central scenario for population growth, economic growth, energy prices and climate change. Sensitivity analysis with respect to other scenarios has been described in the sector chapters (Part II).

11.2 The energy–transport nexus

Power generation and transport are the world's largest contributors to CO_2 emissions at 40% and 24%, respectively. Due to the transport sector's ~98% dependence on fossil fuels, mass deployment of battery electric vehicles (BEVs) has become an important carbon reduction strategy. However, high carbon reduction benefits depends on global average electricity carbon intensity reaching <100 gCO_2/kWh by 2050 from a current average 460–550 gCO_2/kWh (IEA, 2010). Most regions in the world do not currently generate sufficient low CO_2 electricity to capture the emissions reduction potential of BEVs. There is also uncertainty on additional generation capacity requirements for large BEV fleets. While national studies indicate that additional capacity is not required over the next five to ten years, long-term requirements are far more difficult to predict and will depend on the future grid mix.

Impacts on peak demand are also not well understood, especially when accounting for future growth of household energy demand. While BEVs have the potential to manage peak load, that assumes smart grids and meters are also in place (Bishop et al., 2013) thus increasing sectoral interdependencies between transport, power generation and household demand (Baruah et al., 2014). Decarbonising the transport sector through mass electric vehicle deployment will require sufficient low carbon electricity over the long term. This depends upon investment into additional electricity generating capacity, with major differences in grid carbon intensity depending on different investment strategies. Power investment decisions taken over the next ten years will therefore lock in CO_2 emissions for the next forty to fifty years. However, most industrialised countries are now entering a new investment cycle in power generation, representing an opportunity to deploy more clean and efficient generation technologies.

A key interdependency therefore lies between transport and its future energy needs, particularly if a substantial increase in electricity generation is needed to power electric vehicles and domestic energy use. While the majority of electric vehicle recharging is likely to take place at home, it also seems likely that substantial battery recharging infrastructure will be needed in the field (e.g. at garages, supermarkets, workplaces, etc.). Moreover, electrification of the transport sector could require large investments in national transmission and local distribution networks. The energy and transport models described in Chapters 4 and 5 of this volume have been used to explore these important issues of interdependence between energy and transport. We focus in particular on the electrification of heat and transport (EHT) strategy that was introduced in Chapter 4. This strategy highlights important interdependencies between the energy (including space heating) and transport sectors in order to meet energy and climate policy objectives in Britain.

11.2.1 Case study of Britain

Britain has developed forecasts for BEV and plug-in hybrid electric vehicle (PHEV) deployment from 2010 to 2030. In the government's business-as-usual forecast, BEVs increase from 3,000 to 500,000 and PHEVs increase from 1,000 to 2.5 million. Over the same period, the high growth forecast implies growth of BEVs from 4,000 to 3.3 million and PHEVs from 1,000 to 7.9 million (DfT, 2008). Reflecting this level of ambition transport strategies were developed in Chapter 5 showing high penetrations of PHEVs and BEVs with a range of <10% to 80% by 2050. The resulting total electricity consumption and peak load from these electric vehicles are then assessed based on three different energy demand strategies including minimum policy intervention (MPI) used as a reference case, local energy and biomass (LEB) representing a low electrification strategy and EHT representing a high electrification strategy (see Chapter 4).

As a consequence, total electricity demand could nearly double from current levels reaching ~690 TWh in 2050 assuming an 80% penetration of electrified vehicles (see Chapter 4, Figure 4.6). That would result in the transport sector's share of electricity consumption increasing to ~12% of the total in 2050 from <1% in 2010. However, the low electrification energy demand strategy (LEB) can potentially suppress electricity demand to 2010 levels, but also assumes a major systems transformation underpinned by rapid and ambitious implementation of efficiency measures, and high penetrations of photovoltaic (PV) and combined heat and power (CHP).

Chapter 4 also assesses the range of peak electricity demand resulting in ~70 and 150 GW for LEB and EHT strategies, respectively. The EHT strategy assumes that during peak hours 20% of BEVs are connected for charging and 10% provide vehicle-to-grid (V2G) demand response. Additional sensitivity analysis also indicates that increasing the V2G connection to 20% can decrease system peak load by ~7% (Baruah et al., 2014). Although the LEB strategy uses 27% less total electricity than the reference case (MPI) due to aggressive efficiency measures and fuel switching, its peak electricity demand is only 7% below MPI's due to high penetrations of electric vehicles and heat pumps.

While the switch to electric vehicles will have a beneficial impact on air quality and CO_2 emissions, the latter will depend on the future fuel mix for electricity generation. To meet increasing electricity demand, supply-side investment strategies are also developed showing a large variation in technology deployment to meet long-term electricity demand including nuclear, carbon capture and storage (CCS) and offshore renewables. Results indicate that the capacity of combined cycle gas turbines (CCGTs) in 2050 is high in all strategies (Chapter 4, Figure 4.11). CCGT capital costs are lower than coal plants, and therefore always more economic at low load factors. In strategies where there is a carbon price floor (currently £16/tCO$_2$ rising to £30/tCO$_2$ by 2020), coal plants are also relatively expensive to operate. CCGT capacity in the high nuclear and offshore strategies is exceptionally high. This is in part to deal with inflexible (nuclear) and variable (wind, wave, etc.) generation and to meet peak demand.

The large amount of capacity in the high electrification (EHT) strategy in comparison with the reference case (MPI) is due to increasing EHT. This results in high annual energy and peak demand requirements almost two times larger than in the reference strategy. Generation capacity is even higher (>300 GW) in the offshore EHT strategy due to the variable output of wind turbines. This means that decarbonising the transport sector along with electrifying domestic energy use via heat pumps requires sufficient low carbon electricity over the long term. This depends upon investment into additional electricity generating capacity, with important differences in generation mix depending on which energy supply strategy is taken.

The power generation investment strategy taken to meet increasing electricity demand from transport has important economic and environmental implications. From 2010 to 2050, cumulative investment in power generation for decarbonising electricity generation from offshore renewables is £1 trillion, around double the cost of nuclear at £500 billion and CCS at £600 billion. However, the highest level of carbon emissions reduction is achieved with nuclear (50 gCO$_2$/kWh), followed by offshore renewables (90 gCO$_2$/kWh) and CCS (105 gCO$_2$/kWh). This indicates the potential trade-offs between strategies to achieve low carbon grid intensity, while meeting future electricity demand.

11.2.2 Key messages

The three strategies (MPI, EHT and LEB) that have been adopted in this assessment are contrasting, posting a stark choice about the direction in which the energy infrastructure system should be taken. A high electrification (EHT) strategy provides large energy and carbon reduction opportunities, both by increasing efficiency by fuel switching and greater potential for supply decarbonisation. However, EHT requires new infrastructure and the security of supply from generation diversity can decrease with high dependence on the electricity system. Large-scale electrification of transport can have major infrastructure implications. For example, the transport sector would require new electric fuelling infrastructures, the electricity grid would have to be upgraded (smarter grid, new transmission lines and distribution reinforcements) and additional generation capacity would be needed. With decreasing gas use, alternative usage or decommissioning of existing gas infrastructure has to be addressed. Even aggressive demand reduction, if coupled with moderate electrification of transport, mean that the requirement for electricity infrastructure will be at least as high as at present.

Moreover, part of the technical and economic rationale for increasing the uptake of BEVs and PHEVs is to balance the grid due to high penetrations of intermittent renewables. This however requires a strategic commitment to investment in renewables, which may have long-term decarbonisation benefits over the as yet unproven CCS and unresolved concerns over safety for nuclear. Total investment will vary depending on national context, but the relative differences between investment strategies will be similar due to the global nature of technology learning curves. Therefore, our methods could be applied outside Britain to infrastructure investment decisions for other nations seeking to decarbonise their energy and transport systems.

11.3 The water–energy nexus

The energy system depends on water resources for power generation, fossil fuel processing and irrigation of biofuel feedstock crop. Hydrological resources are particularly vulnerable to physical constraints, climatic variability and rising population demand. In 2010, water withdrawals for energy production was $\sim$580,000 GL/yr comprising 15% of the world's total water withdrawals and is expected to increase to 20% by 2035. From the total withdrawal $\sim$66,000 GL/yr was used for consumption and this is projected to increase 85% over the next twenty years (IEA, 2013).

Globally, some 80% of electricity production now comes from thermo-electric power stations, the vast majority of which are cooled using water. The amount of water used depends primarily on the cooling system used, whilst the amount of cooling required depends on the thermal efficiency of the plant. Historically, power stations were sited on rivers from which water was abstracted for direct cooling, and returned at a higher temperature. Although abstraction volumes are high, consumption of water is low ($\sim$1%). Although trends towards more efficient power plants with advanced cooling systems can reduce total withdrawals, water consumption per unit of electricity produced is likely to increase, particularly with CCS and the trend away from once through cooling systems.

Hydrological variability poses a particular risk to electricity generation that is dependent on water. Hence, power plants (both hydro and thermo-electric) are usually situated on bodies of water with reliable yields. However, increasing demands from population growth and other industries may contribute to water scarcity that will likely be exacerbated by climate change. Reduced runoff, particularly during summer and autumn months and coupled with higher air temperatures could increase the likelihood and severity of droughts and low flows (Byers et al., 2014).

11.3.1 Case study of Britain

Power generation expansion to meet increasing electricity demands from transport and domestic end use also has important interactions with the water sector. In Britain, the majority of thermal power plants use water for cooling, in addition to existing hydro and pumped storage capacity. Currently the electricity sector is responsible for $\sim$40% of non-tidal surface water abstractions in England and Wales (Byers et al., 2014). This figure however includes hydropower and pumped storage. The proportion used for cooling of thermoelectric power stations is closer to 3–5%, approximately 160 GL/yr in 2010 (Byers et al., 2014; Tran et al., 2014).

Using the electricity generation strategies presented above, we now show the impacts on freshwater, tidal and coastal water resources in Britain. For electrical power distribution, a geographical area can be identified as a 'busbar' region. For each busbar region, trajectories of future electricity generation are multiplied by abstraction and consumption water use factors and split by distributions that allocate generation to freshwater, tidal waters or seawater. Figure 11.1 shows unconstrained demands for fresh water abstraction and consumption from each electricity generation strategy aggregated over all power generation sites per region (busbars).

In the reference case (MPI NoCC) both abstraction and consumption of freshwater remains fairly stable through to 2050 due to minor changes in the generation mix. The other four strategies have a sharp reduction in abstractions and consumption towards 2020 due to closure of less efficient capacity from the EU Large Combustion Plant Directive (LCPD) (European Parliament and Council of the European Union, 2001). The high water intensity of CCS equipped generation gradually increases abstraction and consumption. Despite having dropped to a quarter of current levels by the 2020s, water use surpasses current levels at around 2035 and is twice the current level in 2050. With the anticipation that there could be limited freshwater supplies in the future, for coal power to have a future in the generation mix of Britain, its generation will need to be not only low-carbon through the use of CCS, but also sited on tidal or coastal sites where water for cooling is not scarce. The high water intensity of coal power with CCS (approximately double and quadruple that of unabated coal and gas power, respectively), in combination with the national CCS Roadmap strategy (DECC, 2012) that encourages clustering of CCS infrastructure, is likely to increase localised water demands in industrial areas (Byers et al., 2014; DECC, 2012).

For tidal sites, cooling water abstractions are dominated by the use of once through cooling, particularly for CCS and nuclear generation. Hence, those strategies have increasing levels of tidal water abstraction and consumption, with 2050 levels at approximately 55% higher than 2010 (Figure 11.2). In the reference case strategies with and without carbon costs (MPI) tidal water use remains at similar volumes, whilst investment in offshore renewables decreases water abstraction and consumption by 2050 to approximately 15% of the 2010 volumes.

Seawater abstractions, in all cases, are expected to increase (Figure 11.3), due to a variety of reasons, such as freshwater constraints and capital and operational cost efficiencies. There is a seven-fold increase in abstraction of seawater for cooling in the nuclear strategy, compared to changes of −35% to +119% in the other four strategies. Whilst the high usage of sea and tidal water in itself is not problematic, this can be considered a proxy for environmental impact on aquatic ecology due to the heated cooling water discharges.

For the reference case (MPI-NoCC) and high CCS strategy the continued presence of coal power in the generation mix leads to considerable increases in freshwater abstraction and consumption for busbars 9 and 16 – equivalent to the river basin regions of Dee & NW England, and the Thames/London region. For the reference case (MPI-CC), nuclear

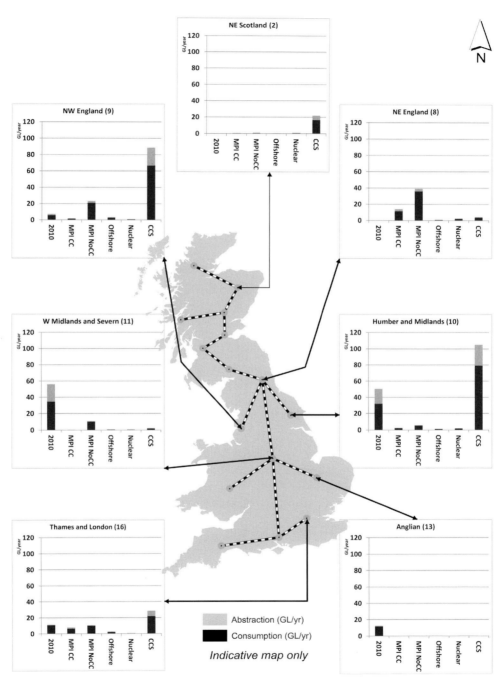

Figure 11.1 Freshwater abstraction and consumption in 2010 and 2050 for each energy strategy disaggregated by region (GL/yr). Note: High offshore (Offshore); high CCS (CCS); high nuclear (Nuclear); MPI with carbon cost (MPI CC); MPI no carbon cost (MPI no CC). Note: Demands for water abstraction and consumption were calculated using the factors listed in Macknick et al. (2012a) and following similar methods to those used in Macknick et al. (2012b) and Byers et al. (2014)

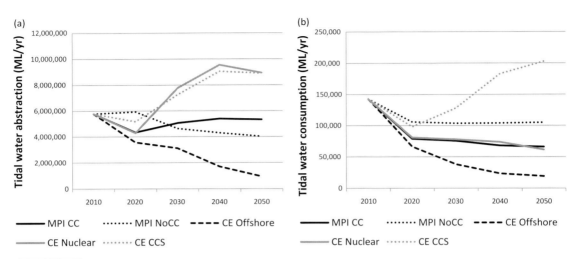

Figure 11.2 Tidal water abstraction and consumption. Due to the use of once-through cooling, the proportion of consumption is much lower

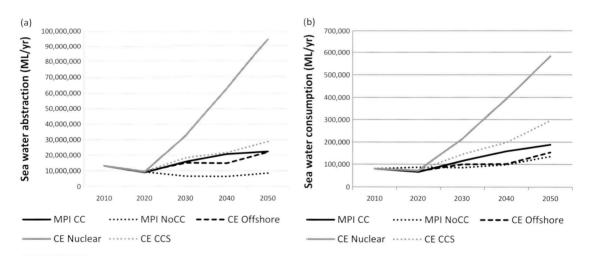

Figure 11.3 Seawater abstractions and consumption. Consumption is very low due to once-through cooling, which results in heated water discharge with possible ecological impacts

and offshore renewable strategies, freshwater abstraction and consumption is reduced significantly in all the regions, besides the demand for development (MPI-CC) in the Thames region which sees a reduction followed by an increase approaching 2050.

Our assessment of regional water resources considered what available flows were available in the largest rivers at a Q_{95} level. A Q_{95} is the fifth percentile flow and indicator of low flows – it represents the flow level exceeded 95% of the time for the duration of the historical flow record. We have assumed that the proportion of licensed abstractions for the

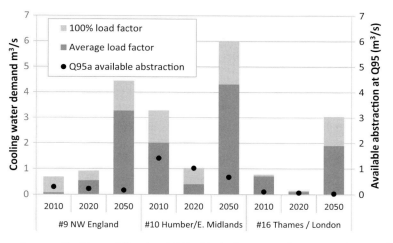

Figure 11.4 Abstractions under normal load factors (LF) and at 100% load for busbars 9, 10 and 16, compared against the adjusted available Q95 flow (Q95a), for the CCS strategy in 2010 and 2050

sector does not greatly exceed proportions that are currently held by the sector. Furthermore, levels of capacity permitted on freshwater in each busbar was limited to a maximum of 0–40% of the whole busbar capacity in 2050, depending on the capacity type and busbar, whilst the rest was allocated to tidal and sea water. The penetration of water-conserving hybrid cooling on freshwater in 2050 was 30% for coal and gas-based technologies, with an additional 10% of air-cooled capacity. The assessment was made with respect to both historical Q_{95} flow levels, as well as Q_{95} for the 2020s and 2050s medium emissions climate scenarios from UKCP09 (Byers et al., 2015).

Considered on an instantaneous basis (in m^3/s), the CCS strategy has levels of generation capacity allocated to busbars 9, 10 and 16 (northwest England, Humber/E. Midlands and Thames/London, respectively), whose abstractions would likely exceed the available water at Q_{95} in 2050 (Figure 11.4). This is the case both at average load factor and if 100% of capacity are simultaneously abstracting. Based on the proposed abstraction reform, capacity would likely have to reduce abstraction at flows below this point unless greater allocations are secured from other license holders. Currently, almost all power stations hold 'unconstrained licenses', hence the indication that the current capacity is already over abstracting at Q_{95}. The MPI-NoCC strategy also indicates lower levels of over-abstraction in busbars 9 and 16, but to a much lesser degree than the CE CCS strategy.

In terms of the impacted capacity due to low flows in 2050, power generation capacities of 6.2, 6.1 and 3.2 GW would potentially face restrictions in busbars 9, 10 and 16, respectively. This would represent 4%, 4% and 2% of total installed thermo-electric capacity in 2050 (167 GW), or cumulatively 10% if all regions were simultaneously experiencing the same low flows.

Excessive abstractions for these two strategies may be mitigated through a variety of measures, including, higher penetrations of dry and hybrid cooling, higher allocations to tidal and sea water sources, or development of capacity in other regions. Alternative cooling water sources may also be obtained, such as treated wastewater, a practice increasingly common in the US (Veil, 2007). Increased use of combined heat and power (CHP) would also reduce cooling demands.

11.3.2 Key messages

For Britain, freshwater abstractions and consumption for energy generation cooling can be expected to decrease, primarily due to the closure of current coal capacity. The Large Combustion Plant Directive and decommissioning of current coal and nuclear capacity puts Britain on a sustainable, low-water trajectory for electricity generation. However, freshwater abstractions and consumption will significantly increase if CCS-equipped generation capacity is aggressively pursued. Conversely, a commitment to offshore renewables is the best alternative for minimising all types of water use and impacts on freshwater aquatic environments. Strategies which minimise freshwater consumption but have higher levels of tidal and seawater use may also be good alternatives (reference case and nuclear), if they can operate within acceptable local environmental constraints. The analysis assumed regional limits of capacity development on freshwater sources, in tandem with increased penetration of hybrid cooling. Without these constraints it is expected that demands for freshwater could possibly exceed available resources at low flows in certain regions (busbars 8–16), regardless of the expected impacts of climate change on hydrological resources. Only the most water efficient generation capacity should be permitted to use freshwater for cooling, whilst more water intensive technologies should be limited to using tidal and seawater.

It is expected by consumers, industry, the government and the regulators that the electricity sector maintains extremely high levels of service reliability, especially when considered as a unit or conglomeration of units and networks. Reliability of individual power stations is not expected to be as high, as there is built in redundancy to account for sudden loss of generation units or periodic maintenance cycles. Nonetheless, the spatial nature of water scarcity means that scarcity in a region or catchment may affect several generation units, and if there is a lack of cooling water, generation may need to be reduced or stopped completely. Similarly, due to the consumptive nature of thermo-electric generation, it is in the water regulator's interest to limit the levels of generation capacity that is developed in a river basin or catchment, in order to prevent over abstraction and maintain a reliable resource of water for an appropriate number of users. Over abstraction may lead to shortages not only for the energy sector but also for other sectors, and may result in environmental damage and breach of environmental regulations, such as the EU Water Framework Directive (Byers et al., 2014).

Energy and water decisions have historically been made independently of each other. Water planners typically assume that they have sufficient energy, while energy planners assume that they have the water that is needed to maintain operations. Both are likely to use different strategic planning, for example, private companies acting under market forces dictate the location of energy infrastructure, while water infrastructure is often located using public interest criteria. A mismatch in planning objectives by different actors can prevent the beneficial siting and combining of technologies. Here we show how policy and investment strategies are needed to encourage the water and energy sectors to move toward integrated resource management, particularly if climate and environmental objectives are to be met.

11.4 Global implications

Planning for sectoral interdependencies is important because infrastructure investment decisions taken over the next ten years will lock in scarce economic resources, environmental impacts and CO_2 emissions for the next forty to fifty years or longer. Industry investment and government planning will have to account for interactions between sectors and the underlying drivers of how this will change over time including consumer markets, technological performance and physical infrastructure build. Global investment into energy infrastructure and technology alone will be in the trillions of US dollars over the next twenty years (IEA, 2010). Massive additional investment will also have to be made to decarbonise the energy and transport sectors through high levels of electrification. The rapid diffusion of advanced energy and transport technologies will not be realised without sustained industry investment and government support. For instance, the climate mitigation and energy security benefits of large electric vehicle fleets can only be realised if there is adequate investment into low carbon electricity supply. Moreover, investment into charging and related infrastructure will have to precede large-scale vehicle commercialisation to increase consumer confidence in new unproven technologies. Yet the diffusion of new electric drive technologies will be shaped by a host of factors including socio-economic and demographic trends towards urban centres, which is not likely to be uniform across regions or cities. Policy formulation will therefore require new stakeholder relationships between local governments and the power and automobile sectors representing different interests that will overlap for the first time.

Our results also have global implications for water and energy policy interactions, particularly regarding CCS. Globally, 80% of electricity production comes from thermoelectric power stations, half of which are coal (IEA, 2009). In primary energy terms, the usage of natural gas, coal and uranium are all expected to double by 2050, as will associated water use if the issue is not addressed more thoroughly (World Energy Council, 2010). CCS remains promising, but has not been applied at operational scales. If CCS technology becomes widely available, take up in countries highly dependent on coal (such as China and India), will see large increases in water consumption. Power stations are often developed relatively close to population centres, yet these are also areas that usually have a high incidence of water scarcity. CCS technology effectively doubles the water-use intensity of electricity generation, although simultaneously reducing emissions by 80–90%. CCS technology is potentially a widely deployable technology that could rapidly reduce global emissions from electricity production, particularly in developing and rapidly industrialising countries. The water impacts of continued coal lock-in however, need to be studied in greater detail to ensure that risks to water resources are managed.

According to Pan et al. (2012), just under a quarter of China's water consumption is for industrial uses, more than half of which is for coal-related activities, including extraction, processing and coal-fired electricity generation. Similar quantities of wastewater are produced, in addition to aquifer pollution, linking coal to not only high consumption, but also significant quality impacts on supply. In recent years China and the US have taken regulatory steps to address the levels of cooling water used in electricity production and industrial facilities. More coordinated action will be required for the development of CCS clusters at industrial facilities that would result in concentrated demands for water. Siting of power stations on the coast has efficiency benefits, but also ecological impacts from heated water discharges and additional risk management required for coastal flooding and sea level rise. Furthermore, not all countries have coasts and there is sometimes competition for coastal land.

The Thirsty Energy programme of the World Bank is working with governments to move towards better cross-sectoral planning between energy and water (Rodriguez et al., 2013). Having identified developing countries as most vulnerable to water–energy pressures, the initiative aims to ensure that in providing reliable electricity to the 2.5 billion people currently without, pressures on water resources are not unduly increased for the 2.8 billion people living in areas of high water stress. Overall, a far greater number of studies are available for hydroelectric power and biomass cultivation, the impacts of which are more complex and highly politicised. However, as water resources become increasingly constrained, what is certain is that there is a growing impetus for more water-efficiency throughout the energy sector. Not only can this reduce scarcity risks and increase resilience to climate change impacts, it tackles access to water and energy challenges that are featuring prominently in the agendas of the Post-2015 and Sustainable Development Goal movements.

11.5 Conclusions

Society will continue to demand environmental improvements with important implications for infrastructure interdependency. The long-term impacts of climate change will also amplify interdependency risks, and should be considered in mitigation and adaptation measures. Of primary importance is the continued provision of essential energy, water and transport services while minimising harmful environmental impacts. However, the challenges that arise from increased interdependency will threaten the provision of those services, and potentially exacerbate environmental impacts.

It is therefore essential to take a long-term and cross-sectoral view in planning for current and future infrastructure systems. New infrastructure often has a long lifetime of fifty to one hundred years, thus current investments define the infrastructure of the future. Whilst a long-term view helps ensure meeting current and future service demand, anticipating demand is challenging due to the high degree of uncertainty in the long term. Moreover, infrastructure provision can encourage patterns of development and land use that become irreversible. Choices about technologies lock in patterns of behaviour and economic activity, and complex sector interdependencies can intensify future uncertainty. Hence, when predicting future demand for a given infrastructure sector, the demands from other sectors must be considered, for instance transportation services to provide fuel for the energy sector, or energy requirements for the water sector. Thus, evaluating the demand for a given sector in the long term requires a coordinated planning effort across infrastructure sectors to balance these interdependencies. Given the long-term nature of infrastructure provision and complex inter-sectoral interdependencies, it is therefore necessary to define broad cross sector strategies while accounting for sector specific interdependencies.

References

Baruah, P., N. Eyre, M. Qadrdan, M. Chaudry, S. Blainey, J. W. Hall, N. Jenkins and M. Tran (2014). "Energy system impacts from heat and transport electrification." *Proceedings of Institution of Civil Engineers – Energy* 167(3): 139–151.

Bishop, J. D. K., C. J. Axon, D. Bonilla, M. Tran, D. Banister and M. D. McCulloch (2013). "Evaluating the impact of V2G services on the degradation of batteries in PHEV and EV." *Applied Energy* 111(Nov): 206–218.

Byers, E. A., J. W. Hall and J. M. Amezaga (2014). "Electricity generation and cooling water use: UK pathways to 2050." *Global Environmental Change* 25: 16–30.

Byers, E. A., M. Qadrdan, A. Leathard, D. Alderson, J. W. Hall, J. M. Amezaga, M. Tran, C. G. Kilsby and M. Chaudry (2015). Cooling water for Great Britain's future electricity supply. *Proceedings ICE – Energy* 168(3), 188–204.

DECC (2012). *CCS Roadmap*. London, UK, Department of Energy and Climate Change.

Defra (2009). *Estimated abstractions from all sources (except tidal), by purpose and environment agency region: 1995–2008*. London, UK, England and Wales, Department for Environment, Food and Rural Affairs.

DfT (2008). *Investigation into the scope for the transport sector to switch to electric vehicles and plug-in hybrid vehicles*. London, UK, Department for Transport.

European Parliament and Council of the European Union (2001). Council Directive 2001/80/EC on Large Combustion Plants. Official Journal of the European Communities. Brussels, Belgium.

Francis, R. and B. Bekera (2014). "A metric and frameworks for resilience analysis of engineered and infrastructure systems." *Reliability Engineering & System Safety* 121: 90–103.

IEA (2009). *Key World Energy Statistics 2009*. Paris, France, International Energy Agency.

IEA (2010). *Energy technology perspectives 2010: scenarios and strategies to 2050 executive summary*. Paris, France, International Energy Agency.

IEA (2013). *World energy outlook 2013*. Paris, France, International Energy Agency.

IPCC (2014). Climate change 2014: mitigation of climate change. *Contribution of Working Group III to the fifth assessment report of the Intergovernmental Panel on Climate Change*. New York, NY, USA, IPCC.

Macknick, J., R. Newmark, G. Heath and K. C. Hallett (2012a). "Operational water consumption and withdrawal factors for electricity generating technologies: a review of existing literature." *Environmental Research Letters* 7(4): 045802.

Macknick, J., S. Sattler, K. Averyt, S. Clemmer and J. Rogers (2012b). "The water implications of generating electricity: water use across the United States based on different electricity pathways through 2050." *Environmental Research Letters* 7(4): 045803.

Ouyang, M. (2014). "Review on modeling and simulation of interdependent critical infrastructure systems." *Reliability Engineering & System Safety* 121(Jan): 43–60.

Pan, L., P. Liu, L. Ma and Z. Li (2012). "A supply chain based assessment of water issues
in the coal industry in China." *Energy Policy* 48(Sep): 93–102.

Rodriguez, D. J., A. Delgado, P. DeLaquil and A. Sohns (2013). *Thirsty energy.* Water
papers. Washington DC, USA, World Bank.

Tran, M., J. Hall, A. Hickford, R. Nicholls, D. Alderson, S. Barr, P. Baruah, R. Beavan, M.
Birkin, S. Blainey, E. Byers, M. Chaudry, T. Curtis, R. Ebrahimy, N. Eyre, R. Hiteva, N.
Jenkins, C. Jones, C. Kilsby, A. Leathard, L. Manning, A. Otto, E. Oughton, W. Powrie,
J. Preston, M. Qadrdan, C. Thoung, P. Tyler, J. Watson, G. Watson and C. Zuo (2014*).
National infrastructure assessment: analysis of options for infrastructure provision in
Great Britain, Interim results.* Environmental Change Institute, University of Oxford,
UK.

Veil, J. A. (2007). Use of reclaimed water for power plant cooling. Report by Argonne
National Laboratory for the US Department of Energy. Chicago, IL, USA.

Williams, J. H., A. DeBenedictis, R. Ghanadan, A. Mahone, J. Moore, W. R. Morrow, S.
Price and M. S. Torn (2012). "The technology path to deep greenhouse gas emissions
cuts by 2050: the pivotal role of electricity." *Science* 335(6064): 53–59.

World Energy Council (2010). *Water for energy.* London, UK.

Analysing the risks of failure of interdependent infrastructure networks

RAGHAV PANT, SCOTT THACKER, JIM W. HALL, STUART BARR,
DAVID ALDERSON, SCOTT KELLY

12.1 Introduction

Sustainability of national infrastructure systems (energy, transport, water, waste and information and communication technologies (ICT)) is contingent on several factors, one of which is reducing the risks associated with infrastructure failure when the system is subject to shocks, be they man-made or natural in origin. Under normal operating conditions infrastructure systems have capacity limitations, contain components nearing the end of their useful life, and face constant stresses in response to technological, demographic, social and lifestyle changes. These changing demands placed on infrastructure systems have been the subject of most of this book. This chapter deals with the more extreme, and fortunately less frequent, situation in which infrastructure systems are subject to conditions that can lead to cessation of service provision. The inter-connectedness of infrastructure system-of-systems means that widespread and cascading failures are possible, leading to major disruption to society and the economy. Avoiding, as far as possible, such damaging events, and planning to cope during those events and enable recovery, is therefore an important aspect of public policy regarding infrastructure provision. If the process of building long-term sustainable, adaptive and resilient infrastructure systems is to take place efficiently then it will need to be underpinned by an approach to understanding infrastructure risks under a range adverse impacts and infrastructure configurations.

The starting point for managing the risks of infrastructure failure, and building resilience to shocks is to understand the performance of infrastructure system-of-systems. What are the hazards to which these systems are exposed? What are the points of vulnerability? What are the potential consequences should failure occur? What are the actions that could be taken to reduce the risk of failure, promote graceful rather than catastrophic failure, enable coping during disruptive events and promote recovery to an enhanced state of resilience? How should potential actions to enhance resilience be prioritised?

This chapter addresses the problem of understanding risk of failure in national infrastructure system-of-systems. It thereby provides the basis for targeting remedial actions. By quantifying risks in economic terms, we provide methodology that can help to make the case for investment in resilience. Focusing on the risks of infrastructure failure and the implied negative economic and social consequences, this chapter introduces a risk assessment methodology to assess failures in interdependent infrastructures. The purpose of analysis presented here is to develop an approach that complements long-term system simulation (see Chapter 2) and informs policy decisions by adding a risk perspective.

In this chapter infrastructure risk is mainly understood in context of intensifying impacts of large-scale spatial impacts, with focus on weather-related hazards, non-weather related natural hazards (earthquakes, tsunamis) and accidental man-made hazards. By focusing on large-scale hazard events and large-scale infrastructure networks, the objectives of the risk assessment in this chapter are to answer following set of questions:

1. Where are the key vulnerabilities in national infrastructures?
2. What are implications of interdependence within and across national infrastructures?
3. What are the potential consequences of catastrophic national infrastructure failure for people and the economy?

To demonstrate methods and distil key lessons, case study results for the national infrastructure system in Britain are presented. Compared to various hazards, accidents and attacks, extreme weather related disasters constitute the biggest threat to Britain (Cabinet Office, 2010), prompting need for a national scale risk analysis that allows central governments to understand severity of risks and target interventions (Defra, 2012). There are other developed countries that face infrastructure risks similar to Britain. Methodological and modelling insights from the current risk assessment can therefore serve as a framework for addressing similar challenges recognised by policy planners and researchers worldwide (National Research Council, 2009).

12.2 The role of critical infrastructure in the economy and society

Infrastructure that is of particular importance in delivering goods and services allowing the national economy to operate efficiently, maintain social well-being and contribute to the goal of achieving long-term prosperity is known as critical infrastructure (CI). The term 'critical infrastructure' is generally used to highlight the importance of national infrastructure during natural disasters, disruptive events and malevolent man-made attacks. The European Union has defined CI as (European Commission, 2006):

> ... physical and information technology facilities, networks, services and assets which, if disrupted or destroyed, have a serious impact on health, safety, security or economic well-being of citizens or effective functioning of governments.

In the U.K., Critical National Infrastructure (CNI) is described as (Cabinet Office, 2010):

> ... key elements of national infrastructure which are crucial to continued delivery of essential services to the UK. Without these key elements essential services could not be delivered and the UK could suffer serious consequences, including severe economic damage, grave social disruption or even large scale loss of life.

During their lifetime, CIs are exposed to a multitude of disruptive influences broadly classified under: (i) natural hazards – which include, amongst others, extreme weather

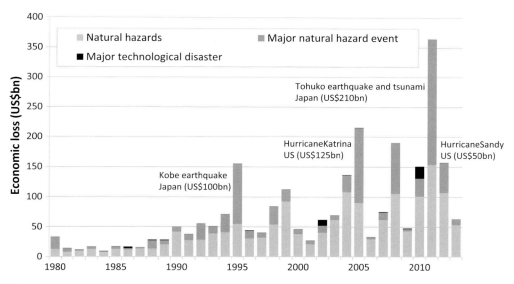

Figure 12.1   Annual estimates of global economic impacts of major natural hazards and technological disasters from 1980 to the 2013 (Source: CRED, 2014)

related hazards (flooding, windstorms, severe winter snows and ice storms, heat-waves and droughts), soil related geohazards (coastal erosion, landslides and embankment failures and subsidence), earthquakes and tsunamis, volcanic eruptions; and (ii) man-made hazards – which include accidental events (industrial accidents and wildfires near urban areas) and deliberate events (malevolent terrorist attacks, cyber-attacks, public disorder and disruptive industrial actions).

In recent years, there has been an increase in the frequency and magnitude of natural hazard events worldwide highlighting the risks imposed by such events on CI. Figure 12.1 shows annual worldwide estimated economic impacts, calculated in US$ billions (nominal value for the year of the event) occurring as a result of natural hazards and man-made accidents between 1980 and 2013 (September 2013) (CRED, 2014). These economic impacts consist of direct costs (e.g. damage to infrastructure, crops and housing) and indirect costs (e.g. loss of revenue, unemployment and market destabilisation). The frequency and cost of single multi-billion dollar natural disasters has been rising over the last decade, and economically the most severe disasters have taken place in developed countries, with substantial built infrastructure. The Intergovernmental Panel on Climate Change (IPCC) recognises, while factoring uncertainties, that in future climate change scenarios there will be an increase in severity and frequency of extreme weather related events (IPCC, 2012), which could result in larger disasters.

While man-made malevolent attacks can have a significant effect on CI functionality in the developed world such attacks are comparatively unusual and mostly locally concentrated. Man-made malevolent attacks can potentially trigger wider and different types of responses compared to natural hazards and accidents, and therefore require different risk perspectives.

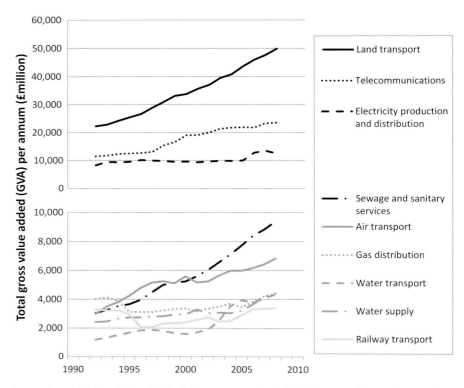

Figure 12.2 Total gross value added from 1990 to 2012 by infrastructure services in the U.K. economy (Source: ONS, 2012)

Having recognised that in long-term risk assessments of CI, extreme weather-driven failures play an increasing role, it is important to develop risk methodologies that incorporate the spatial extent and impacts of such events.

The five CIs considered in this book (energy, transport, water and wastewater, solid waste and ICT) are known to need protection against adverse impacts, which have potential to cause fatalities and economic losses. In most cases, in the developed world, societies are getting better prepared to avoid fatalities, with improved emergency planning measures. However, with growing complexity of infrastructure networks and interdependency between networks, it is not clear whether this growing attention is adequate to deal with the threat that infrastructure failure poses to society and the economy. In 2008 total contribution of the five national infrastructure sectors to Gross Value Added (GVA) in the U.K. economy was 9.2%, with land transport (which includes all commercial land transport activities plus sale of fuel and motor-vehicle distribution and repair) contributing the largest share followed by telecommunications and then electricity production and distribution. This is shown in Figure 12.2, which also highlights absolute contribution towards GVA from each infrastructure sector increased between 1992 and 2008.

Even more important is the dependence of all economic activities on infrastructure as a factor of production, and the particular role that infrastructure has in enabling trade, communication and innovation. Infrastructure represents a critical service within the economic

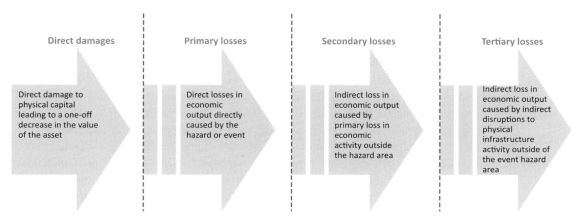

Figure 12.3   Primary, secondary and tertiary mechanisms of economic loss flows

supply chain and when disrupted, further economic transactions relying on that infrastructure might not be possible leading to additional economic losses far beyond normal transactions captured through key-linkage estimates. As an example, disruption of electricity infrastructure may lead to a loss in power throughout the country, completely halting the ability of manufacturers to produce goods and regular businesses to operate computer and ICT equipment. In order to arrive at more accurate estimations of total effects caused by infrastructure disruption it is necessary to estimate both indirect economic effects from loss in production, and the physical indirect effects caused by cascading failure from physical infrastructure systems on the rest of the economy.

Economic impacts from infrastructure failures can be assessed in a multi-step process involving physical damage losses, and direct and indirect economic losses within and beyond the hazard impact zones (see Figure 12.3). Such multi-step analysis goes beyond pure economic linkages between economic sectors and attempts to model physical dependencies between sectors as well as the economic linkages.

12.3 Risk analysis framework

Risk assessment of large-scale CIs covers various aspects of analysis including: (i) characterising spatial properties and probability of occurrence of extreme hazards; (ii) identifying uncertainties in CI responses; (iii) characterising critical assets and their interdependent failures within CIs; and (iv) quantifying vulnerabilities of CIs through interoperation of physical and economic consequences of failures.

Modelling infrastructure risk analysis has evolved from complex systems (Brown, 2007) to system-of-systems (SoS) (Eusgeld et al., 2011) analysis. Since infrastructures are comprised of multiple systems and have expanding cross-sectoral influences, risks are not confined within a single infrastructure. SoS research has focused on infrastructure topology models that account for interdependencies (Pederson et al., 2006) for system vulnerability analysis (Johansson and Hassel, 2010) and cascading failure analysis (Dueñas-Osorio and Vemuru, 2009; Zio and Sansavini, 2011).

Most infrastructure risk methodologies are limited to analysis of networks with similar characteristics confined to regional scales (Holden et al., 2013). For a detailed review of state-of-the-art risk assessment frameworks see Giannopoulos et al. (2012). National-scale risk perspective extends the application and definition of traditional risk analysis frameworks, but there is still a gap in methodologies that combine multiple interdependent networks at such scales. In this respect, the methodology presented in this chapter aims to develop a demonstrable national-scale SoS risk framework that quantifies likelihood and consequences of large-scale infrastructure network failures. The current methodology is useful for understanding catastrophic risks to interdependent networks that propagate towards inter-linked macroeconomic sectors. It is also useful for identifying key infrastructure vulnerabilities based on concentration of network assets and their usages.

12.3.1 Spatial networks and their properties

CIs are large-scale spatially distributed systems with complex interactions within and across their components (assets). CI interdependencies, which denote reciprocal dependence between two assets establishing unidirectional and bidirectional connectivity between systems, are central to their ability to operate and deliver services. In a technologically advanced world, infrastructures are becoming more interdependent and also redundant (backup) infrastructure systems are adding robustness which ultimately improves tolerance to adverse impacts. But interdependencies also trigger failure cascades across systems amplifying disruption impacts.

Broadly, CI interdependencies can be classified into four categories determined by type of infrastructure interactions (Rinaldi et al., 2001): (i) physical – when there are physical flows of resources between infrastructure subsystems; (ii) geographic – when spatial proximity between infrastructure subsystems results in their exposure to similar local environment effects; (iii) cyber – when there is information flow transmitted between ICT systems and other infrastructure subsystems; and (iv) logical – other than physical, geographic and cyber interdependencies, which mainly include human and economic interactions. The challenge in CI analysis lies in translating these different types of interdependencies to a manageable level of complexity that allows us to identify key vulnerabilities and predict risk propagation.

While there are numerous ways in which we can construct interdependent infrastructure models, there is growing evidence and supporting data that suggest collection of critical infrastructure assets are most suitably modelled as networks of nodes and edges (Lewis, 2006). For example, electricity grid is resolved into substations (nodes) and power cables (edges), transport systems are comprised of intersections or stations (nodes) and roads or railway lines (edges), water systems are made up of pumping stations (nodes) and pipes (edges), while telecom-masts (nodes) and telephone cables (edges) are building blocks of telecommunications systems. Network topology, which is the arrangement of nodes and edges, captures essential interdependencies and also indicates flow directionality across important infrastructure components. There are valuable infrastructure characteristics that can be inferred through network topological properties. Network centrality metrics, which measure relative importance of nodes and edges, explain influence of some infrastructure nodes and edges in generating and facilitating resource flows. Infrastructure networks are generally characterised by sparsely connected nodes arranged around some dense clusters of highly connected nodes. If nodes in these clusters are vulnerable to failure then network metrics can help predict propagation effects of failures towards other nodes.

For spatial risk calculations we define a 2D space comprised of points that enclose a boundary, given as $\mathbf{x} = \{\mathbf{x}_1, \ldots, \mathbf{x}_d\}$ (multi-polygon geometry), where a point $\mathbf{x}_k = (x_{k1}, x_{k2})$ denotes spatial coordinates (easting (or longitude) x_{k1} and northing (or latitude) x_{k2}). An infrastructure network is defined as the collection of assets, where an asset is a node or edge. a_i denotes the attributes assigned to a single asset, which contains, amongst other things, spatial geometry. A node's spatial attribute is given as a point geometry, while an edge's spatial attribute is given as a line or multi-line geometry and contains end coordinates that intersect with single nodes.

From asset geo-locations and connectivity rules derived through interdependence relationships we can build appropriate network topology. Two nodes are considered interdependent because one affects the other during normal operations and during failure. We derive network topology from a combination of physical and geographic interdependencies to represent an edge's function in facilitating flows of resources across nodes. Topological network flows occur from source nodes that generate or provide resources, towards sink nodes that deliver resources for final consumption. Based on directionality of flows we are interested in building physical edges (such as roads, railway lines, electricity cables, pipelines, etc.), most of which are not essentially straight lines but are rather curved lines on the ground, which is important for understanding spatial hazard impacts. When physical nature of the edge is not known, we derive straight-line distances between node pairs that transfer resources. Figure 12.4a shows an example interdependent spatial and topological network over the 2D map of Britain. Complete attribute assignment for a single asset comprises its geometry, functional characteristic (node or edge), infrastructure type and behaviour (source or intermediary or sink).

By assigning source-sink characteristics to nodes we are interested in establishing functional pathways for resource flows. A functional pathway includes all nodes and edges that need to be traversed for a resource to reach from source to sink, and shows type of resource being transferred. Multiple functional pathways can exist between same source-sink pairs because infrastructure networks are highly robust and redundant. While building networks we assign values of customers served (over an influence area) by different assets and try to estimate their distributions across different functional pathways. Some pathways have greater flow throughputs than others depending upon amount of resources generated at sources and consumed at sinks, which generates weighted-flow networks. Since sink nodes provide services for final consumption, outputs generated at these nodes measure customers served by networks. Figure 12.4b shows this representation of magnitudes of flows through network edges (creating weighted networks) and demands at sink nodes.

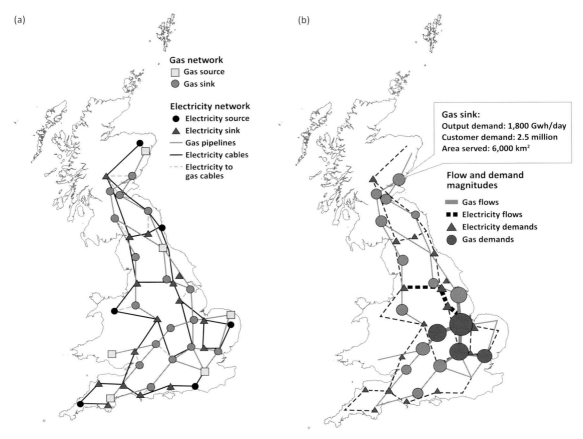

Figure 12.4 Overview of example networks built for risk analysis showing: (a) a topological electricity–gas interdependent network with the different assets (nodes and edges) and their characteristics (source/sink); (b) network flows and demand magnitude estimates derived from source-sink functional pathways which also determine the spatial areas and customers served by assets

12.3.2 Network criticality

By building detailed network representations and developing disruption models we can identify, quantify and prioritise locations and assets in infrastructures based on their network impacts. In particular, we are interested in assessing an infrastructure's criticality hotspot which is defined as a geographical location where there is concentration of CI, measured according to number of customers directly or indirectly dependent on infrastructures in that location. National level hotspots emerge from collection of several clusters of infrastructure critical hotspots.

A method for spatial hotspot analysis is proposed in which a spatially continuous surface of disruptions is generated from discrete measures to produce estimates of disruption density per-infrastructure type for any location in space. A spatial lattice (grid) across the area of interest is generated and at individual points in the lattice a spatial density estimate is derived using a Gaussian weighted intensity Kernel Density Estimation (KDE) approach, whose formulation is given as

$$g\left(\mathbf{x}_i\right) = \begin{cases} \sum_{j=1}^{b} \left\{ \left[W_j I_j\right] \dfrac{1}{\sqrt{2\pi}} \exp\left(-\dfrac{d_{ij}^2}{2h^2}\right)\right\}, & \text{if} \quad 0 < d_{ij} < r \\ 0, & \text{otherwise} \end{cases} \tag{12.1}$$

where $g\left(\mathbf{x}_i\right)$ is density at lattice location $\mathbf{x}_i$, d_{ij} is the distance of the infrastructure asset j from the location $\mathbf{x}_i$, r is the upper bound of search radius for assets around location $\mathbf{x}_i$, W_j is weight associated with infrastructure asset j, I_j is intensity (customer demand value) associated with infrastructure j, h is bandwidth parameter that controls the smoothening applied to the density estimation and $\frac{1}{\sqrt{2\pi}}\exp\left(-\frac{d_{ij}^2}{2h^2}\right)$ is the Gaussian kernel applied to point i that employs distance $d_{ij} \forall j \leq h$.

From KDE values one can recognise statistically significant spatial hotspots of infrastructure criticality via Getis and Ord Gi^* spatial hotspot statistical test (ESRI, 2014), which gives a z-score for each infrastructure asset. Large positive z-scores indicate a spatially weighted summed magnitude that is further from the spatially weighted average of the infrastructure sector and hence tending towards a hotspot. Equally large negative z-scores indicate a spatially weighted summed magnitude that is further from spatially weighted average of the infrastructure sector but tending towards a 'cold' spot (a local neighbourhood where values are lower than average for the entire spatial domain). Considering each individual test to be a part of a larger composite test of hotspots we can also generate a single overall composite hotspot surface.

Identifying network criticalities allows us to assess key vulnerabilities of infrastructures, which forms part of risk assessment albeit without hazard impact. The process of analysis of infrastructure criticality hotspots involves mapping spatial density of criticality for each infrastructure network separately, and combining to map density of infrastructure criticality 'hotspots'.

12.3.3 Network reliability

For risk calculations we formulate reliability, which is the measure of probability of failure studied at network level. To estimate network reliability we associate a function r_i with each asset a_i that defines its functional state. $r_i = 0$ denotes a_i is in 'failed' state and $r_i = 1$ denotes a_i being in 'non-failed' state. Failure refers to inability of the asset to maintain or provide its required functionality in the network. In the current analysis we are only interested in binary asset states. Assuming the entire network contains b assets we can assemble the binary network state vector $\mathbf{r} = \{r_1, \ldots, r_b\}$, whose elements are either 0 or 1 describing which assets have failed and which have not failed.

We are interested in network reliability with respect to extreme spatial hazards that are derived from continuous hazard fields over the 2D space. We assume a continuous spatial hazard generated by a multivariate spatially coherent distribution $f_{\mathbf{Y}}(\mathbf{y})$, from which a sample hazard $\mathbf{y} = \{y_1, \ldots, y_d\}$ can be generated. Here $y_k = y(\mathbf{x}_k)$ denotes hazard measure at point $\mathbf{x}_k$. When subjected to the extreme hazard, if state vector $\mathbf{r}$ contains at least one element whose value is 0, it indicates a failed network state. Network reliability, given in Equation 12.2, is calculated as the likelihood probability that the network is in a failure state $\mathbf{r}$ when hazard exceeds a threshold value $\mathbf{y}_{tr}$, and depends upon joint states of assets and the hazard probability density function $f_{\mathbf{Y}}(\mathbf{y})$ over the entire network.

$$P(\mathbf{r}) = \int\limits_{\mathbf{y}_{tr}}^{\infty} \mathbb{P}[\mathbf{r}|\mathbf{y}] f_{\mathbf{Y}}(\mathbf{y}) d\mathbf{y} = \int\limits_{\mathbf{y}_{tr}}^{\infty} \prod_{i=1}^{b} \mathbb{P}[r_i|\mathbf{y}] f_{\mathbf{Y}}(\mathbf{y}) d\mathbf{y}. \qquad (12.2)$$

We need to include all possible failure combinations of assets that result in network failure, which at maximum comprises possible $2^b - 1$ failure combinations, but if network properties are known then number of scenarios can be narrowed down to a few. We assume that there are p failure combinations (or 'system states') that contribute to overall network failure. We define the vector $\mathbf{r}^j = \{r_1^j, \ldots, r_b^j\}$ to represent jth failure combination and the set $\bar{\mathbf{r}} = \{\mathbf{r}^1, \ldots, \mathbf{r}^p\}$ as the collection of p failure combinations that contribute to overall network failure. The set $P(\bar{\mathbf{r}}) = \{P(\mathbf{r}^1), \ldots, P(\mathbf{r}^p)\}$ shows the different probabilities of failures. Asset failure probabilities $\mathbb{P}[r_i|\mathbf{y}]$ are based on failure processes that occur in the network. Type of network failures we are interested in quantifying are called common cause failures and cascading failures (Dueñas-Osorio and Vemuru, 2009). Common cause failures occur due to extreme spatial hazards event impacts, which affect several network assets simultaneously. Ability of an individual asset to withstand hazard loading is given by a fragility function $L(r_i|y_i) = \mathbb{P}[r_i = 0|y_i]$ that quantifies conditional probability of failure of the asset due to hazard loading on it. Fragility functions are derived from structural properties of assets and for any spatial asset we measure fragility with respect to maximum hazard intersecting it over its spatial extent. Cascading failures generally follow common cause failures as they signify failure propagation. In the current work failure cascades are governed by topological interdependence only and do not incorporate capacity degradations and flow redistributions across networks. To quantify cascading failure effects we propose a network path-centrality metric $\beta_i = 1 - \frac{P_i^f}{P_i}$, which denotes conditional probability of failure of an asset a_i due to its connectivity based on the ratio of functional pathways through it after (P_i^f) and before (P_i) common cause failures. $\mathbb{P}[r_i|\mathbf{y}]$ is a function of $L(r_i|y_i)$ and β_i (Pant et al., 2013).

12.3.4 Network disruption estimation

Outcomes of network failures in this work are quantified in terms of customers disrupted due to interruption of network services. Assuming network state given by $\mathbf{r}$ contains at least one 0 element number of customers served following failure $(S(\mathbf{r}))$, is the union of individual asset customers $(S(r_i))$ and is given by $S(\mathbf{r}) = \bigcup_{i=1}^{b} S(r_i)$. Assuming all sink nodes serve unique customer areas, a simplified expression for post-disruption customer disruptions is $S(\mathbf{r}) = \sum_{s=1}^{\varsigma} S(r_s)$ which gives sum of service deliveries at the sink nodes $s = \{1, \ldots, \varsigma\}$ following disruption.

In the current analysis, we assume that the network state is static and disruptions measured over some appropriate timeframe give difference between welfare (S) that would have accrued had the network continued to function and welfare ($S(\mathbf{r})$) in the case when the network fails. Hence network disruption is given as $\Delta S = S - S(\mathbf{r})$.

12.3.5 Economic loss estimation

Economic losses due to network failures are aggregated and non-spatial. Economic losses due to network failure are twofold producing: (i) direct economic losses – which are a function of costs incurred due to aggregated asset damages and service disruptions and (ii) indirect economic losses – which are a function of direct economic losses and result from propagation of economic losses to the macroeconomic system of industry sectors of which infrastructure networks are a part.

We use input-output modelling to translate disruption estimates to overall economic loss estimates, because it captures both direct and indirect economic losses. Macroeconomic input-output modelling (Leontief, 1951) highlights the degree of importance of an individual infrastructure-based industry (sector) within an economy, quantifying breadth of that sector's contribution and dependence upon all other sectors. When applied to a macroeconomic system comprised of n sectors (each producing a 'homogenous' output) Leontief economic input-output model is given as $q_i = \sum_{j=1}^{n} a_{ij} q_j + c_i$. Here q_i represents economic output of a sector i, $a_{ij} \in [0, 1]$ is the technological coefficient denoting input of sector i required per unit of production in sector j, and c_i represents final demand for sector i output going into private consumption, private investment, government expenditure and exports. Expressed in vector form $\mathbf{q} = \mathbf{Aq} + \mathbf{c}$ the model essentially shows how sector outputs ($\mathbf{q} \in \mathbb{R}^{n \times 1}$) are distributed to meet inter-industry economic flows (intermediate demand sales) ($\mathbf{Aq} \in \mathbb{R}^{n \times n} \times \mathbb{R}^{n \times 1}$) and final demand sales ($\mathbf{c} \in \mathbb{R}^{n \times 1}$). Matrix $\mathbf{A}$, which represents structure of the economy and effects on inter-industry relationships, is quantifiable through vast data sources (Yamano and Ahmad, 2006) and national level statistics (ONS, 2009) making Leontief's model widely relevant and applicable. Derived from input-output model, matrix $\mathbf{L} = [l_{ij}] = [\mathbf{I} - \mathbf{A}]^{-1} \in \mathbb{R}^{n \times n}$, called Leontief inverse, captures total (direct plus indirect) impacts of change in final demands. Another matrix $\mathbf{G} = [g_{ij}] = [\mathbf{I} - \mathbf{B}]^{-1} \in \mathbb{R}^{n \times n}$ where $\mathbf{B} = [\mathrm{diag}(\mathbf{x})]^{-1} \mathbf{A} [\mathrm{diag}(\mathbf{x})]$, called Ghosh inverse (Ghosh, 1958), captures total impacts of change in income also known as the supply side model.

From the input-output model we can derive backward and forward linkages, which are relative indicators that compare 'keyness' of each sector when that sector is compared to other economic sectors. Total backward linkages, $BL_j = \sum_{i=1}^{n} l_{ij}$, represent all input requirements ad infinitum to produce one extra unit for sector j, and total forward linkages, $FL_i = \sum_{j=1}^{n} g_{ij}$, represent the change in output from sector i resulting from one unit change in input from sector i. With knowledge of relative contribution provided by different economic sectors to a single sector's overall 'keyness' it is possible to identify potential for 'dependency risks'. Using this framework a value less than one indicates low interdependencies with the rest of the economy while a value greater than one indicates high interdependency. Figure 12.5 represents linkages between the different infrastructure networks and the rest of the U.K. economy and Table 12.1 shows the linkage values,

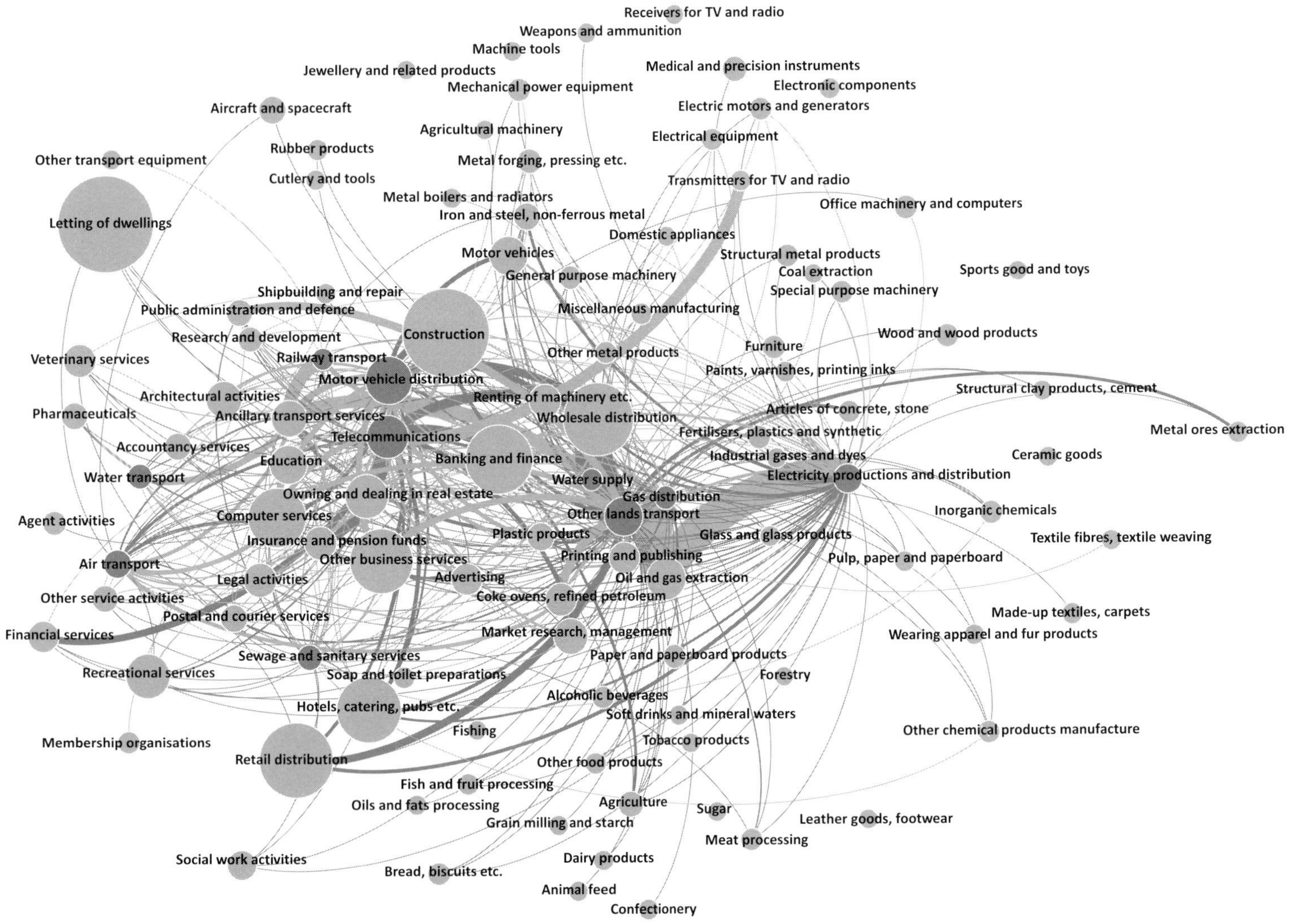

Figure 12.5 Economic input-output linkages between infrastructures (energy, transport, water, waste and ICT) (in red) and other sectors (in green) of the U.K. economy. Bubble sizes reflect the contribution to value added and line thicknesses represent absolute value of trade

Table 12.1 Economic linkages for different infrastructure types

		Total Forward Linkages	
		Low (<1)	High (>1)
Total backward linkages	Low (<1)	Water supply Water transport Air transport	Telecommunications Sewerage and sanitary services
	High (>1)	Railway transport	Electricity production Gas distribution Land transport Ancillary transport

highlighting that several key CIs have high forward or backward linkages or both making them high impact sectors.

In the input-output model $\mathbf{q} = \mathbf{A}\mathbf{q} + \mathbf{c}$, explained previously, infrastructures are part of a macroeconomic system of sectors. In order to capture initial effect of disruption, sectors primarily supported by infrastructure are separately identified in the $\mathbf{c}$ vector. Assuming infrastructure-based sectors are given by subscript $l = \{1, \ldots, m\} \subset \{1, \ldots, n\}$ within the list of sectors, we investigate two economic effects of disruptions:

1. Damage to physical infrastructure – this may be quantified in terms of capital costs incurred when assets are damaged (or cost of reinstating assets) and for an asset in infrastructure-based sector l is written as $K_l(r_i) \equiv K_l(r_i = 0)$. It is reasonable to assume that total cost of reinstating this infrastructure network when it is in failure state set $\mathbf{r}$ is given by adding individual asset damages, that is, $K_l(\mathbf{r}) = \sum_{i=1}^{b} \max[0, 1 - r_i] K_l(r_i)$.
2. Loss of service – in order to convert services to economic costs we assume that the ratio between economic loss ($\Delta c_l(\mathbf{r})$) and cost of pre-disruption service delivery (c_l) is same as the ratio between amount of service disrupted ($\Delta S_l(\mathbf{r})$) and pre-disruption service delivery (S_l), that is, $\Delta c_l(\mathbf{r}) = \frac{\Delta S_l(\mathbf{r})}{S_l} c_l$.

Once we have estimated demand losses for infrastructure-based sectors we can form the input-output demand loss vector $\Delta \mathbf{c}$ and by solving the system $\Delta \mathbf{q}(\mathbf{r}) = [\mathbf{I} - \mathbf{A}]^{-1}[\Delta \mathbf{c}(\mathbf{r})]$ we obtain total economic loss effect (direct plus indirect) of infrastructure disruption on the circular economic flow over a period of one year. The expression for economic losses incurred from the network in failure state $\mathbf{r}$ is given in Equation 12.3.

$$D(\mathbf{r}) = \sum_{l=1}^{m} K_l(\mathbf{r}) + \sum_{i=1}^{n} q_i(\mathbf{r}). \tag{12.3}$$

Similar to reliability estimation, overall network economic loss due to different failure states can be assembled into the set $D(\bar{\mathbf{r}}) = \{D(\mathbf{r}^1), \ldots, D(\mathbf{r}^p)\}$.

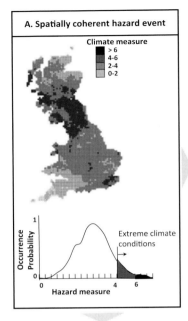

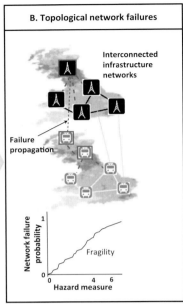

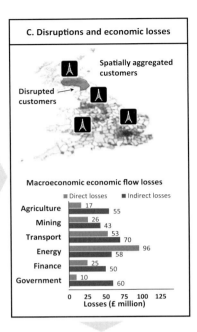

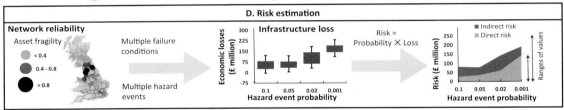

Figure 12.6   Spatial risk methodology framework illustrating the different steps in building the risk calculations

12.3.6 Integrating to produce spatial risk analysis

Infrastructure risk is the product of the network reliability and the aggregated economic loss and is given as $R(\mathbf{r}) = D(\mathbf{r}) P(\mathbf{r})$, when the network is in failure state $\mathbf{r}$. The risks due to different failure outcomes are collected in the set $R(\bar{\mathbf{r}}) = \{R(\mathbf{r}^{1}), \ldots, R(\mathbf{r}^{p})\}$. Figure 12.6 illustrates the risk methodology framework developed through Sections 12.3.1–12.3.5, showing different components and highlights models that need to be built and combined for a coherent risk analysis.

1. Component A shows a spatially coherent hazard event selected from a hazard event set via Monte Carlo sampling.
2. Component B shows interconnected national scale infrastructure networks (separated for visual clarity) failing from a hazard event based on fragility functions and connectivity.
3. Component C illustrates spatial customer disruption calculations and economic loss estimations. Customer disruptions at detailed spatially disaggregated scales are

aggregated and converted to direct macroeconomic loss estimates, which are then used for calculating indirect and overall macroeconomic losses.

4. Component D shows risk estimation that follows from hazard event generation, network reliability and loss calculations. For a particular infrastructure network exposed to hazard loading we can identify assets that have high fragilities and are thus candidates for failure. For each hazard event we generate multiple mechanisms via Monte Carlo simulations and further repeat the process over different hazard events. Different outcomes for each hazard and across different hazards give ranges of possible economic losses and eventually risk outcomes.

Figure 12.6 demonstrates a workflow where network failures in Component B are dependent upon Component A hazard event results, and similarly Component C calculations follow from Component B failure analysis. Component D integrates risk calculations and is dependent upon the execution of previous three components.

12.4 Case study demonstration

Britain has been subjected to major infrastructure disruptions due to large-scale climate hazards (Pitt, 2008), many of which have been identified in the National Risk Register (NRR) as likely to occur over the next five years (Cabinet Office, 2013). Here we use Britain's infrastructure networks to illustrate how risk analysis can be conducted at a national scale and to demonstrate the insights that the analysis provides.

12.4.1 Interdependent infrastructures syntheses

In the present analysis we explore dependence of other major infrastructure assets and networks on electricity. Such a choice is governed by practical reasons because: (i) every other infrastructure requires electricity to operate; (ii) electricity has high linkage effects over the entire economy; (iii) electricity infrastructure is prone to major disruptions that have widespread effects (Pitt, 2008); and (iv) the U.K. plans to invest the most in energy infrastructure in the long term (HM Treasury, 2011).

As listed in Table 12.2, we have assembled point assets belonging to different infrastructure types and also built national-scale electricity infrastructure networks for England and Wales and major railway and road networks at a national scale. The data sources for assembling infrastructures are detailed in Appendix E (available online at itrc.org.uk).

We have constructed a hierarchical representation of the electricity network for England and Wales, consisting of an integrated electricity power generation and transmission network with synthetic sub-transmission and distribution networks. A geographic electricity transmission network consists of nodes that represent functionally important substations (400 kV and 275 kV) and power generators, and their connecting edges representing overhead lines or underground cables. We formulated synthetic representations of distribution network, forming lower levels of the hierarchy; stepping down electricity to 132 kV, 33 kV

Table 12.2 List of asset types for which population attribution has been completed		
Asset name	Sector	Asset type
Electricity transmission and distribution	Energy	Network
Gas transmission take-off points	Energy	Point assets
Wastewater treatment works	Wastewater	Point assets
Water towers	Water	Point assets
Telco masts	Telecommunications	Point assets
Railway stations and tracks	Transport	Network
Major road junctions and roadways	Transport	Network

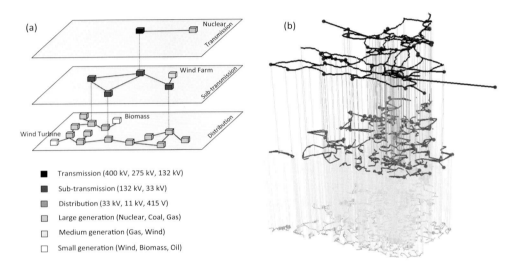

Figure 12.7 (a) provides an abstract representation of the electricity network hierarchy, highlighting connectivity between levels based on operational voltages and the mapping of different generator types based on capacity classifications; (b) provides a pictorial representation of the electricity hierarchy for England and Wales, consisting of an integrated electricity power generation and transmission network with synthetic sub-transmission, distribution networks

and 11 kV substations. The synthetic network structures were tested with some available data found in the northwest region of England. Electricity generation facilities (nuclear, coal, gas and wind) with operational capacities are also embedded at various levels of the hierarchy. These result in three classes of generation and three connectivity types: (i) large connected to the transmission system; (ii) medium connected to 132 kV hierarchy level where operational capacity is greater than 20 MW; and (iii) small connected to 33 kV hierarchy level where operational capacity is less than 20 MW. An abstract representation of the hierarchy alongside a map view for England and Wales is given in Figures 12.7a and 12.7b, respectively.

We have mapped spatial dependencies that represent physical connections between dependent assets and supporting electricity assets. We first identify the level of hierarchy

Table 12.3 Mapping between dependent assets and support electricity assets in the hierarchy

Asset type	Level of connection to electricity hierarchy
Railway stations	11 kV/415 V
Wastewater treatment works	11 kV/415 V
Water towers	11 kV/415V
Telco masts	11 kV/415V

where assets that have a dependency on electricity connect. These mappings, detailed in Table 12.3, have been found correct for majority of cases. For each dependent asset we find the geographically nearest electricity substation of correct operational voltage and build a dependency edge as a straight line.

12.4.2 Customer assignments for utility networks

We have assigned population values to utility assets, which include electricity distribution substations, gas, water, wastewater and telecommunication masts-related assets. Assignment of population values is a two-part process comprising: (i) deriving infrastructure asset footprints to estimate spatial area of influence around each distribution level asset and (ii) assigning customer values to each distribution level asset based on a spatial union of asset footprint with census derived population estimates. We assume that resources and services delivered by distribution level assets are consumed locally and by all members of population.

Asset footprints are formulated by constructing disjoint voronoi polygons where each point within the polygon is closer to its corresponding point asset. This method assumes each asset provides an equally weighted service and assets influence only the closest space around them. Figures 12.8a–h explain the process of customer assignments and provide results derived for wastewater treatment works, gas off-take points, telecommunication masts and water towers, respectively.

By transferring assigned non-electricity customers their supporting electricity assets via corresponding dependency link, we perform a location allocation assignment to distribute them along paths in the electricity hierarchy based on source capacities and source-sink flow strengths. Figure 12.9 illustrates the process of customer assignments across the electricity hierarchy.

12.4.3 Transport network assignments and disruptions

For the railway network we develop origin-destination trip assignments to estimate passenger-trips, which give throughputs across stations and tracks. It is assumed that once trips are assigned along particular routes there is no rerouting in case of disruptions. Hence flow rerouting, congestion and other trip reassignment mechanisms are not considered here. Using data (see Appendix E – available online at itrc.org.uk) we take into account

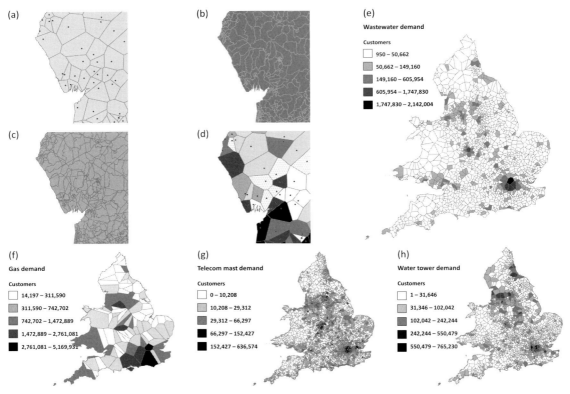

Figure 12.8 Stages in assigning customers to assets: (a) introduces a set of asset footprints; (b) provides a view onto ward level population data; (c) represents the union of both the asset footprint and the bounded population data; (d) gives customer estimates transferred to asset footprints; (e), (f), (g) and (h) provide customer assignments derived for wastewater treatment works, gas off-take points, telecom masts and water towers, respectively

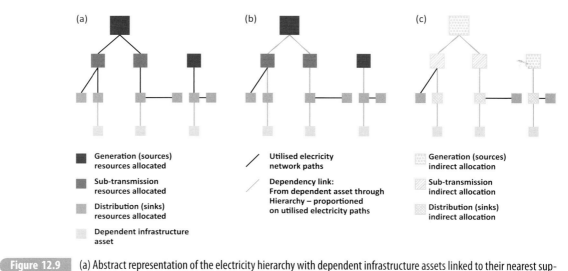

Figure 12.9 (a) Abstract representation of the electricity hierarchy with dependent infrastructure assets linked to their nearest supporting electricity asset in space; (b) the distribution of non-electricity customers along paths, proportioned by sink-source electricity flow along that path and (c) the attribution of the sum non-electricity customers at node assets in the network

station entries, interchanges and exits, and train frequencies along routes and build origin–destination (O–D) daily trip assignments of passengers in the railway network. Figures 12.10a and 12.10b represent the flows of all passengers through stations and along routes (metrics for assessing their criticality) for the national railway network.

We also build road networks with volume of traffic along specific length sections of the road network. In particular, we concentrate on the major road network deriving passenger traffic flow information from data. Figure 12.10c shows the major road network with flow statistics along the links.

12.4.4 Criticality hotspot results and discussion

Spatial hotspots of infrastructure criticality across England and Wales, were estimated by constructing a 1-km sampled spatial lattice over which Gaussian KDE surfaces were derived in terms of customer intensity, using the Equation 12.1 formulation with I_j set to customer demand associated with a particular infrastructure asset j, $W_j = 1$ for all infrastructure types, and a bandwidth (h) set to 5 km by experimentation. KDE surfaces were generated for all infrastructure sectors considered in Section 12.4.2. The z-scores were calculated for each 1-km sampled lattice location in an infrastructure sector, and for each location the spatially weighted summed magnitude of customer demand within a defined spatial neighbourhood was calculated. The spatially weighted summed magnitude of customer disruption for a location was compared to the spatially weighted average of the infrastructure sector for the entire spatial domain under consideration (i.e. average of the entire data-sets) was converted to a z-score.

Hotspot results are presented for four analysis cases:

1. KDE estimates for integrated electricity demand, combining demand for electricity and dependent demand for non-transport infrastructure assets (Figure 12.11a). Overall results suggest a relatively small number of key geographical locations play a major role in satisfying electricity demand within England and Wales, particularly with respect to dependent demand associated with other non-transport infrastructure types.
2. KDE estimates for rail network throughputs (Figure 12.11b), which show there are only a small number noticeable hotspots of passenger flows concentrated regionally around London. Flows radially propagate along mainline rail connections into rest of the network.
3. KDE estimates for road passenger flows (Figure 12.11c), which show that, unlike the rail network, larger more significant regions of flows are found around several regional concentrations.
4. The composite z-score analysis (combination of customer demand for assets, electricity demand relating to assets and transport passenger flows, Figure 12.11d) shows that London is a major focus of asset demand and has a spatially continuous hotspot with multiple peaks that covers a large spatial extent. There are other composite hotspot locations that have substantial spatial influence and customer disruption impacts.

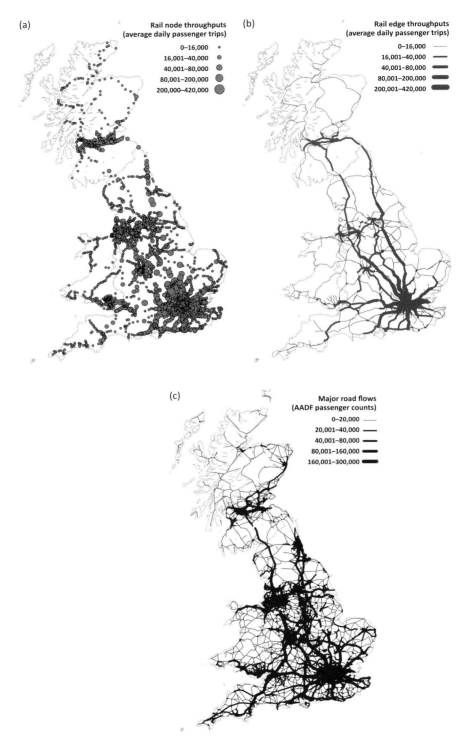

Figure 12.10 Results of the transport network analysis showing estimates of: (a) node criticality, which represent the passengers throughputs at each railway network node (station or junction); (b) edge criticality, which represent the passenger throughputs at each railway network edge (track section); (c) an illustration of the average annual daily flows (passenger numbers) along links on the major road network from Great Britain. This also indicates the criticality of the each road section

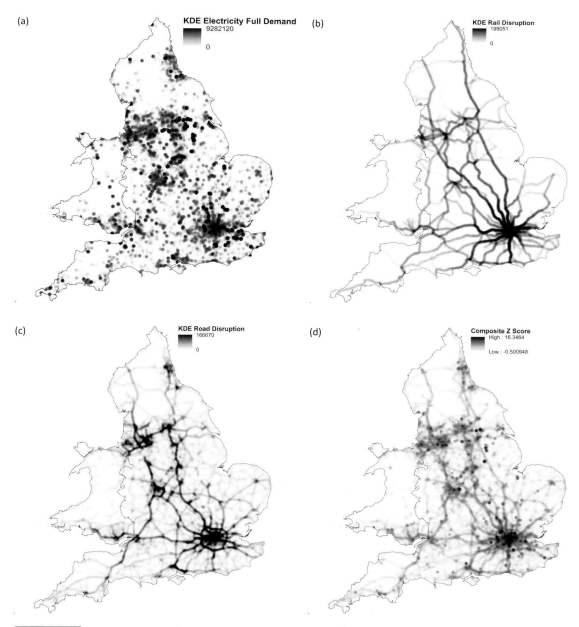

Figure 12.11 (a) Kernel density estimated full electricity customer demand including GSP, gas, telecoms, water and water treatment assets dependency; (b) kernel density estimated rail disruption (stations and track); (c) kernel density estimated road network disruption on the basis of passengers and (d) composite z-scores of population demand and disruption of assets, electric network, rail and road

12.4.5 Flood risk analysis results and discussion

Following from the hotspot analysis, we perform a flood risk assessment of some of the key assets in the electricity transmission network and their dependent infrastructure point assets. We have used the National Flood Risk Assessment (NaFRA) flood likelihood map for England and Wales, which provides information on the estimated likelihood of flooding to areas of land within the flood plain of an extreme flood (0.1% or 1 in 1,000 chance of fluvial and/or tidal flooding in any year). The likelihood of flooding is estimated based on assessments undertaken for 85 river catchments and coastal cells (50 m × 50 m), taking in account the probability that the flood defences will overtop or breach, and the distance of the impact cell from the river or the sea. The results of the analysis are presented for three flood likelihood risk categories as: (i) low – the chance of flooding each year is 0.5% (1 in 200) or less; (ii) moderate – the chance of flooding in any year is 1.3% (1 in 75) or less but greater than 0.5%; and (iii) significant – the chance of flooding in any year is greater than 1.3%. Figure 12.12 shows an example map with areas for different flood likelihoods and some electricity substations that intersect with the different areas, indicating the level of hazard exposure.

The purpose of the flood risk analysis presented here is to highlight the likelihood of the hazard exposure and the potential ranges of worst-case impacts resulting from failures of key infrastructure assets. It is recognised here that other information (hazard temporal and spatial probabilities, asset fragilities, interdependent failure probabilities) is required for capturing all uncertainties in risk estimation. The current analysis provides a useful template for identifying the high-level risks to infrastructures. At this level the comparative levels of systemic risk estimation is useful for informing national-level planning and resource management.

Figure 12.13 provides the ranges of possible customer disruptions over different flood likelihood scenarios. Here we account for customer disruptions due to failures of individual electricity substations and some other infrastructure assets dependent upon these substations, arranged in ascending order of cumulative disruptions. The results highlight the impact of one infrastructure (electricity) on others, as there are several instances of much larger potential customer disruptions for other assets when they are inoperable due to electricity failures. Amongst the dependent infrastructures railway operations are most affected by electricity failure, because individual rail assets (stations) are used for cross-country travel leading to a much wider network impact. Failures in utility based sectors can be confined to smaller geographic areas with fewer customers.

Following the estimates of customer disruptions, we explore the economic losses that could result from such disruptions only. Figure 12.14 gives the ranges of economic loss estimates per day at the macroeconomic scale (for Britain), generated from the economic model discussed in Section 12.3.5. Here we have considered 38 industry sectors that represent the British economy and used the national level input-output tables representing commerce for these industries in the year 2009 (ONS, 2009). The results show both the direct and indirect economic losses per day in the entire economy. In the case of the infrastructure sectors (highlighted in grey) these losses are primarily direct losses (with a few indirect losses due to the closed loop nature of the IO model), while in the case of other

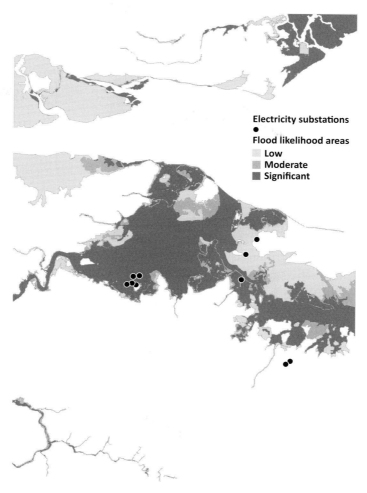

Electricity substations
●
Flood likelihood areas
  Low
  Moderate
  Significant

Figure 12.12 Example of flood Likelihood map with intersecting electricity substations

sectors the losses are exclusively indirect losses. The three most economically impacted sectors in descending order are telecommunications, air transport and electricity, which is different from the order to customer disruptions between these sectors. Also, sectors such as business services and real estate, postal and courier and coal, gas and mining have comparatively significant indirect losses to the directly affected sectors. Hence from the economic analysis we are able to see some levels of direct and indirect impacts that are not intuitively predictable. The ranges of potential total, direct and indirect losses (rounded to the nearest million) are also shown in Figure 12.14.

From the probabilities of the hazard event, the ranges of the flood likelihood estimates and the total economic losses we can produce risk scores for each instance of electricity asset failures. This is shown in Figure 12.15, where each risk score is estimated by taking the product of the event probability (1/1,000), the likelihood of flooding (these are taken

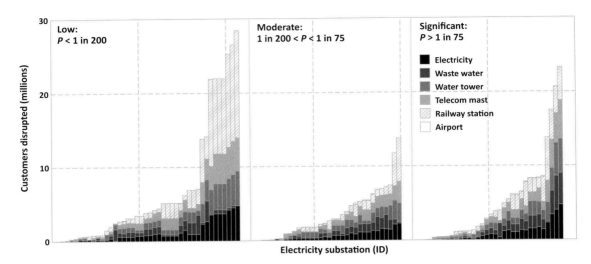

Figure 12.13 Ranges of potential customer disruptions due to electricity substation exposure to different flood likelihood events

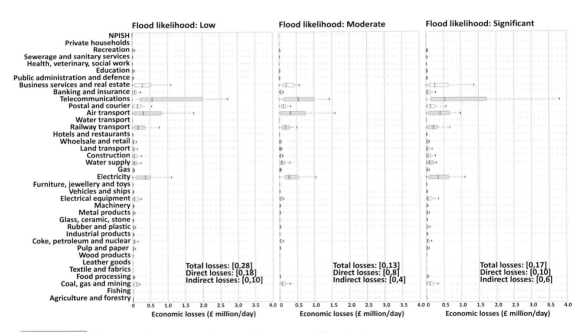

Figure 12.14 The ranges of macroeconomic economic losses across different industry sectors

as ranges: Low [10^{-10}, 1/200]; Moderate [1/200, 1/75]; Significant [1/75, 1]); and the total economic loss estimates (in £ million/day). We can use these risk scores to set up a scale of asset criticality where assets having risk scores within a range are considered most critical.

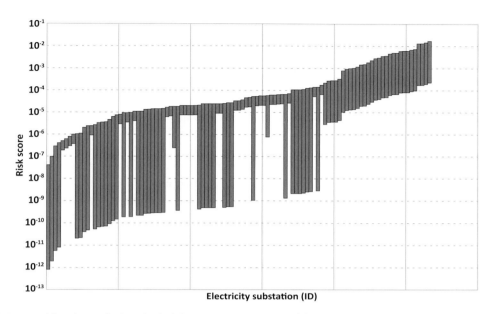

Figure 12.15 Ranges of the risk score for the individual electricity transmission asset failures

12.5 Conclusions

This chapter has presented methodology for national scale analysis of potential risks to interdependent infrastructure. We have proposed a system-of-systems risk methodology that provides a step-by-step quantification of national infrastructure reliability, disruption and risk due to large-scale spatially coherent probabilistic hazard events. A major step in this analysis has been the construction of interdependent national-scale spatial networks and connecting those networks with spatial data on use of infrastructure services. The methodology for assessing infrastructure spatial criticality, measures hotspot locations where large numbers of customers are both directly and indirectly affected by infrastructure failure. In calculating risk we take into account multiple infrastructure failure states when shocked and for each state estimate: (i) failure probabilities of individual assets due to hazard impacts and network interdependence towards estimating overall network reliability and (ii) network disruptions in terms of metrics of influence on population and economy. This analysis is useful for strengthening our understanding of the full scale of risks across interdependent networks and their indirect economic impacts. The framework is also useful for identifying key vulnerabilities in networks, which should be the target for action to reduce risks.

The risk calculations here consider only infrastructure network elements survival or failure aggregated over a timeframe representing the duration of the impact. The dynamics of failure, system flow redistribution and evolving economics are not considered. Since the current risk methodology considered detailed interdependent network representations and flow pathways it has scope to include network dynamics of load redistributions and recoveries. These ideas can be expanded in future work on process-based infrastructure risk estimation.

Even though the current analysis considers present day states of infrastructure the methodology provides a template for long-term risk planning. For a developed country like Britain the majority of core infrastructure has already been built with most future changes resulting in incremental changes to network architecture by adding or removing a few assets. The network synthesis methods developed in this analysis are able to accommodate such network changes.

As highlighted in the risk results there are wider economic failures and disruptions beyond the disrupted infrastructures. These effects need to be linked to sector level strategy performance and cross-sector impacts need to be incorporated in long-term infrastructure provision, so that cascading failure can be anticipated and contained. Such analysis benefits policy makers and infrastructure service providers who are making long-term resource allocation decisions and investments to strengthen infrastructure resilience.

References

Brown, T. (2007). "Multiple modeling approaches and insights for critical infrastructure protection." *NATO Security Through Science Series D – Information and Communication Security* 13: 23.

Cabinet Office (2010). *Strategic framework and policy statement on improving the resilience of critical infrastrucuture to disruption from natural hazards*. London, UK.

Cabinet Office (2013). *National risk register of civil emergencies*. London, UK.

CRED (2014). EM-DAT: the international disaster database, Centre for Research on the Epidemiology of Disasters.

Defra (2012). UK Climate change risk assessment: government report. London, UK, Department for Environment, Food and Rural Affairs.

Dueñas-Osorio, L. and S. M. Vemuru (2009). "Cascading failures in complex infrastructure systems." *Structural Safety* 31(2): 157–167.

ESRI (2014). "Hot spot analysis (Getis-Ord Gi*) (Spatial Statistics)." Retrieved from http://resources.esri.com/help/9.3/arcgisengine/java/gp_toolref/spatial_statistics_tools/hot_spot_analysis_getis_ord_gi_star_spatial_statistics_.htm.

European Commission (2006). Communication from the Commission of 12 December 2006 on a European Programme for Critical Infrastructure Protection. Official Journal C 126 of 7.6.2007. Brussels, Belgium.

Eusgeld, I., C. Nan and S. Dietz (2011). "System-of-systems approach for interdependent critical infrastructures." *Reliability Engineering & System Safety* 96(6): 679–686.

Ghosh, A. (1958). "Input-output approach in an allocation system." *Economica* 25: 58–64.

Giannopoulos, G., R. Filippini and M. Schimmer (2012). Risk assessment methodologies for critical infrastructure protection. Part I: a state of the art. European Commission JRC Technical Notes.

HM Treasury and Infrastructure UK (2011). *National infrastructure plan 2011*. London, UK, HM Treasury.

Holden, R., D. V. Val, R. Burkhard and S. Nodwell (2013). "A network flow model for interdependent infrastructures at the local scale." *Safety Science* 53: 51–60.

IPCC (2012). Managing the risks of extreme events and disasters to advance climate change adaptation. *A special report of Working Groups I and II of the Intergovernmental Panel on Climate Change*. Field, C.B., V. Barros, T.F. Stocker, D. Qin, D.J. Dokken, K.L. Ebi, M.D. Mastrandrea, K.J. Mach, G.-K. Plattner, S.K. Allen, M. tignor and P.M. Midgely (eds.). Cambridge, UK and New York, NY, USA, IPCC, 582 pp.

Johansson, J. and H. Hassel (2010). "An approach for modelling interdependent infrastructures in the context of vulnerability analysis." *Reliability Engineering & System Safety* 95(12): 1335–1344.

Leontief, W. W. (1951). *The structure of American economy, 1919–1939: an empirical application of equilibrium analysis*. New York, NY, USA, Oxford University Press.

Lewis, T. G. (2006). *Critical infrastructure protection in homeland security: defending a networked nation*, Hoboken, NJ, USA, John Wiley & Sons.

National Research Council (2009). *Sustainable critical infrastructure systems: a framework for meeting 21st century imperatives*. Washington DC, USA.

ONS (2009). *United Kingdom national accounts: the blue book*, Basingstoke, UK, Palgramme Macmillan.

ONS (2012). *United Kingdom input-output analytical tables 2005* (2012 edn). Office for National Statistics.

Pant, R., J. W. Hall, S. Barr, D. Alderson and S. Thacker (2013). National scale risk analysis framework for interdependent infrastructure networks due to extreme hazards. Working paper.

Pederson, P., D. Dudenhoeffer, S. Hartley and M. Permann (2006). *Critical infrastructure interdependency modeling: a survey of US and international research*. Idaho, USA, Idaho National Laboratory.

Pitt, M. (2008). "Learning lessons from the 2007 floods." London, UK, Cabinet Office.

Rinaldi, S. M., J. P. Peerenboom and T. K. Kelly (2001). "Identifying, understanding, and analyzing critical infrastructure interdependencies." *Control Systems Magazine, IEEE* 21(6): 11–25.

Yamano, N. and N. Ahmad (2006). The OECD input-output database: 2006 edition. STI Working paper 2006/8. Paris, France, OECD Publishing.

Zio, E. and G. Sansavini (2011). "Modeling interdependent network systems for identifying cascade-safe operating margins." *IEEE Transactions on Reliability* 60(1): 94–101.

Database, simulation modelling and visualisation for national infrastructure assessment

STUART BARR, DAVID ALDERSON, MATTHEW C. IVES, CRAIG ROBSON

13.1 Introduction

As we have seen in the previous chapters of this book, strategic analysis of the long-term performance of national infrastructure system-of-systems involves tracking of decisions and interdependencies through time. The system-of-systems model integrates infrastructure sector models for energy, transport, water, wastewater and solid waste, along with economic and demographic projections. These components are brought together in an uncertainty and decision analysis framework, set out in Chapter 2. The database system (NISMOD-DB) described here provides a central role in structuring, enabling and visualising the results of this analysis. As well as hosting all of the necessary infrastructure data, it hosts the results of each step in the modelling process, manages the information flows and provides an audit trail of the provenance of results.

Good quality data and information lies at the heart of developing long-term robust and economically sustainable infrastructure investment plans (Rinaldi et al., 2001). Infrastructure operators from around the world have recognised the importance of developing data and information management approaches in order to better plan and implement investment decisions (Woodhouse, 2014). Many companies have invested significantly in developing data and information management systems that allow them to capture, update, manage and analyse their infrastructure assets (Haider, 2013). Indeed, the drive to manage infrastructure assets more efficiently has led to several international and national initiatives with regards to infrastructure asset management standards (e.g. PAS 55 and ISO 55000 (Woodhouse, 2014)).

However, while information on the location and state of assets is critically important to infrastructure providers (Amadi-Echendu et al., 2010), the ability to plan long-term infrastructure provision requires such data to be augmented with information on the myriad vulnerabilities, demand/capacity constraints and lifespan renewal issues of an infrastructure system (Rinaldi et al., 2001; RAE, 2011). Moreover, it is now widely acknowledged that in order to understand how an infrastructure system or sector is performing, detailed information and understanding of the dependencies and/or interdependencies that exist between it and other infrastructure sectors is required (Rinaldi et al., 2001; RAE, 2011), which has in turn led to the development of a *system-of-systems* approach to critical infrastructure analysis, modelling and planning (CST, 2009; Otto et al., 2014).

Several large-scale research initiatives, such as the U.S. National Research Council report on Sustainable Critical Infrastructure Systems, the Dutch programmes on Next Generation

Infrastructure and Knowledge for Climate, the Australian Critical Infrastructure Protection Modelling and Analysis (CIPMA) programme and the U.K. Infrastructure Transitions Research Consortium (ITRC) are developing the new suite of infrastructure analysis and modelling tools required to provide a holistic system-of-systems understanding of critical infrastructure. Within such initiatives, and more generally, it has been recognised that the ability to collate, integrate and manage a wide range of diverse infrastructure data at a range of spatial scales and measurement granularity in a cohesive and logical manner is a key requirement.

Nevertheless, relatively little attention has been given to the development of data management systems explicitly designed for and able to handle the wide range of disparate data and relationships required to support a system-of-systems approach to infrastructure analysis and modelling. In this chapter we consider the requirements of such a data management system, and in particular the database framework required to support a system-of-systems approach to infrastructure analysis, modelling and planning. Using this framework, we present a U.K. national scale spatial database framework for infrastructure systems analysis called NISMOD-DB and demonstrate how this is used to support the modelling of the long-term capacity/demand requirements of interdependent infrastructure systems (Chapter 2) and also undertake interdependent infrastructure network risk and vulnerability analysis and modelling (Chapter 12).

13.2 Structuring a national infrastructure database system

A logical approach to the design of a high-level infrastructure system-of-systems data management system is to consider infrastructure to be organised in a hierarchical manner. At its lowest level of granularity it comprises individual assets (Hudson et al., 1997; Nastasie et al., 2010), where at the holistic system-of-systems representation it is an integration of multiple infrastructure systems and sectors in terms of their physical, operational and strategic characteristics (Pederson et al., 2006). These two extremes are connected via assets forming infrastructure networks, which collectively form infrastructure sectors/systems, which themselves aggregate to form the overall holistic infrastructure system (Figure 13.1). Thus, any system-of-systems infrastructure data management system must be able to represent the key characteristics of each of these different components of the hierarchy; namely, assets, networks, sectors/systems and system-of-systems.

Within such a system-of-systems hierarchical perspective of infrastructure it is important to be able to explicitly represent collections of infrastructure components at one level in the hierarchy that can be aggregated to form an integrated component at the next level, such that the 'value chain' (Amadi-Echendu, 2006) is apparent in the infrastructure asset management approach adopted. In many respects, such a perspective of hierarchical organisation naturally leads to the explicit treatment (in terms of representation and management requirements) of the intra-system dependencies and interdependencies that exist. For example, to represent an electricity transmission or local distribution network requires explicit knowledge of not only the individual assets, such as sub-stations and cables, but also knowledge of their topological connectivity.

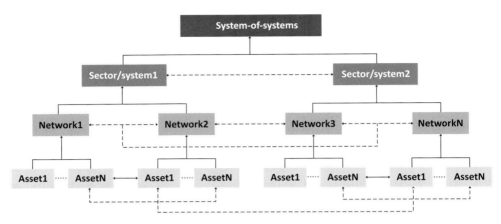

Figure 13.1   The hierarchical representation of different data levels within a system-of-systems representation of infrastructure

However, a system-of-systems approach also requires an explicit treatment of the horizontal within-level-dependencies and interdependencies that exist between infrastructure components. Such dependencies and interdependencies may exist at any level of the hierarchy, from individual assets through to the system/sector level. For example, at the level of individual assets it is critical to know which electricity sub-station is providing power to particular water supply pumping stations. At a more aggregated scale we may wish to know how an infrastructure network system responds to perturbation within any of the other systems they are dependent upon. Equally, it is desirable to understand how performance in one infrastructure sector may be influenced by that of another sector. Thus, any system-of-systems infrastructure data management approach, and related database architecture, must be able to represent explicitly the key asset, network, system and sector dependencies and interdependencies present (Figure 13.1).

In data management terms, different infrastructure components (asset, network, system and sector) will require a set of key generic attributes to describe them, both in terms of their intrinsic characteristics but also in terms of their relationships to other components within the infrastructure hierarchy and in terms of their horizontal within-level dependencies and interdependencies. The development of appropriate database management schema that can capture in a generic flexible manner the attributes of infrastructure components from assets to entire integrated systems is critical to developing a robust system-of-systems approach to infrastructure analysis and modelling.

Amadi-Echendu et al. (2010, p. 14) recognise five key characteristics required to describe an asset; namely its spatial, time, measurement, statistical and organisational generality. These five dimensions describe any infrastructure asset in such a way that one is able to retrieve where it is located (spatial), the scale of representation (spatial), its current function, its physical attributes and operational capacity (measurement), its future operational capacity (time), the vulnerabilities and risks it may face (statistical) and its relationships to other components within the infrastructure system both physically and strategically (organisational). The first four of these 'dimensions' describe an asset's intrinsic physical state both at present and potentially into the future, and provide a template for how one may go about building a set of database schema to describe assets through to entire infrastructure systems.

Organisational generality provides a means by which the relationships between infrastructure assets and other infrastructure components (network, system and sector) can be understood, providing a convenient mechanism by which, at least conceptually, both horizontal and vertical dependencies and interdependencies can be handled in data management terms. Vertical organisational relationships allow one to create aggregations of assets to form networks, aggregations of networks to form a system and/or sector representation. Equally, horizontal relationships allow the mapping of the dependencies and interdependencies that exist between individual assets, networks, systems and sectors to be represented. Such a perception of the organisational dimensions provides a naturally intuitive translation to implementation within a (relational) database framework, with the first four dimensions forming the key fields of the relations at different levels of the infrastructure data hierarchy, while the organisational dimension constitutes the vertical and horizontal relationships between these that are required to understand overall system connectivity.

While the five dimensions of infrastructure (assets) provide a convenient way by which to conceptually consider the requirements of system-of-systems infrastructure data management system, it is important to recognise that as one moves further up the infrastructure data hierarchy the nature of both the intrinsic attributes and relationships change. At the level of assets and networks the physical connections, flows, movements and exchanges that exist dominate and connectivity and topology are the key relationships of interest. Thus at such levels the key data representation requirements are the inherent physical spatial links that exist between different assets within networks and the geospatial dependencies/interdependencies that exist between different infrastructure sub-networks and the assets within these.

At the level of sectors and systems, while one would still require the ability to understand the topology and connectivity of the their constituent networks, the ability to represent the geo-temporal capacity, demand and performance characteristics dominate, both within a single sector or system, and also in terms of aggregated sectors and systems. Thus, a system-of-systems data management system must be able to handle the representation of sectors, systems and aggregations of these in a flexible manner, potentially having to account for different zonal administrative geographies and temporal quantisation of individual sectors and systems, as well attribution that is based less on in-situ physical measurements but rather on multiple metrics of performance under defined sets of strategies (see Chapter 2).

As one considers the data representation requirements of a system-of-systems infrastructure data management framework, it becomes clear that at increasing levels of aggregation the database management tools and functionality required become increasingly complex. For example, for individual assets one requires only the ability to encode the four intrinsic components. However, as one starts to build data representations of sectors, systems and their aggregations a wider range of functionality for creating, managing, searching and reporting the database contents are required. The key characteristics of the different infrastructure levels along with their key intrinsic and extrinsic characteristics and required database functionality are presented in Table 13.1.

Table 13.1 Key requirements of a national scale database system (NISMOD-DB) organised by different levels of infrastructure granularity

Infrastructure granularity	Individual assets	Individual network	Individual sector/system	Interdependent networks	Interdependent sectors/systems	System-of-systems cross-sectoral
Example	Pumping station Gas compressor Power station Telecoms mast Rail line	Electricity transmission Water supply network Road network	Energy Waste Water ICT Transport	Electricity/water Water/electricity Rail/electricity Road/electricity	Water and energy Transport and energy Waste and energy	Aggregations of multiple sectors; energy, waste, transport, water
Key characteristics	Function	Spatial and temporal topological dynamics	Spatial and temporal capacity and demand performance	Spatial and temporal topological dynamics	Spatial and temporal capacity and demand performance	Spatial and temporal aggregated capacity and demand performance
Geography	Location – point, polyline, polygon	Spatial network model	Zonal geography	Spatial network-of-networks	Cross product of n-zonal geographies	UK and administrative zones
Generic attributes	Location Function Condition Capacity Performance	Topology Condition Capacity Performance	Capacity Demand Performance	Topology Condition Capacity Performance	Capacity Demand Performance	Capacity Demand Performance
Relationships	None	Topology	None	Topology	Intra-sectoral	N-sectoral
Data management types	Spatial Tabular	Spatial Network topology Tabular	Spatial Tabular	Spatial Network topology Tabular	Spatial Tabular	Tabular
Key tasks	Build geometry Build tables	Build geometry Build tables Build network Verify network Clean network	Build geometry Build tables Clean tables Order tuples Search tuples	Build geometry Build tables Build networks Build spatial interdependencies Verify spatial interdependencies	Build geometry Build tables Clean tables Order tuples Search tuples Build sectoral relationships Metadata provenance Group sectoral	Build geometry Build tables Clean tables Order tuples Search tuples Build sectoral relationships Metadata provenance Group sectoral Aggregate results Calculate metrics

13.3 The NISMOD-DB database system

In response to the requirement for a system-of-systems approach to infrastructure data management a bespoke database management system has been developed for the U.K. called NISMOD-DB (National Infrastructure Modelling Database). NISMOD-DB aims to collate, manage and provide infrastructure data for the entire U.K. across the different hierarchical levels presented in the previous section; ranging from complete geospatial inventories of assets within a particular infrastructure network, through to aggregated metrics pertaining to the performance of multiple interdependent infrastructure sectors. To date NISMOD-DB has been primarily used to support the data management requirements of a suite of infrastructure models ranging from interdependent infrastructure network risk analysis (NISMOD-RV) through to high-level long-term capacity demand modelling between multiple infrastructure sectors (NISMOD-LP) (Figure 13.2).

NISMOD-DB has been designed and implemented in a modular manner and thus comprises a suite of database management modules which constitute the overall database management system (Figure 13.2). These modules are:

i. The physical database: a *PostgreSQL* RDBMS, along with its spatial extension *PostGIS*. *PostgreSQL/PostGIS* allows the full functionality of a relational database management system to be employed with the ability via the *PostGIS* extension to represent spatial data layers in the form of geometry tables (Obe and Hsu, 2011).

ii. NISMOD-DB Scripts: a suite of scripts that have been developed for the day-to-day management of the data in NISMOD-DB. Generic scripts for data loading, cleaning, relation initialisation, relationship mapping and basic queries have been developed as standard SQL scripts. Advanced management and analysis scripts have been developed using the PostgreSQL procedural programming language pgSQL.

iii. NISMOD-DB Networks: a set of specialist software that has been developed for the construction, storage, management and analysis of spatial infrastructure network models including interdependent infrastructure network models. The software is written in Python and employs the Python network module NetworkX which is widely used in network complexity analysis.

iv. NISMOD-DB Vis: comprises a suite of software tools that allow users to interact with the data and information in the NISMOD-DB database and visualise data-sets and result-sets. Visualisation tools have been written using a variety of software packages including Highcharts JS, OpenLayers, jQuery, Flot and D3js and are coupled directly to the database store using GeoJSON.

Within the *PostgreSQL/PostGIS* database of NISMOD-DB relations (tables) are organised into logical groups for management purposes on the basis of whether they are (i) used in NISMOD-LP for the long-term capacity/demand modelling (Chapter 2); (ii) standard relations (tables) for the NISMOD-RV risks and vulnerability modelling and analysis or (iii) spatio-topological infrastructure network models or interdependent network-of-networks models used in the NISMOD-RV modelling (Figure 13.2). Although the database

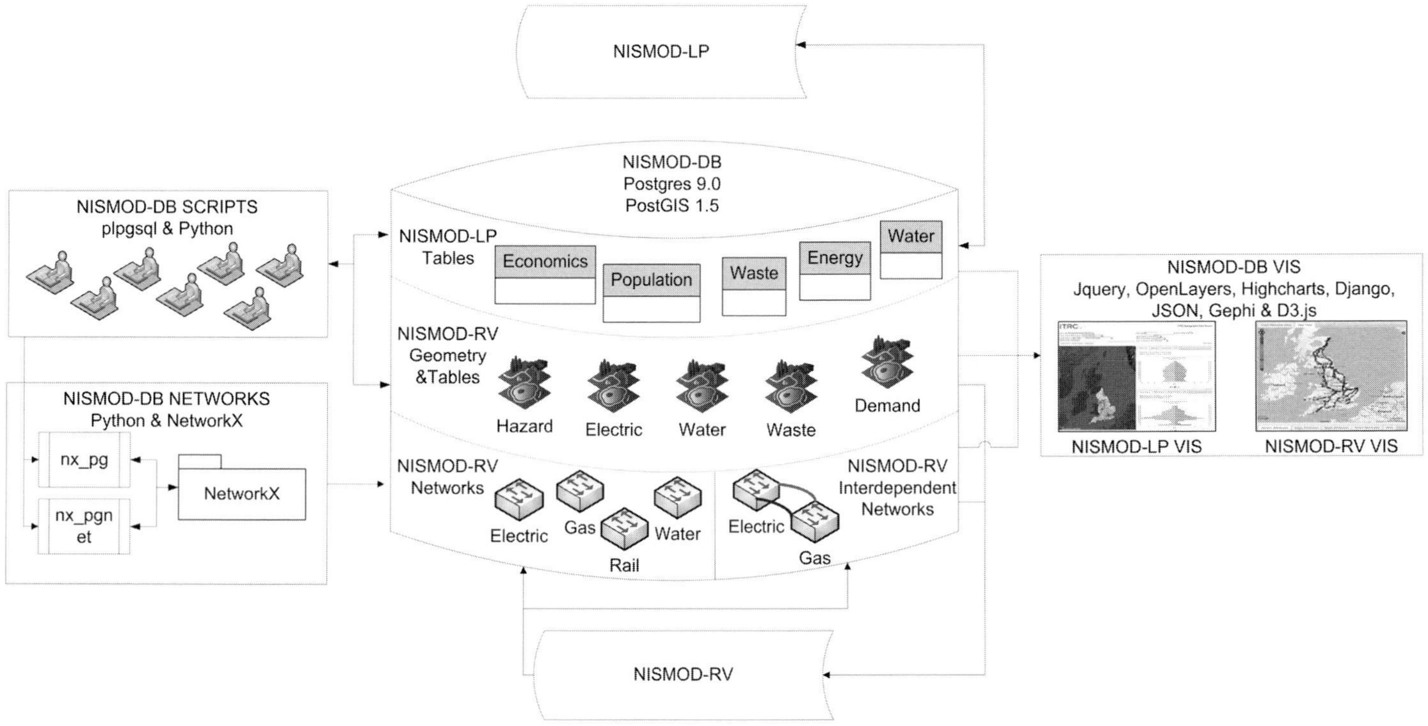

Figure 13.2 The overall organisation of the NISMOD-DB spatial database, analysis and visualisation framework and its linkages to other modelling software

has been organised into these logical groups, the potential exists, if required, to create relationships between the components within NISMOD-LP relating to the broad-scale performance of an infrastructure sector/system with the underlying physical networks and assets within the NISMOD-RV components. This flexibility, conceptually at least, facilitates analysis of individual assets through to a system-of-systems representation of national infrastructure across multiple spatial scales.

Throughout the design and implementation of NISMOD-DB the retention of and ability to present the geospatial characteristics of the national data-sets stored has been a key consideration. The use of the *PostGIS* extension to *PostgreSQL* provides a relatively straightforward means by which the location and geometry of features can be represented via its addition of a geometry field to the normal relation structure of *PostgreSQL*. This facilitates the representation of heterogeneous administrative geographies of the different sectors and systems represented in the NISMOD-LP section of NISMOD-DB (e.g. census geographies for demographics compared with water resource zones of water companies), as well as the representation of individual assets of an individual infrastructure network. The *PostGIS* extension also provides a wide range of spatial operators and functions, over and above the common management tools of a relational databases system (Obe and Hsu, 2011). These have been used extensively in NISMOD-DB to develop bespoke management, processing and analysis software scripts in NISMOD-DB Scripts in relation to the geospatial characteristics of the infrastructure assets, networks, systems/sectors stored in NISMOD-DB.

13.4 Using the database to enable system-of-systems modelling of long-term infrastructure performance

The database acts as a result-set 'broker' between the sector system models of NISMOD-LP. Model result-sets are written to the database, at which time dependent models can make a request to the database for the result-set fields they require for their own parameterisation. All of the infrastructure system models rely on the results of the demographic and economic models, while, for example, the energy supply model requires results from energy demand model to execute (Figure 4.1).

In order to act as a data conduit the NISMOD-DB tables for NISMOD-LP have been organised by infrastructure sector, alongside the supporting data that are used to estimate demand for infrastructure services; namely, economic, population (demographic), transport, energy supply, energy demand, water supply, wastewater and solid waste. In general, each relation for a particular sector model represents an entire single model run for the time-period of interest. As each model generates spatially zonal (administrative) time-series quantised (e.g. yearly through to decadal increments) capacity and/or demand predictions, each relation has, in addition to its primary key, fields to represent the spatial, temporal and measurement outputs. For example, Figure 13.3 shows an annotated table from NISMOD-DB

	Population character	Year integer	Gender character varying	Category character varying	Location character varying	Value precision	Pop inte	gor_id integer	id [PK] integer
1	A	2010	Persons	Total	London	7811627.123	1	1	81
2	A	2010	Persons	Total	South East	8448470.135	1	2	86
3	A	2010	Persons	Total	East of England	5787151.503	1	3	80
4	A	2010	Persons	Total	South West	5246325.012	1	4	87
5	A	2010	Persons	Total	West Midlands	5461423.726	1	5	89
6	A	2010	Persons	Total	East Midlands	4479772.324	1	6	79
7	A	2010	Persons	Total	Yorkshire and the Humber	5286785.911	1	7	90
8	A	2010	Persons	Total	North West	6933415.294	1	8	83
9	A	2010	Persons	Total	North East	2592934.429	1	9	82
10	A	2010	Persons	Total	Wales	3043932.443	1	10	88
11	A	2010	Persons	Total	Scotland	5212979.612	1	11	85
		Year attribute: temporal dimension			Spatial attribute: geographic dimension	Population attribute: metric dimension			Primary Key: unique record identifier

Figure 13.3 Example table of NISMOD-LP results encoded in the NISMOD-DB database system

of one of the demographic population model runs highlighting the fields representing the primary key, temporal, spatial and measurement aspect of this result-set run.

While the internal structure of individual result-set tables in NISMOD-DB for the infrastructure sector models are relatively straightforward, the tables and relationships required to represent an entire infrastructure sector can result in a complex database structure. This is because many infrastructure sector models involve complex relationships and interactions, with intra-system and network interactions expressed in the modelling outputs; information that needs to be fully represented and encoded in NISMOD-DB. For example, the transport system model produces results for different transport modes including road, rail, air and sea-freight in terms of their individual demand and also their intra-relationships (as a sector). In order to fully express these in the relational format used in NISMOD-DB requires in excess of ninety separate tables. Figure 13.4 shows a generalised entity-relationship model that expresses in an abstract manner the resulting tables and relationships required for the transport sector model.

In addition to individual sector tables and relationships, we have developed in NISMOD-DB a bespoke metadata high-level set of tables and relationships, along with associated database scripts that allow us to manage all NISMOD-LP data and results, undertake provenance management of the result-sets generated by NISMOD-LP and also generate the required reporting tables, relationships and metrics required for subsequent reporting via the NISMOD-DB Vis dashboard tools (Section 13.6). Figure 13.5 shows the overall structure of the metadata components of NISMOD-DB in relation to the main model tables and functions of NISMOD-LP.

The management and reporting tables, relationships and scripts allow authorised users to group arbitrary sub-sets of model runs across multiple strategies into bespoke result-set packages which are stored persistently in the database, such that they can be edited, audited, queried, analysed and if required exported/downloaded from the database. Suitably

Rail Zone

Transport Input Tables: RailZone

TR_I_RailZoneElasticities_Base
TR_I_RailZoneElasticities_Run
TR_IO_RailZone
TR_IO_RailZoneElectrificationDates
TR_IO_RailZoneExternalVariables

Transport Output Tables: RailZone

TR_O_RailZoneOutputData

Rail Link

Transport Input Tables: RailLink

TR_I_RailElasticitiesLink_Base
TR_I_RailElasticitiesLink_Run
TR_IO_RailLink
TR_IO_RailLinkElectrificationDates
TR_IO_RailLinkExternalVariables
TR_IO_RailLinkNewCapacity

Transport Output Tables: RailLink

TR_O_RailLinkFuelConsumption
TR_O_RailLinkNewCapacity_Added
TR_O_RailLinkOutputData

Road Link/Flow

Transport Input Tables: Road Link/Flow

TR_I_RoadLinkElasticities_Base
TR_I_RoadLinkElasticities_Run
TR_IO_RoadLink_Hourly
TR_IO_RoadLink_Annual
TR_IO_RoadLinkExternalVariables
TR_IO_RoadLinkNewCapacity
TR_I_VehicleFuelConsumption_Base
TR_I_VehicleFuelConsumption_Run

Transport Output Tables: Road Link/Flow

TR_O_RoadLinkNewCapacity_Added
TR_O_RoadOutputFlows

System and Lookup

System Tables

ISL_Climate_Scenarios
ISL_Constants
ISL_ErrorLog
ISL_Logic_Levels
ISL_MessageLog
ISL_ModelRuns
ISL_ModelVersion
ISL_Parameters_Base
ISL_Parameters_Run
ISL_Socioeconomic_Scenarios
ISL_Strategy_Model_Order_Base
ISL_Strategy_Model_Order_Run
ISL_Strategy_Portfolios
SYS_Sources
SYS_Units
SYS_Packages
SYS_LU_Sectors

Transport Lookup Tables

TR_LU_AirFlow_InitialData
TR_LU_AirFlows
TR_LU_AirNodeCapacityChange
TR_LU_AirNode_InitialData
TR_LU_AirNodes
TR_LU_CommonVariables
TR_LU_Countries
TR_LU_FareE
TR_LU_GORMappingIDS
TR_LU_GORRoadRailZoneMappingIDS
TR_LU_Initial_SeaFreight
TR_LU_LADRoadRailZoneMappingIDS
TR_LU_Modules
TR_LU_RailLinkCapacityChange
TR_LU_RailLinkFlows
TR_LU_RailLink_EVScaling
TR_LU_RailLink_ElectrificationSchemes
TR_LU_RailLink_InitialData
TR_LU_RailLinks
TR_LU_RailZoneCapacityChange
TR_LU_RailZone_EVScaling
TR_LU_RailZone_ElectrificationSchemes
TR_LU_RailZone_InitialData
TR_LU_RailZones
TR_LU_RoadLinkCapacityChange
TR_LU_RoadLink_DailyProfile
TR_LU_RoadLink_FreeFlowSpeeds
TR_LU_RoadLink_InitialData
TR_LU_RoadFlows
TR_LU_RoadZoneCapacityChange
TR_LU_RoadZone_InitialData
TR_LU_RoadZone_GOR_GVA_Proportions
TR_LU_RoadZones
TR_LU_SeaFreightCapacityChange
TR_LU_SeaFreight_InitialData
TR_LU_SeaPorts
TR_LU_VehicleCategoryFuelConsumption

Air Node/Flow

Transport Input Tables: Air

TR_I_AirElasticities_Base
TR_I_AirElasticities_Run
TR_IO_AirFlow
TR_IO_AirFlow_ExternalVariables
TR_IO_AirFuelConsumption
TR_IO_AirNode
TR_IO_AirNodeExternalVariables
TR_IO_AirNodeNewCapacity

Transport Output Tables: Air

TR_O_AirFlowOutputData
TR_O_AirNodeNewCapacity_Added
TR_O_AirNodeOutputData

Sea Freight

Transport Input Tables: SeaFreight

TR_I_SeaFreightElasticities_Base
TR_I_SeaFreightElasticities_Run
TR_IO_SeaFreight

Transport Output Tables: SeaFreight

TR_O_SeaFreightNewCapacity_Added
TR_O_SeaFreightOutputData

Road Zone

Transport Input Tables: RoadZone

TR_I_RoadZoneElasticities_Base
TR_I_RoadZoneElasticities_Run
TR_IO_RoadZone
TR_IO_RoadZoneExternalVariables
TR_IO_RoadZoneNewCapacity

Transport Output Tables: RoadZone

TR_O_RoadZoneOutputData
TR_O_RoadZoneNewCapacity_Added
TR_O_RoadZoneFuelConsumption

Figure 13.4   A generalised entity-relationship model of the transport sector database tables in NISMOD-DB

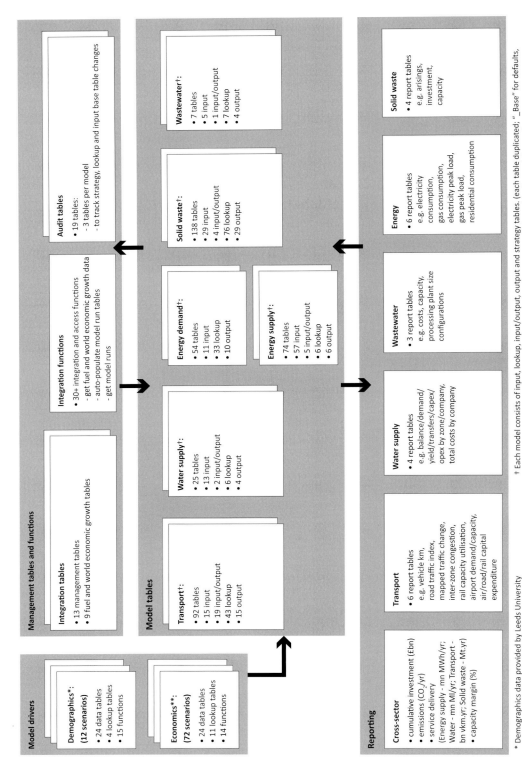

Model drivers

Demographics*:
(12 scenarios)
- 24 data tables
- 4 lookup tables
- 15 functions

Economics**:
(72 scenarios)
- 24 data tables
- 11 lookup tables
- 14 functions

Management tables and functions

Integration tables
- 13 management tables
- 9 fuel and world economic growth tables

Integration functions
- 30+ integration and access functions
  - get fuel and world economic growth data
  - auto-populate model run tables
  - get model runs

Audit tables
- 19 tables:
  - 3 tables per model
  - to track strategy, lookup and input base table changes

Model tables

Transport†:
- 92 tables
- 15 input
- 19 input/output
- 43 lookup
- 15 output

Water supply†:
- 25 tables
- 13 input
- 2 input/output
- 6 lookup
- 4 output

Energy demand†:
- 54 tables
- 11 input
- 33 lookup
- 10 output

Energy supply†:
- 74 tables
- 57 input
- 5 input/output
- 6 lookup
- 6 output

Solid waste†:
- 138 tables
- 29 input
- 4 input/output
- 76 lookup
- 29 output

Wastewater†:
- 7 tables
- 5 input
- 1 input/output
- 7 lookup
- 4 output

Reporting

Cross-sector
- cumulative investment (£bn)
- emissions (CO₂/yr)
- service delivery
  (Energy supply - mn MWh/yr; Transport - bn vkm.yr; Solid waste - Mt.yr)
- capacity margin (%)

Transport
- 6 report tables
  e.g. vehicle km, road traffic index, mapped traffic change, inter-zone congestion, rail capacity utilisation, airport demand/capacity, air/road/rail capital expenditure

Water supply
- 4 report tables
  e.g. balance/demand/ yield/transfers/capex/ opex by zone/company, total costs by company

Wastewater
- 3 report tables
  e.g. costs, capacity, processing plant size configurations

Energy
- 6 report tables
  e.g. electricity consumption, gas consumption, electricity peak load, gas peak load, residential consumption

Solid waste
- 4 report tables
  e.g. arisings, investment, capacity

* Demographics data provided by Leeds University

** Economics data provided by Cambridge Econometrics

† Each model consists of input, lookup, input/output, output and strategy tables. (each table duplicated;
"_Base" for defaults, "_Run" for model execution. Each model also has 1 function to auto-populate tables for model execution.
All information subject to change due to on-going development.

Figure 13.5 The high-level system control metadata tables and relationships developed in NISMOD-DB for the management of NISMOD-LP result-sets

authenticated users can query a package and/or strategy result set and obtain a detailed listing of (i) the model runs that comprise the package and/or strategy; (ii) the endogenous cross-sectorial parameters used for entire model runs; (iii) the individual exogenous sector model parameters and state variables employed; and (iv) the precise result-sets from one individual sectoral model that were used to parameterise any subsequent sectoral model (e.g. the case of the energy demand and supply models).

13.5 Using the database to enable analysis of risks and vulnerability in interdependent infrastructure networks

In addition to providing the underpinning data management and reporting structure for long-term infrastructure planning and decision analysis in NISMOD-LP, the database has also been extensively utilised to encode, manage and analyse physical infrastructure ranging from individual assets through to national scale interdependent network-of-networks representations of infrastructure. Much of the work undertaken at the level of individual assets, collections of assets and their representation as infrastructure networks has been conducted in relation to gaining a better understanding of spatio-temporal vulnerability and risk of infrastructure within the U.K. to a range of hazards (NISMOD-RV).

In order to model, analyse and quantify infrastructure risk it is important that both the hazards of interest and also the assets that may be affected are represented in a spatially explicit manner. Primary data on infrastructure assets are stored as PostGIS spatial geometry tables as either point, polyline or polygonal features along with their corresponding attribution such as function, operating conditions/ranges and where appropriate fragility information. Hazards are predominantly represented as gridded spatial fields that provide measurements of the magnitude of the hazards at specific locations within their spatial domain. Again, data within NISMOD-DB is organised and grouped by infrastructure system; namely, spatial hazards, energy (electric transmission/distribution, gas transmission/distribution), water (supply and waste), transport (road, rail, air, sea (ports)) and solid waste.

Until recently few database management systems have had the ability to represent, manage and process data that is naturally expressed as a topological network. Moreover, the extension to representing spatio-topological network structures and combinations of these to build network-of-networks models does not exist within existing database management systems. However, it is important to represent, manage and process spatial infrastructure networks (e.g. electricity transmission, electricity distribution and water supply) and combine these as interdependent networks (e.g. explicit representation of the dependencies and interdependency between electricity transmission and water supply networks) in order to understand infrastructure network vulnerability and risk.

In NISMOD-DB the standard database schema of *PostgreSQL/PostGIS* has been extended to develop a spatial network database schema that can represent, manage and analyse collections of infrastructure assets as a single networked system (entity). Moreover, the database schema developed allows an explicit encoding of interdependency between different infrastructure networks to be stored. Figure 13.6 shows the new database schema developed in NISMOD-DB. Object-orientated inheritance is used to create an instance of a spatial infrastructure network from the base schema when presented with geometry tables representing infrastructure asset point and polyline features (e.g. an electricity transmission network model from electricity substation points and transmission polylines). Each network comprises a Nodes table that contains both the attributes and geometry of the node data, and Edges and Edge_Geometry tables that store the attributes and geometry of the edge data, respectively (Figure 13.6). Using inheritance a potentially arbitrary number of network models can be stored in NISMOD-DB along with details on the explicit dependency relationships that exist between nodes of one network and the nodes of a second one.

Figure 13.7 shows an example network-of-networks representation for part of the U.K. using the NISMOD interdependent network database schema. Such network representations are created by a series of specialised scripts NISMOD-DB Networks that employ *NetworkX* in order to check for topological consistency when constructing a new network before writing it back to NISMOD-DB. Additional processing scripts have been developed for the national infrastructure networks in NISMOD-DB to perform tasks such as database maintenance and pre- and post-network construction filtering and analysis. For example, modules have been developed for pre- and post-network construction cleaning and filtering to remove or correct systematic geometry or attribute errors that result in incorrect network models being generated. For example, software modules that connect hanging edges to nodes can be used to automatically remedy digitisation errors such as the example shown in Figure 13.8 where the original U.K. National Grid Gas Transmission pipes do not intersect spatially with the gas compressor. The corrected version has been derived in this case by creating completed edges between the pipelines and the compressor node on the basis of a comparison of the gas pipeline and gas node names.

The NISMOD-DB spatial infrastructure network database schema and its associated software modules have been used to represent and analyse a range of infrastructure networks including interdependent infrastructure networks sensitivity to extreme hazards (Pant et al., 2014), understanding customer spatial demands on infrastructure networks (Thacker et al., 2014), infrastructure network modelling in developing nations (Kennedy-Walker et al., 2014) and for the modelling of failures between interdependent networks such as electricity transmission and critical public transport networks such as the U.K. London underground (Barr et al., 2012, 2013; Robson et al., 2015). Figure 13.9 shows the interdependent infrastructure networks modelled by Barr et al. (2013) which employed the NISMOD-DB schema to generate an interdependent network model of electricity transmission and the London underground system. Subsequent cascading failure modelling (Johansson & Hassel, 2010) was used for several failure types (Crucitti et al., 2004; Dueñas-Osorio et al., 2007; Mishkovski et al., 2011) to generate graph metrics (Albert & Barabási, 2002; Boccaletti et al., 2006) of the failure dependency of the London tube network with respect to the electricity transmission (Figure 13.10).

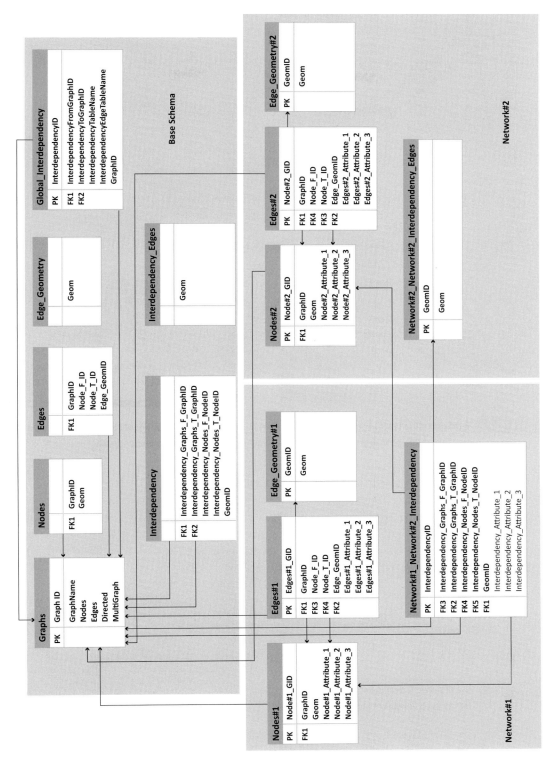

Figure 13.6 Entity relationship diagram showing the database schema developed for the representation of interdependent spatial infrastructure networks

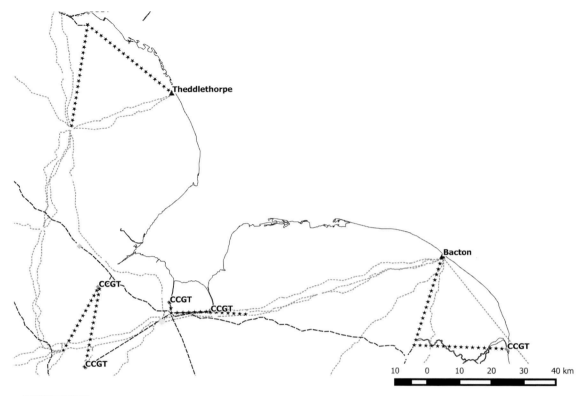

Figure 13.7 A network-of-networks representation of U.K. national grid electricity and gas transmission networks for part of east coast of the U.K. generated using the NISMOD-DB interdependent network database schema

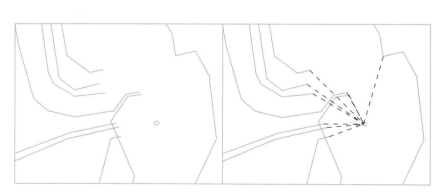

Figure 13.8 Use of network cleaning scripts to construct a topologically valid infrastructure network model for the national gas transmission network of the U.K.

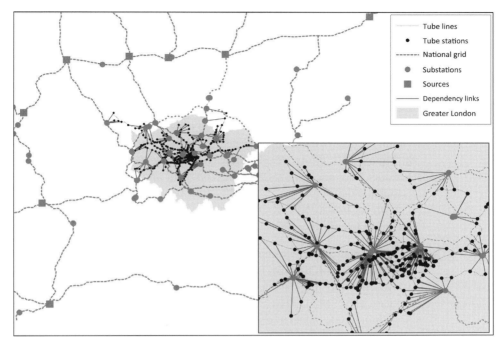

Figure 13.9 The electricity grid for southeast England and the London tube network. Inset: Central London showing spatial dependency edges (mappings) between electricity substations and tube stations

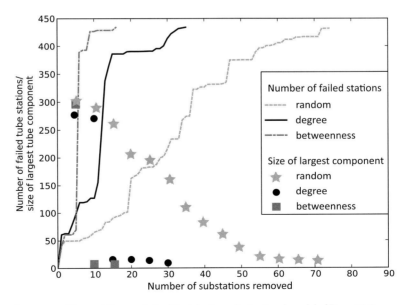

Figure 13.10 Failure metrics generated from a failure analysis of the interdependent network model of Figure 13.9

13.6 Decision support and visualisation

As well as providing database functionality, NISMOD-DB is equipped with visualisation tools to allow users to interact with the often complex high-dimensionality data and result-sets represented within such extended database architectures. Importantly, the type of tools required to achieve this will be different for different infrastructure components within the overall hierarchical system-of-systems representation of infrastructure presented in Figure 13.1. Moreover, in order for such tools to be of utility one must carefully consider how one exposes and presents data and results to different types of end user or stakeholder.

With regards to NISMOD-DB four primary groups of user have been recognised that will need to engage with data and result-sets:

i. **Administrative:** Require access to the finest level of granularity of the data and result-sets in NISMOD-DB. These users will have permission to add, modify and delete schema. Existing database administration tools (e.g. *PGAdmin* for *PostgreSQL* in the case of NISMOD-DB) are essential. However, such users would also benefit from data mining tools linked to visual analytics in order to assist the recognition of data outliers, data blunders and systematic errors within the database tables etc.

ii. **Researcher:** Require access to groups of data and result-sets but not the individual tables within the database. They require the ability to interrogate and explore high-dimensionality data and results in a detailed manner in order to gain new understanding and insights on infrastructure performance currently and into the future. Visualisation and reporting tools based-around data cube analytics that allow a detailed level of interaction to be effectively conveyed both quantitatively and qualitatively are essential for this group.

iii. **Decision maker:** Require the ability to engage with the data and results in a visual manner often in terms of an aggregated, generalised or summarised presentation. Data will need to be linked to sources and results presented in a visual manner with a focus on conveying the key points rather than an exhaustive description of data or result-set. However, such users should still have the ability to explore and query into the details underpinning any high-level abstracted presentations of data and results.

iv. **Public:** Require access to summary result information in the form of easy to understand visualisations accessed through tools, such as wizards, that guide the user through the process of querying the information.

Efforts into the development of NISMOD-DB has focused, to date, on providing tools for decision-makers such that they can engage with the data and results generated from the overall NISMOD infrastructure modelling framework. This has been achieved by developing a series of prototype web-enabled visualisation dashboards that directly link to the NISMOD-DB database. This allows users, via the use of synchronous requests, to directly retrieve requested data/results from the database and then visualise it appropriately within the particular tool in question. Such a *query-retrieve-visualise* approach allows users flexibility in terms of engaging with the NISMOD outputs but at an appropriate level of abstraction and generalisation such that much of the data complexity is hidden from the user.

Two primary NISMOD Vis (the visualisation module of NISMOD-DB) visualisation tool-sets have been developed to date; one for the presentation of the results generated by the long-term capacity/demand modelling of NISMOD-LP and one for the analysis of the data and results of the infrastructure network vulnerability and risk analysis of NISMOD-RV. Both tools have been constructed using a *Django-enabled* web framework along with *OpenLayers* map client, *Google Maps Application Programming Interface (API)* and *Highcharts*.

The visualisation tool-sets for understanding infrastructure long-term capacity/demand are structured around several components: (i) the presentation of the exogenous demographic and economic drivers of the infrastructure sector models both in terms of the disaggregated administrative geographies of the U.K. and also the corresponding population age-distribution (Figures 13.11 and 13.12); (ii) performance of single sectors such as energy supply (Figure 13.13) where the performance of different future strategies of delivery can be explored again in terms of the administrative geography and also key performance metrics; (iii) bi-sectoral performance analysis that investigate the dependencies that exist between two sectors in terms of key metrics such as Figure 13.14, where water usage from energy generation under different energy supply sub-strategies is visualised over time; and (iv) cross-sectoral performance where multiple sectors (energy, transport, water, wastewater, etc.) are represented in terms of their performance for multiple future strategies based on performance metrics of interest such as Figure 13.15, where four possible future strategies are visualised in terms of the cross-sector CO_2 emissions and future investment (Alderson et al., 2014).

The infrastructure networks visualisation tool for NISMOD-RV comprises a web-enabled dashboard that allows multiple dimensions of infrastructure networks encoded using the interdependent network database schema presented in Section 13.2 to be visualised and explored. The dashboard comprises two main view windows; the map view which shows the actual geography and spatial layout of an infrastructure network in terms of its edges and nodes, and the graph/network view which provides a graph-rendering of the topological layout of the infrastructure network. For example, Figure 13.16 shows the two different views for the U.K. national gas transmission network model held in NISMOD-DB.

Within each view it is possible to explore a range of different attributes for any visualised infrastructure network model; generic network attributes such as the name of the network loaded and its rendered style, edge attributes describing the coding, function and layout of the edges (links) within a visualised infrastructure network, node attributes which present the different categories of nodes in the network and corresponding layout feature, and graph attributes which report a range of metrics on the topological structure of the network via a coupling to the NetworkX python modules metric functions. The last of these functions has been widely used to check the topological validity of the network models encoded

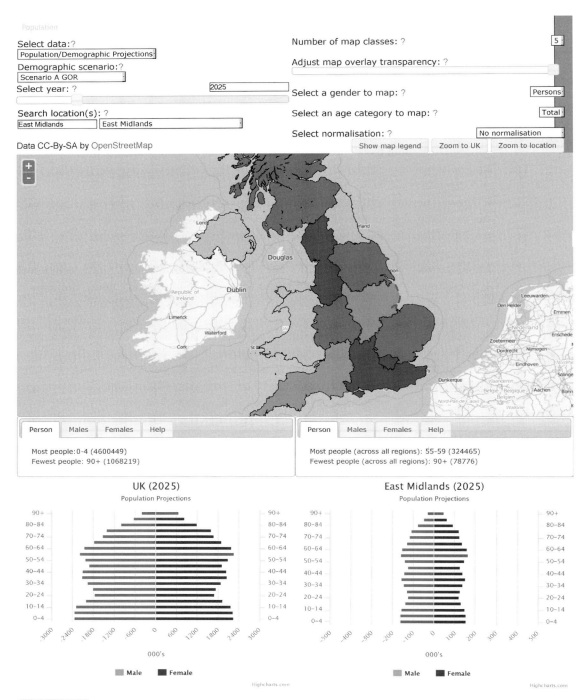

Figure 13.11   NISMOD-DB Vis web-enabled visualisation dashboard for the analysis of demographic exogenous drivers for the NISMOD-LP sector models

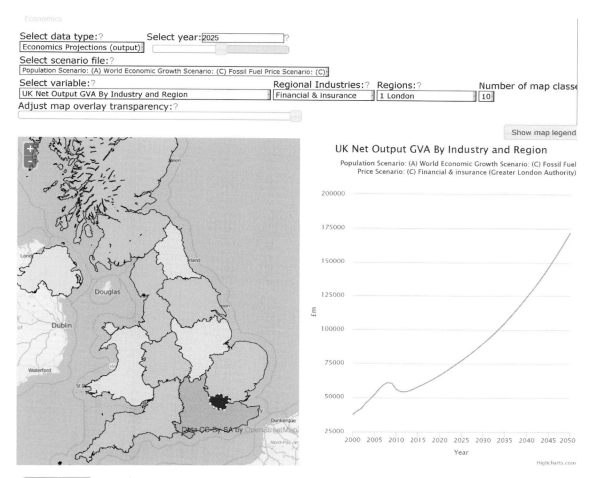

Figure 13.12 NISMOD-DB Vis web-enabled visualisation dashboard for the analysis of demographic exogenous drivers for the NISMOD-LP sector models

using the interdependent network schema, for example, applying constraints that ensure network models are connected (e.g. rail network) or if they are comprised of multiple components, that they do not have any isolated nodes (e.g. a regional electricity distribution network).

The final way a user can visualise and engage with an infrastructure network is via the interactive select–query–report functionality of the dashboard. This allows any feature within an infrastructure network to be interactively selected. This results in a query being sent to NISMOD-DB and the record corresponding to the feature in the infrastructure network model being retrieved and returned to the dashboard, where it is presented to the user.

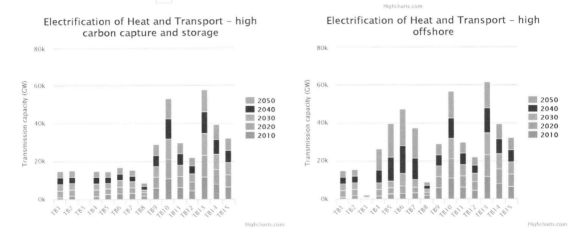

Figure 13.13 NISMOD-DB Vis visualisation of energy supply power transmission network expansion for different energy supply strategies between 2010 and 2050

13.7 Conclusion

In order to manage and plan infrastructure systems at a national scale, suitable analytical tools are required. This has been recognised as being particularly important in relation to an improved understanding of interdependent infrastructure systems (Rinaldi et al., 2001).

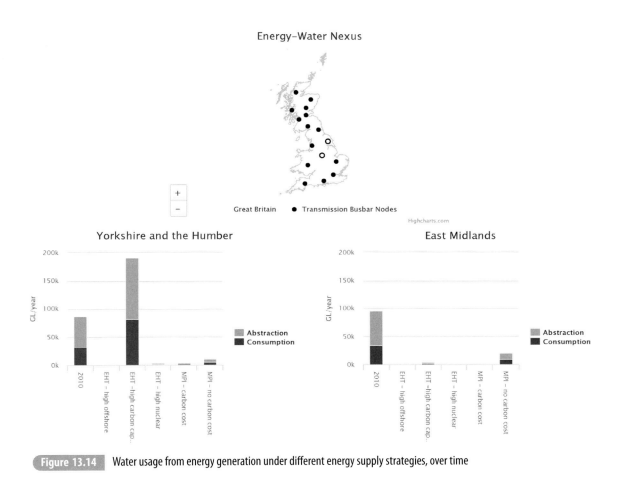

Figure 13.14   Water usage from energy generation under different energy supply strategies, over time

Database systems and associated tools will play a critical role in the future management, analysis and modelling of individual and interdependent infrastructure systems (Rinaldi et al., 2001). Indeed, we argue that it would be practically impossible to conduct efficient, auditable and comprehensive simulation modelling and decision analysis without a well-structured database to marshal the steps in the analysis, host the results and provide an audit trail through the analysis.

   In this chapter, we have described the development of a national infrastructure database management system (NISMOD-DB). This database system offers a flexible environment for the storage of a range of different infrastructure data, both spatial and aspatial. Furthermore, the development of a bespoke interdependent infrastructure network schema opens up the possibility of using the database as the foundation for complex infrastructure simulation modelling.

Policy Option(s):

☑ Minimal Policy Intervention (P-MI)
☑ Long-term Capacity Expansion (P-CE)
☑ Increasing System Efficiency (P-SE)
☑ System Restructuring (P-SR)
☑ All policies

Update

For more information, please follow this link to the interim report results: ITRC Interim Results, 2014
For more information about the UK Infrastructure Transitions Research Consortium (ITRC), please follow this link.

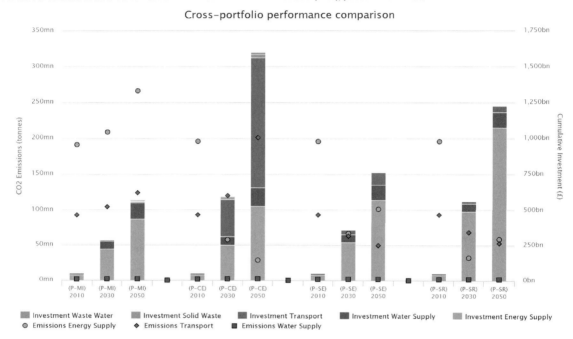

Cross–portfolio performance comparison

Figure 13.15 Multi-sector contributions to CO_2 and investment for multiple cross-sector strategies

Visualisation is an essential aspect of the database functionality to enable modellers and analysts to scrutinise and interpret the results. The NISMOD-DB system provides high-dimensional spatial-temporal data sets which need to be visualised, including via maps and time series plots. The visualisation facilities are also essential to communicate both the concepts and the content of the results to a range of decision-makers and other stakeholders. As decision-makers become more engaged in the analysis, they will need to query the data and will pose new questions. NISMOD-DB provides the facilities to enable this type of interactive engagement in the system-of-systems analysis.

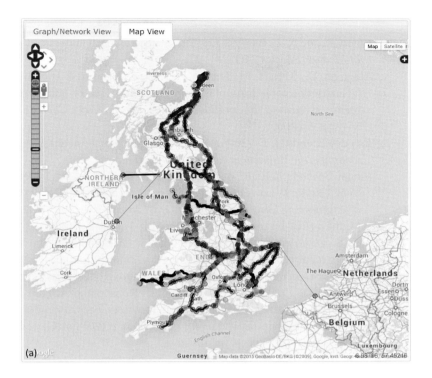

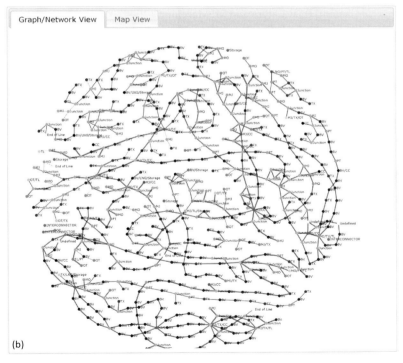

Figure 13.16 Two examples of the infrastructure network visualisation dashboard of NISMOD-DB VIS generated from the U.K. national gas transmission network model held in NISMOD-DB

References

Albert, R. and A.-L. Barabási (2002). "Statistical mechanics of complex networks." *Reviews of Modern Physics* 74(1): 47–97.

Alderson, D., S. Barr, M. Tran, J. W. Hall, A. Otto, A. J. Hickford and E. Byers (2014). Visualisation tools for multi-perspective, cross-sector, long-term infrastructure performance evaluation. *International Symposium for Next Generation Infrastructure 2014*. Vienna, Austria.

Amadi-Echendu, J. E. (2006). New paradigms for physical asset management. *3rd World Congress on Maintenance*. Basel, Switzerland.

Amadi-Echendu, J. E., R. Willett, K. Brown, T. Hope, J. Lee, J. Mathew, N. Vyas and B.-S. Yang (2010). What is Engineering Asset Management? *Definitions, concepts and scope of engineering asset management*. J. E. Amadi-Echendu, K. Brown, R. Willett and J. Mathew (Eds.). London, UK, Springer. 1: 3–16.

Barr, S., D. Alderson, C. Robson, A. Otto, S. Thacker and R. Pant (2013). A national scale infrastructure database and modelling environment for the UK. *International Symposium for Next Generation Infrastructure*. Wollongong, Australia.

Barr, S. L., T. Holderness, D. Alderson, C. Robson and A. Ford (2012). An open source relational database schema and system for the analysis of large scale spatially interdependent infrastructure networks. *4th Open Source GIS Conference (OSGIS)*. Nottingham, UK, Nottingham Geospatial Institute.

Boccaletti, S., V. Latora, Y. Moreno, M. Chavez and D. U. Hwang (2006). "Complex networks: structure and dynamics." *Physics Reports* 424(4–5): 175–308.

Crucitti, P., V. Latora and M. Marchiori (2004). "A topological analysis of the Italian electric power grid." *Physica A: Statistical Mechanics and Its Applications* 338(1–2): 92–97.

CST (2009). *A national infrastructure for the 21st century*. London, UK, Council for Science and Technology.

Dueñas-Osorio, L., J. Craig, B. Goodno and A. Bostrom (2007). "Interdependent response of networked systems." *Journal of Infrastructure Systems* 13(3): 185–194.

Haider, A. (2013). *Information systems for engineering and infrastructure asset management*, Springer Gabler.

Hudson, W. R., R. Hass and W. Uddin (1997). *Infrastructure management: integrating design, construction, maintenance, rehabilitation and renovation*. New York, NY, USA, McGraw-Hill.

Johansson, J. and H. Hassel (2010). "An approach for modelling interdependent infrastructures in the context of vulnerability analysis." *Reliability Engineering & System Safety* 95(12): 1335–1344.

Kennedy-Walker, R., T. Holderness, D. Alderson, B. E. Evans and S. L. Barr (2014). Network modelling for road-based faecal sludge management. *Proceedings of the Institution of Civil Engineers: Municipal Engineer*, Institution of Civil Engineers/Thomas Telford.

Mishkovski, I., M. Biey and L. Kocarev (2011). "Vulnerability of complex networks." *Communications in Nonlinear Science and Numerical Simulation* 16(1): 341–349.

Nastasie, D. L., A. Koronios and A. Haider (2010). Integration through standards – an overview of internal information standards for engineering assets. *Definitions, concepts and scope of engineering asset management.* J. E. Amadi-Echendu, K. Brown, R. Willett and J. Mathew (Eds.), London, Springer. 1: 239–258.

Obe, R. and L. Hsu (2011). *PostGIS in action.* Greenwich, CT, USA, Manning Publications Co.

Otto, A., J. W. Hall, A. J. Hickford, R. J. Nicholls, D. Alderson and S. Barr (2014). "A quantified system-of-systems modelling framework for robust national infrastructure planning." *IEEE Systems Journal* 99: 1–12.

Pant, R., J. Hall, S. Barr and D. Alderson (2014). *Spatial risk analysis of interdependent infrastructures subjected to extreme hazards.* Vulnerability, uncertainty, and risk: quantification, mitigation, and management. M. Beer, S.-K. Au and J. W. Hall (Eds.) ASCE: 677–686.

Pederson, P., D. Dudenhoeffer, S. Hartley and M. Permann (2006). *Critical infrastructure interdependency modeling: a survey of US and international research.* Idaho Falls, ID, USA, Idaho National Laboratory.

RAE (2011). Infrastructure, engineering and climate change adaptation – ensuring services in an uncertain future. Report written by the Royal Academy of Engineering on behalf of Engineering the Future.

Robson, C., S. L. Barr, P. James and A. Ford (2015). Spatially modelling dependent infrastructure networks. *GISRUK 2015.* Leeds, UK.

Rinaldi, S. M., J. P. Peerenboom and T. K. Kelly (2001). "Identifying, understanding, and analyzing critical infrastructure interdependencies." *Control Systems Magazine, IEEE* 21(6): 11–25.

Thacker, S., R. Pant, J. W. Hall, S. L. Barr and D. Alderson (2014). Building interdependent infrastructure networks and customer assignment models for understanding spatial demands. *The future of national infrastructure systems and economic prosperity conference.* Cambridge, UK.

Woodhouse, J. (2014). "Standards in asset management: PAS55 to ISO55000." *Infrastructure Asset Management* 1(3): 57–59.

14 Governance of interdependent infrastructure networks

RALITSA HITEVA, JIM WATSON

14.1 Introduction

Governance of infrastructure relates to the interactions and decision-making amongst multiple actors that result in the delivery, financing and payment for infrastructure services. Taking the system-of-systems approach adopted in this book, this chapter will approach the governance of infrastructure with a particular emphasis upon the challenges raised by interdependence between infrastructure sectors, building on the U.K. and international examples. This will inform a discussion of the governance implications of cross-sector strategies that have been explored in Chapter 10.

Section 14.2 starts by discussing the key public and private stakeholders in the governance of infrastructure interdependencies and their interests. Section 14.3 discusses the emerging interest in the governance of infrastructure interdependencies, in joint infrastructure investment and coordination between sectors, and the stakeholders and processes through which these could be achieved. Section 14.4 discusses two case studies of infrastructure interdependencies (between electricity, ICT and private vehicles and between water and energy). It also reflects on the governance implications of the four cross-sector strategies discussed in Chapter 10. The aim is to consider how governance processes might need to change to address interdependencies between infrastructure sectors in the future. The chapter concludes with a summary in Section 14.5.

14.2 Stakeholders in the governance of infrastructure interdependency

During the twentieth century, the governance of most national infrastructure sectors has moved from a fragmented set of arrangements – with a mix of public and private provision – towards a national, market-led governance model. Principal features of this model are the introduction of competition into the infrastructure system to facilitate greater economic efficiency and reduce prices (Mitchell, 2008). Current U.K. infrastructures include (i) price regulated market driven infrastructures, such as ports and airports that are paid for by consumers; (ii) unregulated market driven consumer paid infrastructure, such as telecommunications and power plants; (iii) government funded infrastructure such as roads;

and (iv) private-public finance initiatives delivered by Local Authorities, such as waste (CST, 2009).

The evolution of institutions, rules, regulations and ownership arrangements has been accompanied by an increasingly diverse set of actors. The most prominent governance actors are national government departments, national economic regulators and national environmental regulators. The European Commission has played an increasingly important role, especially in environmental regulation and in pushing for more open markets and competition. Governance at the local scale has also become more pronounced – for example, through increasing interest of Local Authorities in energy and utility companies.

The state has multiple stakes in the governance of infrastructure – and the governance of interdependencies between infrastructures. These include optimising the role of infrastructures in economic growth, the need to meet public policy goals (e.g. emissions reductions), a potential role for government as an investor and the important security and political roles played by infrastructures (Bouchon, 2006). These three related roles for the state provide the basis for understanding its specific interests in the governance of infrastructure interdependencies.

14.2.1 The state as a guarantor of security

The 'traditional' role of the state as a protector is linked to its monopoly on issues of security. The state and its agents (national authorities and decision makers) are expected to guarantee national interests and to maintain the integrity and continuity of access to services (such as continued energy supply) to citizens within their territory. Developed economies, like the U.K. are also facing what Bouchon (2006) calls an expanding 'spectrum of threats', which include a range of manmade disasters such as terrorism and natural disasters. However, the ability to maintain integrity and continuity of access to services for citizens increasingly depends on more than one critical infrastructure and the interaction between them (Bouchon, 2006; Cabinet Office, 2011; Frontier Economics, 2012). The range of threats to infrastructure security that the state needs to take into account has changed over time, and is expected to change further in the future (Ministry of Defence, 2014). There is increasing global recognition that the awareness of and ability to govern (e.g. identify, protect, adapt and respond to) threats emerging from the growing interdependencies between critical sectors is fast becoming a focal point of guaranteeing security within national borders, whether in the context of continued energy supply or protection from cyber-attacks. This role for the state is challenging not only because of the complexities involved in understanding interdependencies but also because of the need to move away from established patterns of governing sectors independently from each other and develop new ways of working more closely across sectors and with a wider range of actors at multiple scales (such as regional authorities, the EU, local communities and global supply chains).

14.2.2 The state as guarantor of public goods and services

Another key role for the state is making sure that suitable regulatory arrangements and incentives are in place to ensure the provision of public goods and services, such as transportation and access to drinking water (*Public* commonly refers to 'of, related to or serving the economy'). Although the definition of public goods and services has changed over time, it has retained 'efficiency' of provision as a core value. Once produced, public goods and services can be consumed by an additional consumer at no additional cost, and consumers cannot be excluded from consumption. However, such non-excludable goods are under-produced in the private sector (Holcombe, 1997) and are therefore most commonly guaranteed or provided by the state (Cox, 2009; OFT, 2009). As a result, infrastructure sectors tend to be subject to a range of policies and regulations to ensure access to services and to provide incentives for investment and maintenance.

Helm (2013) argues that U.K. infrastructure has multiple negative externalities (such as carbon emissions) and market failures (such as a tendency for oligopolies), making it unlikely that adequate infrastructure provision can take place without state intervention. The effectiveness of state intervention depends on the systemic interaction of a complex network of actors, often in rigid systems characterised by lock-in of particular technologies, power relations and methods of governance, supported and reinforced by interacting systems of rules, institutions and actors within social networks (Geels, 2004; Weber and Hemmelskamp, 2005).

The extent to which states recognise the possibility of using infrastructure interdependencies for the provision of public goods and services (e.g. by reducing the cost of infrastructure development and maintenance through grouping different utility projects in the same area) varies significantly. While in the U.K. the role of the state so far seems to be limited to identifying and removing (mainly) financial barriers to joined-up infrastructure projects in principal (Morgan, 2013; U.K. Regulators Network (UKRN), 2015), the USA and Canada have adopted a proactive approach to facilitating such projects taking place. One such proactive initiative is the use of Envista – a web-based infrastructure project coordination application – by the USA Department of Public Works, for better coordination of projects among municipalities and utility service providers. Envista enables counties to identify opportunities to optimise the construction planning process, save on costs and reduce the environmental and economic impact of operations (Allegheny County, 2014).

14.2.3 The state as an environmental protector and low carbon leader

The state also plays a significant role in environmental protection by producing legislation, guidelines and recommendations for reducing the impact of infrastructure development and maintenance on the environment. This includes making Environmental Impact Assessments integral to acquiring planning permission, and monitoring and enforcing the implementation of environmental policies. Over the past decade states have adopted the principles of sustainable development to varying degrees, and have made sustainability a consideration in political, social and economic activities. Although the level of ambition of government policies is often contested, developed states are often perceived as leaders in setting the agenda and targets (such as energy efficiency and emissions reductions) for low carbon transitions and climate change adaptation. The U.K.'s Climate Change Act (2008), which incorporates legally binding targets for emissions in the medium and longer term is a particularly good example of strong environmental protection legislation.

Interdependencies between sectors pose additional challenges for designing and implementing effective environmental policies, and may require trade-offs to be made between sectors. For example, while the proliferation and advancement of ICT use (such as superfast broadband) could reduce the number of daily commutes (and thus reduce associated carbon emissions), it relies on the existence of big data centres that are very energy intensive and could also increase the use of energy in households as more people work from home.

14.2.4 Non-state actors in the governance of infrastructure interdependencies

A range of non-state actors such as owners of infrastructure assets, private infrastructure investors and insurers are also involved in the governance of infrastructure interdependencies. Key concerns for infrastructure providers are revenue generation, profit, maintaining the reliability of the service delivery, business continuity and how to prevent the potential loss of competitiveness of the service delivered (Frontier Economics, 2012). Another key concern is related to perceptions of insufficient political will for significant infrastructure changes, a lack of continuity and clarity of government policy and delays to major state infrastructure projects (Business, Innovation and Skills (BIS), 2013; Public Accounts Committee, 2013; Stothart, 2013).

Global private investment companies, especially pension funds, are looking for more investment opportunities in U.K. infrastructure, enabled by regulatory change and an institutional environment that mitigates upfront financial risks for institutional investors (Jones and Llewellyn, 2013). However, in order for such investments to be realised, the associated risks need to be low enough. In some U.K. sectors, particularly energy, risks are simply too high at present to attract large amounts of capital from these investors (Blyth et al., 2014).

Consumers and citizens are concerned with the cost, availability and quality of a large number of infrastructure services, such as electricity and water supply. Some consumers are also concerned about the ethical and environmental implications of how these goods and services are produced, delivered and used (Eleini, 2010; Consumer Focus Scotland, 2012; OECD, 2012). Going further, a recent report by Green Alliance has noted that due to these and other concerns, citizens should be engaged much more fully in local and national decisions about infrastructure development (Mount, 2015).

Through the granting of planning permissions and their relative degree of independence over spending, Local Authorities are actively involved in the governance of infrastructures. They are playing an increasingly proactive role in facilitating joined-up infrastructure projects, with the objective of improving efficiencies and minimising infrastructure costs. Prominent examples can be found in Canada, such as Strathcona County who developed a multi-purpose utility corridor to accommodate crude oil and natural gas pipelines, water lines, wastewater mains, telecommunication assets and a road network (Strathcona County, 2007). In Europe, the Smart Cities agenda – advocated by EU programmes and the Department for Business, Innovation and Skills (BIS) – has made some cities more proactive in the governance of infrastructure interdependencies. The aim is to provide a good quality of life with a minimal consumption of resources by intelligently joining infrastructures (energy, mobility, transport, communication, etc.) at different levels (building, district and city) through the use of ICT (Elber, 2013). Smart Cities could therefore offer a platform for tackling policy problems, such as increased greenhouse gas emissions mitigation, that fall inherently between sector boundaries (Smart Cities and Communities, 2013).

However, cities suffer from similar difficulties in governing infrastructure interdependencies to states. For example, cities have high levels of technical and operational complexity, which can serve as barriers to integrating physical city infrastructures (Falconer and Mitchell, 2012). It has also been suggested that some city administrations have organisational structures that lack an integrated approach (Marche and McNiven, 2003). In the U.K. context, however, many of the challenges faced by cities are linked to cuts in funding since the 2008 financial crisis rather than a lack of integration per se (Travers, 2012).

The European Union (EU) also plays an important role in infrastructure governance. The EU has made significant contributions to regulating the environmental impacts of infrastructure sectors, the development of more integrated European energy markets, and other strategic initiatives such as the development of European transport corridors. Through the use of Directives, the EU has provided requirements, definitions and standardisation (e.g. for infrastructure corridors) across sectors and Member States in areas were the EU has specific competencies, as well as through the provision of communications from the Commission to the Council and the European Parliament. For example, the European Commission has stated that it will seek to actively pursue synergies and interdependencies in infrastructure development in the transport and energy sectors in its guidelines for trans-European energy networks (European Commission, 2006).

Another way in which the EU governs infrastructure interdependencies is through the creation of overarching programmes and instruments like the Connecting Europe Facility (CEF). This aims to remove barriers to the joint consideration (both, across borders and infrastructures) of the development of critical infrastructures. CEF was established in 2011 as a new integrated long-term instrument for investing in EU infrastructure priorities in Transport, Energy and Telecommunications. It is designed to encourage synergies between programmes and sectors (European Commission, 2011).

This analysis shows that the state has a particularly important role to play. The next section takes a closer look at the different aspects of infrastructure interdependencies, and the development of new approaches to govern them.

14.3 Emerging trends in the governance of infrastructure interdependencies: infrastructure coordination and joined up thinking

The U.K. National Infrastructure Plan 2014 states that significant investment in infrastructure will be required until 2050, with over £460 billion in the pipeline of planned investment over the period to 2020 and beyond (HM Treasury and Infrastructure UK, 2014). Governments in the U.K. and elsewhere that wish to see such a significant increase in infrastructure investment are faced with two main questions: (i) how to deliver the necessary levels of public and private investment and (ii) how to maximise the economic and social impact and provide value for money.

One important way in which the efficiency of infrastructure investment could be improved is by taking a more integrated approach, rather than focusing on each infrastructure sector separately. The rationale for such an integrated approach has traditionally been one of risk management. In common with other governments, the U.K. government has tended to focus on risks to infrastructure reliability (and hence, to national security), and the extent to which these risks are being exacerbated by increasing infrastructure interdependencies.

Whilst this rationale remains strong, the U.K. government is also interested in exploring the range of potential economic gains to be made by different forms of infrastructure coordination. OECD (2006) and Frontier Economics (2012) claim that joint infrastructure investment and coordination can lead to more coordination and information about infrastructure interdependencies. Furthermore, it could also foster greater trust between different stakeholders, especially between public and private actors, and national, regional and local authorities.

14.3.1 Rationales for joint infrastructure investment and coordination

One way of achieving savings and efficiency gains from the construction and/or maintenance of infrastructure is by developing more opportunities for what the Joint Regulators Group terms 'joint infrastructure investment' (Morgan, 2013). Joint infrastructure investment could refer to several different possibilities, one of which is the joint provision (through economies of scope) of several infrastructures. This is thought to increase opportunities for economic growth (OECD, 2006; Frontier Economics, 2012).

With respect to new investment, infrastructure providers could 'roll out' new infrastructure in more than one sector simultaneously (e.g. electricity and broadband) to yield savings. One-off and annual benefits to infrastructure providers could include: (i) unlocking new investment and growth (e.g. by higher surrounding property values); (ii) delivering new capacity at lower costs; (iii) providing infrastructure maintenance at lower cost; and (iv) reducing the need for additional capacity to meet future demand (Frontier Economics, 2012). In some cases, joint infrastructure investment could develop new 'infrastructure corridors'. Corridors are bundles of infrastructure that link two or more urban areas and can include ICT infrastructure, power lines and cables, water pipes, natural gas, electricity and sewage (Priemus and Zonneveld, 2003).

Some of the key barriers to this more co-ordinated approach to infrastructure provision are related to the investment of time of the parties involved, including the potential for delays; technical differences and complexity of design standards during construction; the possibility of higher costs of construction; and more complex procedure for obtaining planning permission. There may also be broader impediments to coordination that are due to conflicting policy initiatives that impact on infrastructure providers (Frontier Economics, 2012). POST (2010) highlights that lack of communication between infrastructure providers in different sectors can make it difficult to identify interdependencies and to target resources efficiently. Many of the barriers to infrastructure coordination exist between sectors irrespective of their ownership or regulatory structure due to a long tradition of operating separately (Morgan, 2013).

Recent experience in the U.K. provides further illustration of some of these barriers (Watson and Rai, 2013). For example, a case study of interdependencies between electricity, ICT and private vehicles highlights the importance of economic regulation. This case study, which focused in particular on the development of smarter electricity networks in the U.K., shows that economic regulation has been a barrier to innovation in the past. More recently, regulation has acted as a catalyst for innovation and integration (e.g. through the Low Carbon Networks Fund, LCNF).

This case study also showed that the three infrastructure sectors involved are characterised by significant differences with respect to their relationships with consumers, their products, levels of competition and innovation, and the nature of policies and regulations that apply to them. For example, while the ICT sector is very dynamic and proactive, with limited economic regulation, and a lot of innovation driven by competition, the electricity sector (particularly electricity distribution) is more reactive, is subject to significant economic regulation, and is characterised by more gradual change. Another big difference is that in the ICT sector customers are willing to pay more for innovation. In the electricity sector, by contrast, companies have traditionally supplied an undifferentiated product and there is a strong political and consumer focus on affordability and keeping prices as low as possible.

These differences are unlikely to be resolved simply through the creation of joint multi stakeholder institutions like the Office of Low Emission Vehicles (created within the Department of Transport, the Department for Business, Innovation & Skills and the Department of Energy & Climate Change) and the Smart Grid Forum (a government-industry body). The development of future incentives and policies for smarter grids will need to take into account the hierarchical and evolutionary relationship between the electricity, ICT and private vehicle sectors in the U.K. Large-scale take-up of electric vehicles is strongly dependent on the progressive further integration (co-evolution) between electricity and ICTs in the next twenty years.

The diffusion of electric vehicles is likely to require a significant increase in electricity generation capacity, and upgrades to the electricity grid (Xenias et al., 2014). Furthermore, to realise the emissions reductions required – for electric vehicles and for the energy system as a whole – the electricity generation sector will need to be largely decarbonised by 2030 (Ekins et al., 2013). Electric vehicles can also provide systemic benefits by helping to integrate intermittent sources of generation – particularly wind power.

To realise these benefits, further co-ordination may be required between the electricity, transport and ICT sectors. This could include the removal of regulatory and investment barriers to cross-sectoral infrastructure investment. It could also mean further regulatory actions to ensure that innovation in smarter electricity networks and learning is sustained beyond the life of current demonstration programmes such as LCNF.

A second example that illustrates the case for more 'proactive' approach to infrastructure interdependencies is the interaction between the electricity and water sectors in the U.K. Once again, differences in economic regulation and policy priorities between these two sectors have been a source of tensions. Water companies have been subject to increasingly stringent environmental regulations at national and EU levels to improve water quality. One consequence of this is that water provision has become more energy and carbon intensive – a trend that runs counter to policy objectives for carbon emissions reduction. In order to balance these pressures, some U.K. water companies have invested in on-site renewable energy projects. However, the incentives for them to do so have been weakened because the costs of such projects cannot be recovered from water consumers – and they do not count towards some of those companies' emissions reduction obligations (Watson and Rai, 2013).

14.3.2 Recent developments in U.K. governance of infrastructure interdependencies

Awareness of the need for a more systematic approach to infrastructure interdependencies has been growing in the EU and U.K. over the past five years (Pitt, 2008; POST, 2010; Cabinet Office, 2011; European Commission, 2011). This includes the need to build in further resilience to risks that could affect more than one infrastructure sector. As the case studies in the previous section indicate, it also includes the need for more proactive information exchange and coordination within (and between) the government and private sectors.

In response, Infrastructure U.K. (part of HM Treasury) was created in 2010 to facilitate the development of a strategic approach to producing and delivering infrastructure and the National Infrastructure Plan (Public Accounts Committee, 2013). In the National Infrastructure Plan 2013, the U.K.'s economic regulators were asked to consider any regulatory barriers that might hinder shared infrastructure projects, shared facilities or revenue sharing across the different sectors that they regulate, as well as ways of encouraging more joined up infrastructure provision (HM Treasury and Infrastructure UK, 2013).

In parallel with this, the government's Cabinet Office (2011) has emphasised the strategic position of economic regulators – particularly in relation to infrastructure resilience. However, its view is that existing governance arrangements should be used to improve cross-sectoral co-operation on resilience. Only if these arrangements are found to be inadequate should any new duties be considered.

Despite these activities, the extent of cross-sectoral governance in the U.K. has been limited so far. The Joint Regulators Group (JRG) established in 2005 brought together economic regulators for energy, water, transport and health. It aimed to facilitate collaborative work, support information exchange between regulators, and to act as a single point of contact for government, investors or other stakeholders interested in cross-cutting regulatory issues (Morgan, 2013). JRG clarified some regulatory complexities such as how coordination between a monopoly network sector and another sector would work. It also provided regulatory guidance, including on the sharing of costs and the identification of a lead regulator for determining coordination arrangements.

However, the Group did not focus significantly on interdependencies issues. For example, although the JRG recognised the value of creating a national database of joint infrastructure investments, it concluded that it was not the responsibility of economic regulators to gather and deliver such information. This is in contrast to some other European countries. For example, maps indicating spare capacities in ducts owned by incumbent infrastructure companies are provided in Portugal (through the CIS portal) and in the Flanders region of Belgium (by Agiv). In each case, the aim is to enable access for other companies and sectors (Koistinen, 2013).

In 2014, the JRG was replaced by the U.K. Regulators Network (UKRN) which has a more explicit remit to understand and address issues that cut across more than one infrastructure sector. Like the JRG, its members include economic regulators for the energy, water and transport sectors. It remains to be seen what the impact of this Network will be.

In addition to these general 'top down' reforms, there have been some more specific attempts to introduce cross-sectoral governance arrangements. One example is OLEV, whose cooperation agenda is focused on energy and climate change, and the introduction of ultra-low emission vehicles. It therefore leaves out wider issues pertaining to land use and environmental impacts other than greenhouse gases and emissions.

14.4 Future evolution of infrastructure governance

This section examines how governance arrangements might need to change further to address key interdependencies in the future. This analysis is carried out in relation to the four strategies that have been explored in Chapter 10. It emphasises that changes in governance should not be seen in isolation from the types of infrastructure investments and technologies that could be envisaged. Rather, changes in governance are likely to be co-dependent with changes in physical infrastructures, technologies, business models and citizen's behaviour.

There are many important features of future governance that are not possible to predict over such long time scales, even if the strategies and their implications are taken as given. For example, the extent to which markets and/or government intervention will be required to deliver each strategy cannot be predetermined. Some government intervention is likely to be required to realise many of the changes to technologies, infrastructures, consumer choices and business models in these strategies. But this does not necessarily mean that governments need to engage in more central planning than they do now. In some cases, market mechanisms may be better.

In some cases, the realisation of these strategies could require much greater decentralisation of power by national government. This could be particularly relevant for the strategies that differ most significantly from today's situation. Further uncertainty stems from the potential for significant constitutional change within the U.K. during the period to 2050. For example, Scottish devolution has already led to differences between infrastructure governance there and the arrangements that apply elsewhere in the U.K. Such constitutional changes would have profound consequences, but their unpredictability means that they are not explicitly discussed here. References to 'national government' should therefore be taken to apply to the U.K. as a whole.

National infrastructure strategies

Each of the national infrastructure strategies analysed in Chapter 10 has different implications for policies, regulations and the wider governance of infrastructure sectors in the U.K.

The *Minimum Intervention (MI)* strategy emphasises the maintenance of current infrastructure systems, with little change in the rates of investment in new infrastructure systems, and little attention to environmental sustainability. This implies a set of governance arrangements that are similar to those that were in place in the immediate post-privatisation period in the U.K. These arrangements placed a primary emphasis on competition, economic efficiency and consumer prices – but did not place significant emphasis on environmental protection, innovation or on interdependencies.

This strategy would therefore imply the 'rolling back' of many of the regulations and policies that are currently in place in the U.K., including climate legislation such as the Climate Change Act 2008 and EU environmental Directives. Publicly funded programmes to support innovation such as LCNF would be discontinued. In addition, there would continue to be little emphasis within governance arrangements on coordination between infrastructures. Some ad hoc co-ordination might continue, but only where there are easy economic gains to be made.

The *Capacity Expansion (CE)* strategy involves a primary emphasis on supply side infrastructure investment and expansion to meet demand. Policies and regulations would therefore need to provide clear incentives for this investment, and for infrastructure providers to take long-term decisions whilst managing any associated financial risks. Like the first strategy, there is little emphasis on sustainability or a more coordinated cross-sectoral approach to infrastructure investment and operation.

This strategy includes particularly high levels of investment in nuclear power. Given the capital-intensive nature of this technology and the past experience of escalating costs, it is likely that a policy framework would be required that substantially reduces financial risks to investors. This could be achieved, for example, through long-term fixed price contracts for power production or arrangements that allow power companies to pass costs to consumers.

The strategy also leads to increasing carbon emissions after 2030, partly as a result of rising transport emissions. This implies that there is a lack of a cross-sector strategy for reducing emissions, and that the Climate Change Act will have been either repealed or significantly modified. The significant rise in transport emissions suggests that more specific regulations for the transport sector (e.g. vehicle fuel efficiency standards) are also likely to be weak.

The *System Efficiency (SE)* strategy focuses primarily on making current infrastructure systems more efficient by achieving higher levels of capacity utilisation, and by reducing end-use demand using a wide range of technological innovations (ICT) and policy interventions. This would require more fundamental policy and regulatory reforms so that infrastructure providers have a primary incentive to improve efficiency, to focus more on the demand side and to improve environmental sustainability. It will also require significant amounts of coordination within and between infrastructure sectors to help realise efficiency gains.

Therefore, this strategy is likely to require a strong overall policy framework for climate change mitigation such as the Climate Change Act. However, in common with the first two strategies, carbon emissions reductions to 2030 as a result of this strategy are partly offset by a subsequent rise between 2030 and 2050. This indicates some weakening of climate legislation or a lack of comprehensiveness that allows reductions in one sector (energy) to be partly offset by increasing emissions in another (transport). Within this policy framework, there will also be a need for a significant carbon price (e.g. through taxation or emissions trading) or for regulations to limit power plant emissions (e.g. an emissions performance standard). This will be required to provide an incentive for carbon capture and storage (CCS) technology deployment in this strategy. The deployment of CCS is also likely to require the economic regulation of the new CO_2 networks that will be developed. Depending on their importance in this strategy, such arrangements might also extend to heat networks.

This strategy requires a radical change in the incentives for infrastructure providers so that they prioritise efficiency throughout infrastructure systems. In the energy sector, policies such as demand side management and integrated resource planning have been used for a long time (e.g. in the US) to provide an incentive for this. However, these approaches have not been applied in liberalised markets. This focus on efficiency, especially on the demand side, would also need to be reflected in any incentives for capacity (e.g. capacity markets). It is also likely to require metering and associated charging structures that reflect the actual *use* of infrastructure services rather than a fixed charge for services such as household waste collection or water supplies. Furthermore, a greater emphasis on standards for end use appliances and technologies would also be expected under this strategy to speed up the introduction of more efficient technologies.

Stronger mechanisms will also be required to realise efficiency gains that depend on action in more than one sector. Economic regulators would need to move beyond recent initiatives like the U.K. Regulators Network towards more comprehensive and proactive collaborative arrangements. These could not only bring together economic regulators and relevant national government departments, but could also include other actors, such as environmental regulators (e.g. the Environment Agency), and different levels of government including Local Authorities. Working across levels of governance in this way will require rebalancing between the better co-ordination by central government and more context specific processes, resources and actions at a local level. A greater emphasis on local governance will also be important since this strategy includes more investment in local energy systems – particularly heating systems that use local biomass resources.

More bottom–up action for this strategy could include the establishment of more multi-agency intermediary organisations such as Resilience Forums at local and regional level. These could bring together utility companies for joint infrastructure investment, and could promote coordination of infrastructure delivery. One existing example is Lincolnshire's Critical Infrastructure and Essential Services Group which facilitates the development of closer relationships and cooperation between infrastructure providers (such as Anglian Water, CE Electric and British Telecom) and Local Authority bodies (such as local drainage boards). It organises regular meetings focusing on topics such as the resilience of critical infrastructure along its coastal strip. These processes are thought to significantly improve the knowledge of infrastructure assets held by national government agencies. They could also help to build trust and facilitate the flow of information between local industry, infrastructure owners and local authorities, through activities such as the development of Information Sharing Protocols (Cabinet Office, 2011).

While such intermediary platforms could provide a starting point for more co-ordination across infrastructure sectors at the local level, they would need to be developed significantly. So far, they have understandably focused on civil contingencies and risks to infrastructures, including risks stemming from interdependencies. Within this strategy, they would need to expand their remits so that they can help to identify cross-sectoral opportunities for economic and resource efficiency, and to highlight situations where there are conflicting or perverse incentives for infrastructure providers.

A further disadvantage of relying on these platforms for infrastructure is that there could be ad hoc coverage, with different levels of engagement in different geographical areas. In the absence of a significant contribution to governance from national institutions, there could be significant gaps in infrastructure development.

The *System Restructuring (SR)* strategy is the most radical approach, aiming to fundamentally redesign the current infrastructure system to improve total system performance. The governance implications are similar to those for the *System Efficiency* strategy, but with much more emphasis on incentives for radical technical change (such as the mass deployment of smarter electricity grids), the widespread application of ICTs and new business models.

For this strategy, a full-scale coordinated approach to governance across sectors will be essential. As with the third strategy, this would involve a more prominent role for 'bottom up' approaches and decentralised action. The strategy emphasises a 'mixed economy' for infrastructure governance in which local, regional, national and international institutions all have roles to play. This is likely to require a second dimension of coordination between these different governance levels. Secondly, it could involve further and 'deeper' integration between different departments and agencies within national government (current examples include Defra, DECC and Ofgem as well as bodies such as Infrastructure UK).

Within this strategy, a more proactive approach is likely to be required to create opportunities and incentives for more coordination between sectors and joined-up infrastructure projects, in particular waste to energy schemes and waste heat from power generation. This could involve rethinking the role of economic regulators and infrastructure agencies like Infrastructure UK to enable the efficiency gains and systemic innovation. It is also likely to require collaboration between the public and private sectors to support demonstration projects (building on recent examples such as LCNF), and learning from these projects, so that the most successful system innovations can be replicated and/or scaled up. Increasing social acceptance of the need for energy system transformation is an important aspect of the more proactive and decentralised approach of this strategy.

At a national level, one option for governance reform for this strategy could be to merge sector regulators to form a single infrastructure regulator that has a greater focus on interdependencies and the overall sustainability and cost effectiveness of infrastructure services. Another would be to retain the current structure of regulators and government departments, and to strengthen bodies like UKRN and Infrastructure UK so that they have the power to identify and implement projects or programmes across sectors. A third option could be to decentralise governance, and to establish more formal regulatory institutions at a local or regional level. This could include a stronger role and new powers for Local Authorities. It also implies a weaker role for national institutions. However, since this system restructuring could deliver regionally dispersed benefits in the short to medium term, the role of the government in rebalancing any potential gaps, and ensuring large-scale benefits from achieving economies of scale may still be important.

The role of Local Authorities and other local governance arrangements could be particularly important under this strategy, not only in meeting demand through heat networks but also in providing innovative initiatives in cities. Under banners such as 'Smart Cities', 'EcoCities' and 'Sustainable Cities', some cities are already engaging in horizontal coordination between a wide range of stakeholders through municipal networks such as the Covenant of Mayors. These networks are creating opportunities for shared learning at the sub-national level which could be supported and expanded further. As indicated earlier, this expansion would require new powers and more resources for cities and other Local Authorities – a reversal of the recent trend of budget cuts.

If integrated within a national governance framework that focuses more on infrastructure interdependencies, local networks would be valuable hubs of knowledge and experience. This would help to address information gaps between different stakeholders; to align objectives between actors in different sectors and governing institutions; to develop capacity for joined-up infrastructure delivery and maintenance; and to help develop methods and tools for monitoring and measuring the performance of relevant agencies and industry.

14.5 Conclusion

This chapter has shown that there are important rationales for the governance of infrastructures to place more emphasis on interdependencies – both from the point of view of managing risks to the provision of infrastructure services and for improving efficiency and sustainability. By exploring the roles of different actors involved in infrastructure governance, the chapter has highlighted the significant difference between *governance* and *government*. Whilst national governments have an important role to play, developing more sustainable infrastructures will continue to require the involvement of multiple actors including government and public agencies at different levels (international, national and sub national), infrastructure providers as well as citizens. Given that many of the possible future trajectories for infrastructures will continue to be controversial, even when overall goals are widely shared, a greater involvement of citizens in priority setting at national and local levels is likely to be increasingly important (Mount, 2015).

Using examples from the U.K., the chapter has shown how governance arrangements are already starting to change to take interdependencies into account more fully. It has also explored what further changes might be required in the future – especially if there is a continuing desire to move beyond 'business as usual'. The chapter's analysis has shown that governance will need to support greater coordination across infrastructure sectors in two of the future strategies explored earlier in Chapter 10: the System Efficiency and the System Restructuring strategies.

In the case of the System Efficiency strategy the state would have to introduce radical changes to sector policies and regulations to realise the efficiency gains and emissions reductions envisaged. It is also likely that national governance arrangements would need to work alongside a variety of decentralised governance 'platforms' and actions. In the case of the System Restructuring strategy, further emphasis will be required at all levels on supporting innovation, particularly through the application of ICTs, to help meet efficiency and sustainability goals. This strategy would require a more radical shift to decentralised governance, alongside strengthening the governance of infrastructure interdependencies at national level.

Irrespective of the future pathway taken by infrastructures, governance will continue to be a crucial part of the picture. It will continue to co-evolve with changes in technologies, behaviours and business models. Furthermore, the overall approach taken to governing infrastructures – including the extent of decentralisation, the relative roles of markets and the policy instruments chosen – will depend heavily on wider political and policy changes at local, national and international levels.

References

Allegheny County (2014). "Utility and infrastructure project coordination." Retrieved from www.alleghenycounty.us/publicworks/utility.aspx.

BIS (2013). *The Smart City market: opportunities for the UK*. BIS Research Paper No. 136. London, UK, Department for Business Innovation & Skills.

Blyth, W., R. McCarthy and R. Gross (2014). Financing the power sector – is the money available? UKERC Energy Strategies Under Uncertainty.

Bouchon, S. (2006). The vulnerability of interdependent critical infrastructure systems: epistemological and conceptual state-of-the-art. EUR 22205 Report. Institute for the Protection and Security of the Citizen.

Cabinet Office (2011). *Keeping the country running: natural hazards and infrastructure*. London, UK, Cabinet Office.

Climate Change Act (2008). *Climate Change Act 2008*. London, UK, HM Government.

Consumer Focus Scotland (2012). Evidence to the Scottish Parliament Infrastructure and Capital Investment Committee: Water Resources (Scotland) Bill.

Cox, K. R. (2009). Public good. *The international encyclopaedia of human geography*. R. Kitchin and N. Thrift (Eds.). London, UK.

CST (2009). *A national infrastructure for the 21st century*. London, UK, Council for Science and Technology.

Ekins, P., I. Keppo, J. Skea, N. Strachan, W. Usher and G. Anandarajah (2013). The UK energy system in 2050: comparing low-carbon, resilient scenarios. UKERC Research Report. London, UK, UKERC.

Elber, U. (2013). *Smart cities – joint programme review*. Brussels, Belgium, European Energy Research Alliance.

Eleini, N. (2010). Public attitudes towards climate change and the impact of transport: 2006, 2007, 2008 and 2009. Summary report for the Department for Transport. London, UK.

European Commission (2006). Decision No 1364/2006/EC of the European Parliament and of the Council, Official Journal of the European Union L262, from 6 September 2006 laying down guidelines for trans-European energy networks and repealing Decision 96/391/EC and Decision No 1229/2003/EC.

European Commission (2011). *Proposal for a regulation of the European Parliament and of the Council establishing the connecting Europe facility*. Brussels, Belgium.

Falconer, G. and S. Mitchell (2012). *Smart City Framework: a systematic process for enabling Smart+Connected communities*. San Jose, CA, USA, CISCO Internet Business Solutions Group.

Frontier Economics (2012). Systemic risks and opportunities in UK infrastructure. A report prepared for HM Treasury and Infrastructure UK. London, UK.

Geels, F. W. (2004). "From sectoral systems of innovation to socio-technical systems: insights about dynamics and change from sociology and institutional theory." *Research Policy* 33(6–7): 897–920.

Helm, D. (2013). "British infrastructure policy and the gradual return of the state." *Oxford Review of Economic Policy* 29(2): 287–306.

HM Treasury and Infrastructure UK (2013). *National infrastructure plan 2013*. London, UK, HM Treasury.

HM Treasury and Infrastructure UK (2014). *National infrastructure plan 2014*. London, UK, HM Treasury.

Holcombe, R. (1997). "A theory of the theory of public goods." *The Review of Austrian Economics* 10(1): 1–22.

Jones, R. and J. Llewellyn (2013). *UK Infrastructure: the challenges for investors and policymakers*. London, UK, Llewellyn Consulting, for Pension Insurance Corporation.

Koistinen, P. (2013). Cost saving measures for broadband roll-out. *Conference on monitoring electronic communications and information society services in the enlargement countries*. Izmir, Turkey.

Marche, S. and J. D. McNiven (2003). "E-Government and E-Governance: the future isn't what it used to be." *Canadian Journal of Administrative Sciences* 20(1): 74–86.

Ministry of Defence (2014). Global strategic trends – out to 2045. *Strategic Trends Programme*. Swindon, UK.

Mitchell, C. (2008). *The political economy of sustainable energy*. Basingstoke, UK, Palgrave Macmillan.

Morgan, I. (2013). Joint Regulators Group (JRG) subgroup report on cross sectoral infrastructure sharing. Joint Regulators Group.

Mount, A. (2015). Opening up infrastructure planning: the need for better public engagement. Report for Green Alliance. London, UK.

OECD (2006). *Infrastructure to 2030 – telecom, land transport, water and electricity*. Paris, France, Organisation for Economic Co-operation and Development.

OECD (2012). *Inclusive green growth: for the future we want*. Paris, France, Organisation for Economic Co-operation and Development.

OFT (2009). *Government in markets: why competition matters – a guide for policy makers*. London, UK, Office of Fair Trading.

Pitt, M. (2008). *Learning lessons from the 2007 floods*. London, UK, Cabinet Office.

POST (2010). Resilience of UK infrastructure. *postnote*. London, UK, Parliamentary Office of Science and Technology.

Priemus, H. and W. Zonneveld (2003). "What are corridors and what are the issues? Introduction to special issue: the governance of corridors." *Journal of Transport Geography* 11(3): 167–177.

Public Accounts Committee (2013). HM Treasury: planning for economic infrastructure. London, UK, Public Accounts Committee, Forty-second report.

Smarter Cities and Communities (2013). Integrated action plan – report process and guidelines. Smart Cities Stakeholder Platform, Finance Working Group Guidance Document.

Stothart, C. (2013). Energy sector now biggest concern as survey finds majority of firms lack faith in infrastructure policy. *Construction News*.

Strathcona County (2007). Strathcona County municipal development plan – bylaw 1–2007 (Chapter 16 Utility Systems).

Travers, T. (2012). *Local government's role in promoting economic growth: removing unnecessary barriers to success*. London, UK, Local Government Association.

UKRN (2015). Innovation in regulated infrastructure sectors – summary report. UK Regulators Network.

Watson, J. and N. Rai (2013). Governance interdependencies between the UK water and electricity sectors. Working Paper. ITRC/University of Sussex.

Weber, M. and J. Hemmelskamp (2005). *Towards environmental innovation systems*. Berlin and New York, Springer.

Xenias, D., C. Axon, N. Balta-Ozkan, L. Cipcigan, P. Connor, R. Davidson, A. Spence, G. Taylor and L. Whitmarsh (2014). Scenarios for the development of smart grids in the UK: literature review. Working Paper. UKERC.

15 The future of national infrastructure

JIM W. HALL, ROBERT J. NICHOLLS, MARTINO TRAN, ADRIAN J. HICKFORD

15.1 A system-of-systems approach

In Part I of this volume, we make the case for taking a systems approach for planning national infrastructure. Analysis of infrastructure systems has conventionally dealt with each infrastructure sector – energy, transport, water, waste, digital communications and others – in isolation. Each of these sectors has their own planning approaches and specific ways of making the business case for investment. The development of investment plans has been based on assumptions that have been developed for each sector: of demand from households and industries, and of the hazards to which the systems may be exposed. Even when cross-sectoral interdependencies are significant, for example, in thermoelectric energy generators' demand for cooling water, these have been dealt with via assumptions based on past experience. These approaches have enabled the development of the elaborate systems that we now witness, which sustain modern societies and economies.

The analysis described in this book is motivated by a hypothesis that the current approaches to planning, design and management of infrastructure systems that have served modern societies well in the past are not fit for purpose as we move into the future. We are already observing increasing convergence of infrastructure services, enabled by technology – information and communications technologies (ICT) in particular. Though it may be convenient to assume so, the demands for infrastructure services are not independent for each infrastructure sector. Ultimately they derive from people and businesses who are demanding multiple infrastructure services in ways that are correlated in space and time and across sectors. By considering infrastructure services in combination rather than in isolation, there are opportunities both to access synergies and to manage the risks of interdependence.

In Part II of this volume, we have considered each infrastructure sector separately, recognising their distinctive characteristics. Energy, transport, water, waste and digital communications are distinctly different systems, making use of specialised technologies that need to be understood if they are to be managed effectively in the future. These chapters have provided an overview of how these systems function and have each used new national infrastructure system models to analyse the prospects for the future under changing patterns of demand. We have explored the range of technological options that might be adopted in the foreseeable future and analysed contrasting strategies for infrastructure provision in these systems. Whilst the legacy of existing infrastructure, particularly in mature economies like Britain, tends to lock in established development pathways, we have been able to identify and test radically different futures for national infrastructure in each of the infrastructure sectors that we have considered. Whilst our approach in Chapters 4–9 was sector-specific, we have sought to adopt a consistent approach to reviewing these infrastructures, enabling us to highlight similarities, whilst fully recognising differences.

In Part III, our distinctive contribution has been to integrate across sectors, exploiting the full benefits of an assessment methodology that is based upon long-term, cross-sectoral strategies for national infrastructure provision. Integrating across sectors into a system-of-systems perspective, reported in Chapter 10, has enabled examination of factors that jointly influence all sectors, such as population and the economy. We have analysed the interdependence between infrastructure sectors through cross-sectoral demands for infrastructure services. The examples presented in Chapter 11 explored how much electricity will be required by the transport sector given different levels of uptake of electric vehicles and rail electrification, and how much water might be required by thermoelectric power plants. They both demonstrate that cross-sectoral dependencies are real and important when considering future demand.

We have also explored, in Chapter 12, the ways in which extreme hazards can impact multiple infrastructure networks, and how resulting failure can propagate across and between these networks. The analysis also showed the economic implications of disrupted service delivery. All of this understanding is necessary if provision of national infrastructure is to avoid being haphazard and incremental, and if systemic vulnerabilities are to be avoided. Without system-of-systems analysis it is hard to demonstrate that the major investments that are now taking place will yield outcomes that are beneficial and sustainable in the long term. Infrastructure investments are long-lived and have the potential to lock in patterns of development, so it is essential to analyse decision pathways under a wide range of possible future conditions.

The strategies and scenarios that we have analysed do not exist in isolation. Infrastructure and the services that infrastructure systems provide are deeply embedded in the economy and society. The delivery of infrastructure is governed by complex arrangements, spanning the public and private sectors and different levels of government. Chapter 14 has therefore analysed the governance context for the provision of infrastructure services, which in its current forms will be challenged by the system-of-systems approach that we have pioneered. Governance arrangements have evolved in different ways for different sectors. Chapter 14 has argued that it would be naïve to suppose that unified cross-sectoral governance arrangements are feasible, or indeed desirable, but management of complex interdependent systems requires flows of information, so that interdependencies (both propitious and malicious) are understood, and so that coordination can be exerted where necessary. The system-of-systems approach that we have developed provides a framework within which information, assumptions and plans for the future can be exchanged. This can inform governance approaches for key system-level issues that may currently fall in the gaps between sectoral silos.

15.2 The emerging toolset

Our approach is based upon systems analysis and simulation modelling. Given the unpredictability of change in all of the systems that we have explored, the use of simulation modelling might be regarded as being naïve. How can we possibly seek to represent all of the relevant processes and interactions? Of course we cannot. But modelling has provided us with a structured framework within which to explore and test system behaviour, recognising that all models are incomplete, but some are useful. If we had *not* developed models, it is difficult to see how we could have kept track of even a fraction of the complexities, interdependencies and uncertainties that we have been able to explore. Simulation modelling has provided us with the potential to explore a wide range of possible strategies for infrastructure provision and a variety of possible futures that they might inhabit. Moreover, simulation has provided the potential to develop and test *adaptive* strategies, in which investments are not pre-determined, but are mobilised in response to changing demands for infrastructure services and environmental conditions. Strategies do not need to be a deterministic blueprint for the future – they can and should contain contingencies and decision rules to ensure requisite system performance under a wide range of possible conditions. Simulation modelling has allowed us to test and refine these decision rules in a way that otherwise would have been impossible.

The National Infrastructure System Model (NISMOD) framework has also allowed us to explicitly represent interdependencies between sectors in our models and track how those evolve. How, for example, could the uptake of electric vehicles influence electricity demand; how might that demand for electricity be satisfied by different generation technologies; and how, in turn, might new thermo-electric generation influence demand for cooling water from rivers whose flows are depleted by human demands and climate change? We have been able to answer these questions by embedding models within an integrated framework which enables interaction between sectors and through space and time.

Importantly, the lessons from the complex systems analysis described in this book have been learnt along the journey rather than at the destination. The modelling process often provides more insights than the model results per se. In order to build models we have had to examine and reflect upon the relevant system behaviours and interactions. We have had to question the results from the analysis and test them against data and analogues. The analysis has led us to seek new empirical data sets and has forced us to reformulate our hypotheses. Hence the process is iterative and the modelling promotes learning.

NISMOD is underpinned by an entirely new database architecture populated by several hundred national infrastructure network data sets. Assembling and cleaning national infrastructure data has been a time-consuming process, and there is much more to do. However, the NISMOD-DB national infrastructure database is now a scalable proof of concept, around which a growing group of infrastructure owners, operators and analysts can congregate. There will always be security concerns about some aspects of national infrastructure data, and often commercial considerations also. Secure approaches to holding and exposing these data are therefore required, to enable modelling and analysis of interdependencies that can help to inform understanding of interdependencies and development of cross-sectoral strategies. Where data gaps have been impossible to fill, we have used statistical methods for network analysis to synthesise missing elements of the network (Thacker et al., 2015).

Visualisation is central to the process of communicating the concept and results of the analysis. Visualisation tools built on top of the NISMOD-DB geospatial database framework have enabled presentation of results at scales of relevance to decision-makers. The combination of database facilities and visualisation tools enables in-depth scrutiny of the results and tracing back to the input conditions that generated the results. This combination of flexible scrutiny and traceability is essential when dealing with complex multi-dimensional data sets.

The notion of a strategic approach to national infrastructure provision is, as these words are written, acquiring more support in Britain. Given the importance of such an exercise, and the widespread support, it is perhaps surprising that before the work described in this book, there was no national-scale, cross-sectoral, long-term analysis of alternative strategies for infrastructure provision. One explanation for this situation, we believe, is that the tools did not exist to construct such an analysis. And until a national infrastructure systems assessment had actually been conducted, it was rather difficult to envisage how it might be done. We recognise that the analysis described in this book is by no means perfect, as we will discuss further in the next section, and there are other approaches to conducting a national infrastructure assessment. Yet the benefit of what we have done is to demonstrate one version of how a national infrastructure assessment might be conducted. We expect our approach to stimulate debate, be improved upon and ultimately superseded, but the first step in disentangling a complex system is often the hardest. We have made that step for infrastructure system-of-systems.

15.3 Limitations and next steps

John Archibald Wheeler said that 'As our island of knowledge grows, so does the shore of our ignorance'. Constructing a system-of-systems analysis for national infrastructure has made us even more aware of the limitations of our understanding of these complex systems. We see major gaps in the inputs, the processes and the outputs of our systems analysis.

The input data that the analysis is dependent upon has profound limitations. The gaps are most striking in the analysis of interdependent infrastructure failure in Chapter 12. Conducting this type of analysis requires a huge amount of data about the spatial characteristics of loads, the fragility of infrastructure elements and the interdependence between infrastructure networks. In some senses it is not surprising that these data do not all exist. Some of the data relate to very rare and practically unrepeatable events. On the other hand, given the impacts of potential failures that we have identified, more effort in data collection is surely merited. We hope that these results will help to motivate further data collection.

The data for characterisation of demand for infrastructure services and the capacity of systems to supply those services is rather more extensive, though it varies to a great extent across sectors. Here the limitation is in parameterisation of future changes, though many of those future uncertainties will be impossible to resolve. This is why we have adopted an approach based upon scenario analysis and the exploration of robustness of possible strategies across the scenarios, which attempt to capture this uncertainty.

The objective of analysing at national scale, spatially and over long timescales, has limited the resolution of the physical processes that are enacted in our system models. There is always a trade-off between model complexity and the number of model runs that are feasible. Spatial-temporal resolution and complexity of process representation increases runtime, and so limits the number of future scenarios and strategies that can be explored. Greater process-based representation, for example, of road transport or the operation of water supply systems, would be desirable, but would also limit our ability to conduct extensive scenario analysis. Inexorable advances in computational capacity means that these limitations will dwindle in the future. In the meantime, our emphasis will be upon using models of intermediate complexity which we can extensively test and explore.

The analysis of infrastructure performance in Part II of this book, which was synthesised in Chapter 10, used a basket of metrics to assess the performance of alternative infrastructure strategies: the amount of infrastructure service provided; the capacity margin; the investment costs associated with the strategy; and the carbon emissions. These represent four attributes of fundamental interest to decision-makers. However, these performance indicators are not all that decision-makers want to know. The reliability and quality of services is also of utmost importance. Metrics of quality will differ from sector to sector – they are very different for rail transport compared to, say, water supply. The amount of carbon emissions is one metric of environmental impact which we have focused upon –and is particularly important because emissions reduction targets can be a driver of transformative change in energy, transport and across the infrastructure sectors. It is also attractive because the same metric can be used across sectors (though each sector raises its own emissions accounting issues). Whilst we have considered other metrics of environmental impact, for example, in the standard of wastewater treatment (Chapter 7), further work will be required to compile these into an integrated set of environmental metrics.

Chapter 3 explored the economic demand for infrastructure services. It raised the fundamental question about the relationship between infrastructure and the economy. In what ways does infrastructure alter the nature and location of economic activity, and more specifically economic growth? In that chapter, we cited some of the literature and presented some new evidence about the relationship between infrastructure and economic growth, but theoretical understanding is incomplete and the empirical evidence tends to be at an aggregate level. That is a significant gap, given the emphasis upon infrastructure investment as a means of promoting growth. The evidence to make the business case for those investments is not as substantial as we would hope. Aside from that public policy concern, we also recognise a dynamic (and spatial) feedback between economic growth, provision of infrastructure systems and the demand for infrastructure services. The analysis reported here has not closed that feedback loop. For a treatment of that dynamic, readers will have to wait for a future edition of this book!

15.4 The global context

Our system-of-systems analysis has been demonstrated as a case study of Britain. The overall framework can, and will, be transferred to other international settings, though the analysis will have to be modified to those contexts. The work is most readily transferrable to other mature economies, like the U.K., with a large stock of existing infrastructure, that are facing major choices about infrastructure replacement, upgrade and carbon emissions reduction. However, the approach is generic and can be transferred more widely.

The need for infrastructure in emerging economies and least developed countries is even more acute. Around US$53 trillion of infrastructure investment will be required globally through to 2030, equivalent to an annual 2.5% of global GDP (OECD, 2012). If that investment is to be employed wisely, in ways that will not be regretted in the future, because of missed opportunities or locked in harmful impacts, then it is essential that decisions are taken with an understanding of future changes and risks. The principles that have been developed in this book urgently need to be applied to inform those investment decisions: looking across sectors; understanding demand for infrastructure services; explicit representation of uncertainty through scenario analysis; analysing infrastructure needs and flows at an appropriate spatial resolution; and developing long-term adaptive cross-sectoral strategies. These challenges are even more complex in the rapidly changing context of emerging economies. Hence, the need for systematic analysis of future pathways for infrastructure provision is even more pressing in those contexts.

15.5 Lessons for the future

The timescales over which infrastructure will exist are daunting. Many assets have lifetimes of fifty years or more, and infrastructure investments can lock in patterns of development more or less irreversibly, as shown by the legacy of historic investments such as transport links. Over those timescales there are very large uncertainties about changes in population, lifestyles, behaviour and technology. Because these changes depend on innovations and human choices that have not yet happened, they are inherently unpredictable. Yet if regrettable investment choices are to be avoided, it is necessary to explore the range of possible futures that infrastructure systems may inhabit. By developing and applying methodology and models for long-term scenario analysis we have found that meaningful insights do emerge, which can guide such decisions.

The long life of infrastructure and buildings, which we have already alluded to, provides some fairly solid points of reference for the future. Even in the long term, the future is not entirely uncertain. New homes and towns will be built, cities will grow and possibly also decline in size, but the main urban agglomerations endure. The analysis of regional demography in Chapter 3 has demonstrated that the spatial patterns of development in

Britain seem to be enduring through trends of population growth. Indeed it is difficult to construct a plausible regional economic scenario that does not see the southeast of England as the continued pole of demographic and economic activity. That may be a discouraging fact for policies aimed at 'rebalancing' Britain's economy, but it provides some confidence for infrastructure planners that there are some recognisable trends even in a very uncertain future.

The interplay between lifestyles, behaviour and technology represents the greatest uncertainty facing the future of national infrastructure. Road transport is a case in point. The analysis of road transport demand presented in Chapter 5 foresees increases in demand for road transport, moderated mostly by the effect of congestion eventually suppressing demand. Yet that chapter noted the role of technology in possibly substituting (or at least changing patterns of demand) for road transport. We noted a decline in the number of young people with driving licences. We also commented upon literature suggesting that there may be a limit to the number of hours in a day that most people are willing to spend in a car. These technological, economic and lifestyle factors may be combining to reach, or pass, 'peak car', be that defined in terms of car ownership or (of more relevance) car usage. In water, solid waste and energy we have also observed patterns of demand decoupling from economic growth. By one account, global emissions of carbon dioxide from the energy sector did not increase in 2014, the first time there was a halt or reduction in emissions that was not tied to an economic downturn (IEA, 2015). Even in energy demand, a decoupling from economic growth may be under way.

The coupling with population growth is more enduring. Demand for all infrastructure services is strongly influenced by population, and in the case of sewage treatment, demand is a direct multiplier of population. Our analysis of population projections in Chapter 3 reveals a wide range of possibilities in terms of population growth, which are influenced to a large extent by the amount of in- and out-migration to the U.K. Those uncertainties exist worldwide. In emerging economies, the rate of migration to urban areas is another extremely important factor. Nonetheless, despite these uncertainties, there are plausible bounds to the range of future population, which provide a basis for evaluation of strategies for infrastructure provision.

Our analysis of coupled energy and transport systems (Chapters 4, 5, 10 and 11), has revealed a wide range of possible energy futures, demonstrating that major choices do exist. These choices are particularly striking in the energy sector, because of the need to replace ageing and polluting plant, and the rapid innovation in supply and demand. As David Mackay has said, there are many possible energy strategies that could be adopted, but 'Make sure your policies include a plan that adds up' (Mackay, 2009). The analysis we have developed is based on quantified modelling of infrastructure systems and demand for infrastructure services, so all of our strategies 'add up'. Some of the strategies meet the U.K.'s carbon targets and some do not. With the right sequence of choices, the costs of meeting carbon targets are only slightly more than least cost strategies that do not meet the carbon targets. Decarbonisation of infrastructure systems is achievable and affordable, but it requires sustained effort at a large number of points in infrastructure systems and end-user demand.

15.6 A blueprint for national infrastructure assessment

The notion that infrastructure plans and investments should be based on a long-term view of infrastructure needs and performance is an intuitively appealing concept. However, when one delves into the governance of each infrastructure sector, as we have done in Chapter 14 of this volume, one finds apparently intractable complexity of ownership, financing, regulation and decision-making. Investment decisions emerge from the interplay of a wide range of different actors, who are each subject to multi-dimensional drivers and constraints. No approach to systems analysis and assessment will be able to incorporate all of this complexity, and the system-of-systems approach we have presented in this book definitely does not. So what purpose does national infrastructure assessment serve, and what have we learnt about national infrastructure assessment in practice?

National infrastructure assessment helps to construct the 'big picture', or indeed alternative 'big pictures' of how infrastructure systems function and perform today and into the future.

Infrastructures are network systems. Their performance emerges as a result of interactions across the network. Investments in one asset yield benefit through the role that asset has in the overall network. The benefits of infrastructure investment need to be appraised in terms of performance at the system-scale. Whilst each infrastructure sector has developed its own methodologies for planning and appraisal, no individual sector is responsible for the system-of-systems. Yet each sector needs to understand interdependencies in order to be able to cope with future demands from other sectors and manage the risks of interdependence. In addition, government needs to take an overview of the system-of-systems and ensure appropriate regulation and governance across diverse and interacting infrastructures.

A national infrastructure assessment helps to explore and evaluate alternative long-term pathways of infrastructure development. There is a reasonable consensus about the attributes that a modern infrastructure system should deliver to society and the economy: efficient, reliable and affordable services that are environmentally benign. The challenge is to demonstrate the extent to which proposed policies, plans and investments will contribute to that outcome. It may be possible to demonstrate the marginal effect of an individual investment, but that does not help to build confidence that the desired outcomes, for example, in terms of security of supply of water or energy, accessibility to markets or carbon emissions reduction will add up in the long run. Through scenario analysis and simulation, we have demonstrated how national infrastructure assessment can test the effectiveness of alternative strategies for infrastructure provision, and provide evidence of performance in the long run.

Of course none of the strategies we have tested will be precisely adopted in the future. However, we have tested a range of alternative directions of travel, which seek to span the range of future possibilities. We have demonstrated the implications of these diverse strategies in terms of metrics that are of interest to decision-makers. We have demonstrated that the future can be meaningfully analysed as alternative pathways that will lead to foreseeable outcomes. Evidence about those outcomes can be used to compare strategies and make decisions about what to do.

Given the scale of uncertainties, it is often desirable to keep options open and, sometimes, to delay decisions. On the other hand, there are choices that it would be unwise to delay; keeping options open can incur unacceptable costs. So decisions have to be made, and we are most likely to get them right if they are supported by rigorous analysis of possible outcomes, risks and uncertainties.

The complexity of national infrastructure assessment is possibly the greatest barrier to it being applied in practice to inform infrastructure policy, planning and investment decisions. The principle of having a clearly articulated and evaluated direction of travel, which acknowledges uncertainty and seeks to avoid conflicts across sectors and to manage unacceptable risks, is hard to argue with. The challenge is devising an infrastructure plan in practice, and then assembling the evidence about the long-term performance that is needed to choose between alternative plans. The complexity is potentially daunting and the methodology has not been well developed until now. This book has demonstrated one version of how national infrastructure assessment can be carried out, from a quantified system-of-systems perspective. In doing so, we hope to have built confidence that national infrastructure assessment is feasible and informative. Without the work described in this book, the notion of national infrastructure assessment would have been an unproven hypothesis. What we have done can doubtless be improved upon, but we have set out concepts and frameworks that we hope will endure and provide a foundation for further developments.

References

IEA (2015). "Global energy-related emissions of carbon dioxide stalled in 2014." Retrieved from www.iea.org/newsroomandevents/news/2015/march/global-energy-related-emissions-of-carbon-dioxide-stalled-in-2014.html.

Mackay, D. J. C. (2009). *Sustainable energy – without the hot air*. Cambridge, UK, UIT.

OECD (2012). Strategic transport infrastructure needs to 2030. Paris, France, Organisation for Economic Co-operation and Development.

Thacker, S., J. W. Hall and R. Pant (2016). "A methodology for the synthesis of hierarchical infrastructure networks." *IEEE Systems Journal*. In press.

Index